PHYSICOCHEMICAL METHODS FOR WATER AND WASTEWATER TREATMENT

Other Related Pergamon Titles of Interest

Books

ALBAIGES:
Analytical Techniques in Environmental Chemistry

HENSTOCK & BIDDULPH:
Solid Waste as a Resource

HUTZINGER et al:
Aquatic Pollutants - Transformation and Biological Effects

MOO-YOUNG:
Waste Treatment and Utilization - Theory and Practice of Waste Management

ZOETEMAN:
Sensory Assessment of Water Quality

Journals

Conservation & Recycling

Progress in Water Technology

Water Research

Water Supply and Management

Full details of all the above titles and a free specimen copy of any journal available on request from your nearest Pergamon office.

PHYSICOCHEMICAL METHODS FOR WATER AND WASTEWATER TREATMENT

Proceedings of the Second International Conference,
Lublin, June 1979

Edited by

LUCJAN PAWLOWSKI

Maria Curie-Sklodowska University, Lublin, Poland

PERGAMON PRESS

OXFORD · NEW YORK · TORONTO · SYDNEY · PARIS · FRANKFURT

U.K.	Pergamon Press Ltd., Headington Hill Hall, Oxford OX3 0BW, England
U.S.A.	Pergamon Press Inc., Maxwell House, Fairview Park, Elmsford, New York 10523, U.S.A.
CANADA	Pergamon of Canada, Suite 104, 150 Consumers Road, Willowdale, Ontario M2J 1P9, Canada
AUSTRALIA	Pergamon Press (Aust.) Pty. Ltd., P.O. Box 544, Potts Point, N.S.W. 2011, Australia
FRANCE	Pergamon Press SARL, 24 rue des Ecoles, 75240 Paris, Cedex 05, France
FEDERAL REPUBLIC OF GERMANY	Pergamon Press GmbH, 6242 Kronberg-Taunus, Pferdstrasse 1, Federal Republic of Germany

Copyright © 1980 Pergamon Press Ltd.

All Rights Reserved. No part of this publication may be reproduced, stored in a retrieval system or transmitted in any form or by any means: electronic, electrostatic, magnetic tape, mechanical, photocopying, recording or otherwise, without permission in writing from the publishers.

First edition 1980

British Library Cataloguing in Publication Data

Physicochemical methods for water and wastewater treatment.
1. Water - Purification - Congresses
I. Pawlowski, Lucjan
628.1'6   TD430   79-42661
ISBN 0-08-024013-5

In order to make this volume available as economically and as rapidly as possible the authors' typescripts have been reproduced in their original forms. This method has its typographical and in some cases linguistic limitations but it is hoped that they in no way distract the reader.

Printed in Great Britain by A. Wheaton & Co. Ltd, Exeter

CONTENTS

Foreword 1
L. Pawlowski

Pilot Plant Development of Ozone Disinfection 3
R. L. Bunch

Nitrogen Fertilizer Manufacturing Water Effluents
and Control Technology 15

Development of Anion Exchangers for Silica Removal:
History and Present Trends 27
C. Calmon

Demineralization of Brackish Waters by an Ion
Exchange/Electrodialysis Process 41
G. Boari, G. Tiravanti

Ultra-Pure Water at the Service of Technology 55
M. P. Steiner

Electrochemical Method for Phosphorus Precipitation 65
P. Ennet, M. Hannus, H. Molder

Ammonium Phosphate Recovery from Urban Sewages
by Selective Ion Exchange 73
L. Liberti, N. Limoni, R. Passino, D. Petruzzelli

Utilization and Desalination of Saline Mine Waters 87
J. Kepiński

Ion Selective Electrodes in Water and Wastewater
Analysis 99
K. Sykut

Purification of Wastewaters from a Pig Farm by Means
of Electrolysis 107
Z. Drabent, J. Dziejowski, L. Smoczyński

New Coagulants for Treatment of Piggery Wastewater
M. Rutkowski, H. Pielichowski, T. Lanowy, M. Korczak
113

The Ozonization of Dihydroxybenzenes in the Model
Solution
D. Leszczyńska, A. L. Kowal
123

Studies of the Influence of the Coagulation Process
on Certain Pesticides in Water
W. Sztark
133

A. Comparison of Packed-Bed and Expanded-Bed
Adsorption System
S. W. Hermanowicz, M. Roman
141

Electrochemical Oxidation of Sodium Salt of Cetyl
Sulphate Under Galvanostatic Conditions
Z. Gorzka, A. Jóźwiak, B. Kozlowska
153

Electrochemical Oxidation of "Rokaphenol N-6"
Z. Gorzka, K. Jasińska, A. Socha
163

Neutralization of Industrial Wastes Containing Detergents
by Means of Catalytic Oxidation Method
Z. Gorzka, M. Kaźmierczak
175

Removal of Organic Solvents from Wastewaters by Means
of Physico-Chemical Methods
T. Jaroszyński
185

Polyurethane Resins as Oil Sorbents in Wastewater
Treatment
E. Gomólka, J. Rybka
193

The Effects of the Thickening of Various Sewage Deposits
Subjected to the Influence of Ultrasonic Field
J. Bień, E. Kowalska, E. Zielewicz
205

Problems of Choice of Sorptive Filters on the Basis
of BIOT Number Quantities
A. Grossman, B. Kucharski, W. Kusznik
211

The Chemico-Physical Method of Wastewater Treatment
from Glulam Wood Structures
Z. Nieweglowska, B. Bartkiewicz
221

Method for Recovery of Water and Vanadium Compounds
from Wastewater
L. Zagulski, L. Pawlowski, A. Cichocki
229

Ion Exchange Method for Water, Ammonia and Nitrates
Recovery from Nitrogen Industry Wastewater
J. Barcicki, L. Pawlowski, A. Cichocki, L. Zagulski
237

Membrane Separation of the Product in the Process
of Biological Transformation
A. Poranek, T. Winnicki, J. Wiśniewski
245

Contents

Dialytic Processes in Water and Wastewater Treatment T. Winnicki, A. Mika-Gibala, G. Błażejewska	255
Catalytic Oxidation of Inorganic Sulphur Compounds in Reducing Solutions A. L. Kowal, E. M. Klocek	265
Simulation of the Flocculation Process in Filter Beds A. L. Kowal, J. Maćkiewicz	275
Treatment of Wastewater from Hydraulic Transport of Ash in Power Plants B. Dziegielewski, A. M. Dziubek, A. L. Kowal	283
Estimation of Conventional Methods for the Treatment of Urban Storm Water M. Świderska-Bróż	289
The Recovery of Ammonium Nitrate from Fertilizer Factory Wastes L. D. Roland	299
Experiences with Primary Water Purification and Waste Water Treatment Plants in Nuclear Power Plants with Pressurized-Water and Boiling-Water Reactors - New Developments in this Field N. Buser and J. P. Ghysels	313
Author Index	323
Subject Index	325

FOREWORD

There is a strong need for efficient treatment processes which can detoxify, destroy or apply resource recovery principles to wastewater. Much work has recently been done to evaluate the feasibility of applying physicochemical treatment techniques, such as chemical coagulation, filtration, ion exchange and activated-carbon adsorption, directly to raw wastewaters or primary effluents to eliminate entirely the need for biological processes. Chemical coagulation and filtration are used to remove suspended matter from raw wastewater, whereas activated carbon is used to adsorb the remaining soluble organics. Phosphorus removal normally occurs with chemical coagulation. If nitrogen removal is also required, physicochemical processes such as ion exchange and breakpoint chlorination are adaptable.

Energy crisis has turned attention to shortage of not restorable resources. Their economical use has become important for the whole world, therefore recovery of chemicals from wastewater has to be used on larger scale. For these purposes physicochemical unit processes are the most useful. Some of these unit processes are commonly used for industrial wastewater treatment, while others require further R and D efforts before they will become commercially attractive.

For successful contribution of chemists to improve the quality of our environment a forum for exchange of scientific and technological information is needed. It was provided for Polish specialists by the conference organized in 1976, on Physicochemical Methods for Water and Wastewater Treatment.

Because of great interest in these problems of chemists and professional water and wastewater technologists, we have organized the Second International Conference related to the same problem. Papers selected from those presented on the last Conference have been published in this book.

We will organize a third Conference in 1981, proceedings of which we are going to publish in similar form. I will be grateful to readers for suggestions related to the coming Conference as well as to the book. They will be useful to improve our future work.

Lublin
August 8, 1979

Lucjan PAWŁOWSKI

PILOT PLANT DEVELOPMENT OF OZONE DISINFECTION

Robert L. Bunch

*Municipal Environmental Research Laboratory,
U.S. Environmental Protection Agency, Cincinnati, Ohio 45268,
U.S.A.*

ABSTRACT

The adverse environmental effects of residual chlorine in chlorinated wastewater prompted a search for alternative disinfection agents that would be environmentally acceptable. Ozone is used for disinfecting potable water in Europe but because of its cost it has not been used for disinfecting treated wastewater. One of the main items influencing the cost is the ozone contactor. This study presents the data from the comparison of three contactors that are commercially available. They are the positive pressure injector, a packed column and a bubble diffuser. The results show that the bubble diffuser gave the highest ozone utilization and total coliform reduction. A highly significant log-log correlation between coliform reduction and ozone utilization was noted.

KEYWORDS

Ozone; disinfection; ozone dissolution; wastewater ozonation; microorganism reduction; gas contactors.

INTRODUCTION

The disinfection of treated wastewater before discharge to surface waters and of sewage sludges before landspreading is a controversial procedure. Disinfection of treated wastewaters and sewage sludges is not routinely practiced in Europe to the extent that it is in North America. In the Western European countries, disinfection of wastewater is generally employed only to protect public bathing facilities, shellfish areas and wastes from hospitals, clinics and large microbiological laboratories. Reviewing some of the historical events and recent studies might aid in understanding the rationale for the present disinfection requirements in North America.

The first use of chlorine in wastewater treatment was for the control of odors. This dates back to 1854 when it was used to deodorize the London sewage. It was first used to disinfect sewage in North America in 1893 at Brewster, New York. This application was for the protection of the Croton watershed, which is a part of the New York City water supply. The literature indicates an active interest in wastewater disinfection beginning about 1945. Up to that time the main interest in chlorine was for odor control, hydrogen sulfide destruction, and prevention of septicity. Most of the sewage treatment plants practicing disinfection during that time belonged to the U. S. Armed Forces. It was a matter of policy that sewage effluents had to be chlorinated at all Army bases in the United States during World War II. Up until 1972, the disinfection requirements in the USA varied widely from state to state. Seasonal disinfection of wastewater was practiced in many states, while no disinfection was required for certain waters where the possibility of human contact was remote.

The U. S. Federal Water Pollution Control Act Amendments of 1972, Public Law 92-500, clearly established the responsibility of the Environmental Protection Agency (EPA) for reducing and controlling the pollution of navigable waters. The Act established as a National goal that the water quality should provide for the protection of fish, shellfish and wildlife and for recreation in and on the water by July 1, 1983. The principal mechanism used to achieve the objectives of Public Law 92-500 was by implementing effluent standards for industrial and municipal wastewater dischargers. Included in the effluent standards were biochemical oxygen demand, suspended solids, pH, and fecal coliform bacteria. The limit on fecal coliform bacteria was set at 200 per 100 cm^3. This requirement generally necessitated the use of separate, continuous disinfection processes in addition to the biological or physical-chemical processes for BOD and suspended solids removal.

On July 26, 1976, the EPA deleted the fecal coliform requirement from the effluent standard. Reliance on each state to set disinfection requirements for municipal wastewater treatment plants based on site-specific water quality standards is now the practice. Thus, disinfection will still be required by state regulation, but no uniform Federal requirement has to be achieved. Some of the reasons for the change in the regulation are: (1) toxic effects on the aquatic environment of residual chlorine, (2) halogenated organic compounds are potentially carcinogenic, (3) continuous disinfection of wastewater is not needed in certain locations, (4) potential dangers associated with the use of elemental chlorine, and (5) the amendment of the effluent standards provide flexibility in disinfection requirements. Cost was not considered as a main factor, because public health and protection of aquatic environment are the overriding considerations affecting any disinfection requirements. The unnecessary and excessive use of disinfectants, however, is wasteful both in terms of available financial resources and energy.

The adverse environmental effects of residual chlorine in chlorinated wastewaters has prompted EPA to search for alternative disinfection agents which would satisfy the disinfection requirements and be environmentally acceptable. Ozone has been used for more than half a century as a potable water disinfectant in Europe, particularly in France. Presently, ozone is used at 577 water treatment plants in France, 247 in West Germany, and 135 in Switzerland. It has been only in the last five years that thought has been given to using it to disinfect wastewater. The use of ozone for disinfection of treated wastewater requires better engineering controls than for treating drinking water. Drinking water usually has a constant ozone demand, and the flow to be treated is usually constant. The ozone demand of wastewater is quite variable as is the flow. It is much more difficult to pace the flow of ozone to the wastewater demand and flow than chlorine. With chlorine there is a ready supply to meet large variations in demand and the developed instrumentation to control the residual rather precisely. Another difference between ozonizing drinking water and treated wastewater is the amount of ozone needed. For drinking water the dose of ozone for adequate disinfection is usually less than 1 mg/dm^3 while typical requirements for treated wastewater are 5-10 mg/dm^3 of ozone. This will require better dissolution of the ozone. The disinfection efficiency of ozone depends to a large extent on the quality of the wastewater to be disinfected. Ozone is a potent oxidizing agent, and its reaction with oxidizable material is non-selective. The demand exerted by organic matter in effluents can have a marked influence on the disinfection efficiency and reliability of ozone. Care must be exercised in making certain that the ozone produced is utilized in the most efficient manner; otherwise, the operating costs of ozonation may be needlessly high due to excessive use of energy resources. One of the main items influencing the cost is the contactor.

Private industries concerned with supplying ozone generation factilities have as their objective the most cost-effective package. The package normally includes air preparation, ozone generation, and ozone contacting. Every conceivable liquid-gas contacting device is marketed with ozone systems. The U. S. Environmental Protection Agency has initiated at its pilot plant at the Robert A. Taft Laboratory, Cincinnati, Ohio, a project to determine the advantages and disadvantages as well as the economics associated with the contactor. This study has completed evaluating three generic types of contactors commercially available. It is intended to evaluate more in the near future. The three contactors evaluated were: (1) positive pressure injector (PPI), (2) a packed column (PC), and (3) a bubble diffuser (BD).

CONTACTOR DESIGN CONSIDERATIONS

Before describing the pilot plant studies on contactors, a brief review of some of the significant process variables in ozone disinfection might be beneficial. It has been shown by several researchers that the initial rate of inactivation of microorganisms is very rapid in the presence of ozone residual. The reaction time of ozone is much more rapid than chlorine in the disinfection process; therefore, the contact time as a design parameter is not as important as in chlorination. In the presence of a measurable ozone residual, a contact time in the order of 2-5 minutes appears to be adequate for treated wastewater.

In USA wastewaters, a dose of 5-10 mg/dm^3 of ozone is required to meet the effluent standard of 200 fecal coliforms/100 cm^3. Little attempt has been made by researchers to correlate the degree of inactivation of microorganisms with respect to ozone residual in an ozonated effluent. This has been due primarily to difficulties in determining low concentrations of ozone in the presence of organic matter, color, and turbidity in wastewater effluents. The amount of residual ozone obtainable in secondary wastewater effluent depends upon the nature of the organic matter present in the effluent, dosage applied, and the mode of the ozone application. An improved practice would be to correlate the effectiveness of disinfection with ozone residual rather than the dosage applied. Analytical techniques for detecting low levels of ozone in wastewater will have to be improved greatly if this is to be done. The type of microorganisms and the density of microorganisms is also a significant variable. Various microorganisms have varying degrees of resistance toward ozone. For example, E. coli are more susceptible to inactivation by ozone than poliovirus. Because microorganisms exert an ozone demand, a higher density of microorganisms would exert a greater demand and thus reduce the effectiveness of disinfection for a given dosage of applied ozone. In order to obtain the desired inactivation, the ozone residual must be maintained at a level sufficient to inactivate the highest possible density of the particular microorganisms in question.

Mixing is another parameter which affects the ozone residual. Mixing should be sufficient to completely mix the contents of the reactor and to promote the transfer of ozone into the solution. However, mixing should be kept to a minimum to prevent ozone loss.

OZONE CONTACTORS

Fig. 1 is a schematic diagram of the packed column contactor. This is a 230 mm diameter glass column packed with 3.1 m of 12 mm ceramic saddles. A teflon redistributor plate is located midway in the column to redirect the liquid towards the center of the column. Filtered secondary effluent enters the top of the column and exits at the bottom. Mean residence time of the secondary effluent was approximately 35 seconds (measured by a dye study) at a liquid flow rate of 75 dm^3/min and a gas flow rate of 40 dm^3/min. Ozone, injected at the bottom of the column, flowed upward countercurrent to the secondary effluent and exited at the top. A gas sample tap was located on the exhaust line. Effluent from the packed column was directed to two 200-dm^3 covered holding tanks connected in series to allow an additional five minutes contact time with the ozone residual. Liquid sample taps were located as shown in Fig. 1.

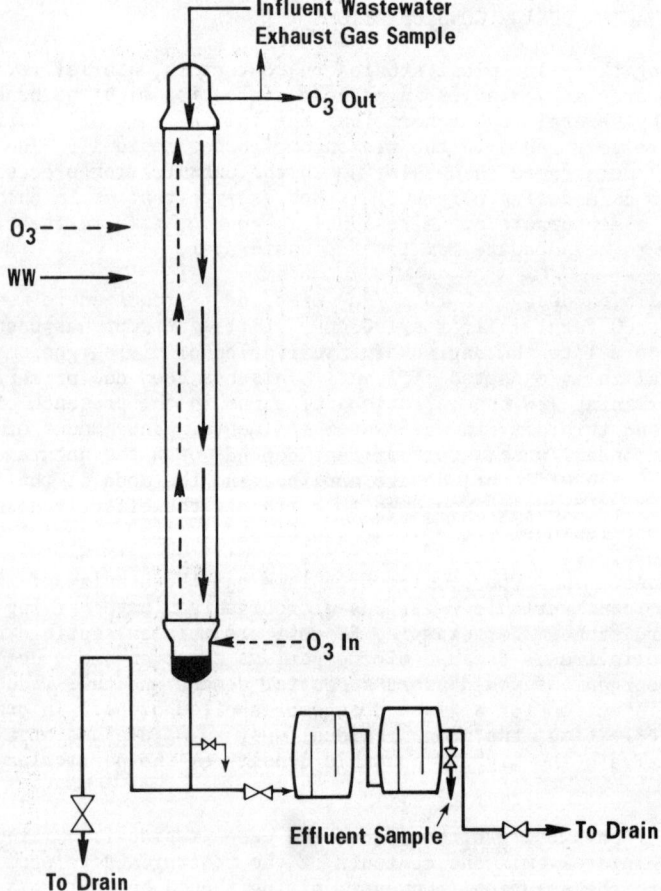

Fig. 1. Packed column ozone contactor.

Fig. 2 is a schematic diagram of the positive pressure injector. Ozone enriched gas was injected into the influent liquid flow in a specially designed "tee-mixing chamber" under a slight positive pressure of 50 to 60 kilopascal. The gas-liquid mixture flowed concurrently downward through an inner cylinder. Upon reaching the bottom of the inner cylinder, the mixture reversed direction and flowed upwards through a larger cylinder. At the top, the gas and liquid separated, the gas flowing to the exhaust gas line and the liquid flowing downwards through a second concentric cylinder, finally emerging at the bottom of a drain line. Mean liquid residence time in the system was approximately 15 seconds (measured by a dye study) at a liquid flow rate of 75 dm^3/min and gas flow rate of 40 dm^3/min.

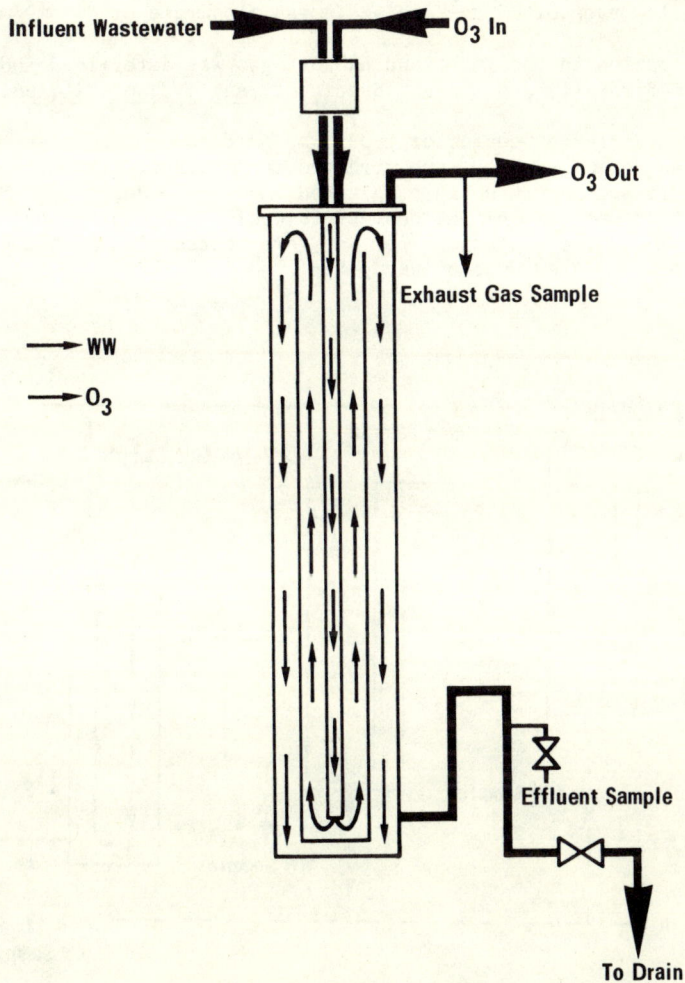

Fig. 2. Pressure injector contactor.

Fig. 3 is a schematic diagram of the bubble diffuser contactor. It consisted of three aluminum columns, each 3.7 m high and 300 mm in diameter, connected in series by PVC piping. The three columns were staggered vertically so that secondary effluent, which has been pumped to the top of the first column, could flow by gravity to each of the two successive columns. The ozone enriched gas was injected through domed ceramic diffusers (Norton Chemical Process Products Division) located at the bottom of each column and connected in parallel to a common header pipe. The first column received 50 percent of the total gas flow, while the other two columns received equal fractions of the remaining 50 percent. Gas and liquid sample taps were located as shown in Fig. 3. Since the bubble diffuser was operated as a system, treated effluent samples were collected at the sample tap of the last column only, while exhaust gas samples were collected after the exhaust gas from each column had been combined into a common line (Fig. 3). Mean liquid residence time in the bubble diffuser contactor, measured by a dye study, was 9.4 minutes at a liquid flow rate of 75 dm^3/min and a gas flow rate of 40 dm^3/min.

Ozone concentration in the inlet and exhaust gas was determined iodometrically by the method of Birdsall, Jenkins and Spadinger, <u>Anal. Chem.</u>, <u>24</u>, 662 (1952).

Testing of the different contactor types was performed using a statistical split plot design where whole plots were arranged in a balanced incomplete block design. The whole plots were arranged in a balanced incomplete design due to the physical restriction that only two of the contactors could be set up in parallel at any given time. Thus, two contactors per day were tested with both contactors receiving the same applied dose at any given time. The order of the dose levels

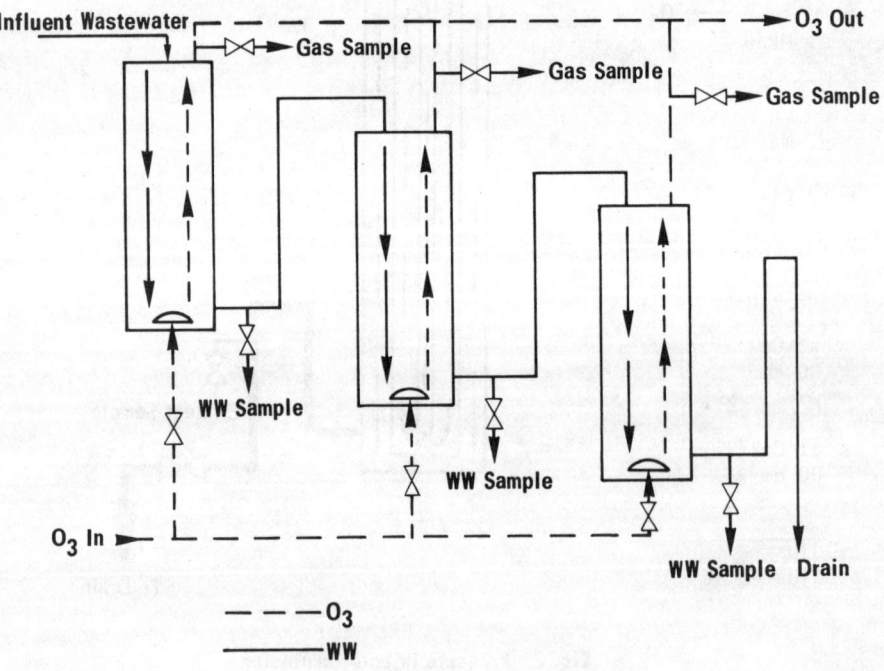

Fig. 3. Bubble diffuser ozone contactor.

was balanced so that each dose level was used first, second, and third in a day the same number of times for each contactor. Randomization consisted of randomizing the order in which the pairs of contactors were compared.

SOURCE OF WASTEWATER

Secondary effluent was obtained from the conventional activated sludge pilot plant at the Robert A. Taft Laboratory, Cincinnati, Ohio. Wastewater entering the plant was of municipal origin. Effluent from the final clarifier was filtered through a multi-media pressure filter (Baker Filtration Company, Model HRC-30D) and then split equally between two of the contactors for parallel evaluation. In Table 1, the minimum, mean, and maximum values of the physical-chemical and bacteriological characteristics of the filtered pilot plant effluent prior to ozonation are presented. Quality was typical of a well-treated secondary effluent. Note that the total coliform density was approximately four times the fecal coliform density.

TABLE 1 Summary of Filtered Pilot Plant Effluent Characteristics Prior to Ozonation (June 22 to July 21, 1978)

Parameter	Mean	Minimum	Maximum
TCOD, mg/dm^3	35	22	46
SCOD, mg/dm^3	31	22	38
TOC, mg/dm^3	11.5	7.7	16.5
TSS, mg/dm^3	3.8	1.4	7.2
NH$_4$-N, mg/dm^3	13.4	7.7	19.2
TKN, mg/dm^3	17.1	10.0	28.7
Turbidity, JTU	2.2	0.6	3.8
pH	-	7.2	8.3
Temperature, °C	22	20	23
Total Coliforms/ 100 cm^3*	2.6 x 10^6 (geometric mean)	2.4 x 10^5	7.9 x 10^6
Fecal Coliforms/ 100 cm^3**	6.5 x 10^5 (geometric mean)	1.4 x 10^5	2.0 x 10^6

* Adjusted for 87% colony verification
**Adjusted for 95% colony verification

RESULTS

The first set of experiments evaluated the percent ozone utilization as a function of applied ozone dose. The applied ozone dose to each contactor was varied by increasing the gas flow rate to liquid flow rate while keeping the inlet ozone concentration constant. The secondary effluent flow to each contactor was 75 dm^3/min and the gas flow rates were 25, 50 and 75 dm^3/min. The concentration of ozone in the inlet gas was maintained at 10.0 mg/dm^3 gas (approximately 0.8 weight percent).

The results, as shown in Table 2, indicated that overall percent ozone utilization was highest in the bubble diffuser, followed by the packed column, and then the positive pressure injector. As the dose was increased, percent ozone utilization in all contactors decreased. In Table 3, the total coliform log reduction (TCLR)

TABLE 2 Mean Percent Ozone Utilization as a Function of Applied Ozone Dosage (Increasing Q_G/Q_L, constant Y_1)*

Dose mg/dm^3	Mean percent ozone utilization (% U) (standard deviation)			Mean % U, all contactors
	Positive pressure injector	Packed column	Bubble diffuser	
3.3	77 (2)	85 (4)	90 (2)	84 (6)
6.7	53 (3)	58 (2)	86 (2)	66 (15)
10.0	41 (3)	46 (3)	81 (3)	56 (18)
Mean % U, All doses	57 (16)	63 (17)	86 (5)	

TABLE 3 Mean Total Coliform Log Reduction (TCLR) as a Function of Applied Ozone Dose (Increasing Q_G/Q_L, Constant Y_1)*

Dose mg/dm^3	Mean total coliform log reduction (standard deviation)			Mean TCLR, all contactors
	Positive pressure injector	Packed column	Bubble diffuser	
3.3	2.90 (0.36)	2.85 (0.36)	3.12 (0.10)	2.96 (0.31)
6.7	2.94 (0.44)	3.13 (0.10)	3.56 (0.25)	3.21 (0.38)
10.0	3.04 (0.34)	3.31 (0.11)	4.52 (0.55)	3.62 (0.75)

*Q_G = carrier gas flow rate
Q_L = liquid flow rate
Y_1 = concentration of ozone in the carrier gas

in the above experiments are summarized. The data reveal that overall mean total coliform log reduction was higher in the bubble diffuser than in either the packed column or the positive pressure injector. The overall mean total coliform log reduction in the latter two contactors was not significantly different. The results of fecal coliform reduction were similar to the total coliform reductions.

The second set of experiments was run to determine the change in percent ozone utilization and microorganism reduction when the applied dose was varied by changing the ozone concentration and maintaining a constant gas to liquid flow ratio. A gas flow rate of 37.5 dm^3/min and a liquid flow rate of 75 dm^3/min to each contactor was used. The maximum ozone concentration attainable with the ozone generator used at this gas flow rate was 15 mg/dm^3 gas (approximately 1.25 weight percent). The three concentration levels selected were 3.0, 9.0 and 15.0 mg/dm^3 gas which corresponded to an applied dose of 1.5, 4.5 and 7.5 mg/dm^3.

In Table 4 the mean percent ozone utilization data per contactor at each dose level are summarized. The results indicated that overall percent ozone utilization was higher in the bubble diffuser than in the packed column or the positive pressure injector while percent utilization in the latter two contactors was similar. The rate of decrease in percent ozone utilization at higher doses was substantially less in the bubble diffuser than in the other two contactors. The total coliform log reduction data are shown in Table 5. The overall total coliform log reduction in all contactors was similar. Although the overall utilization of ozone in the bubble column was higher, only 0.5 mg/dm^3 more ozone was absorbed than in the other two contactors. This increase was not sufficient to cause a marked increase in total coliform log reduction. The fecal coliform reduction shows a similar pattern.

When the total and fecal coliform log reductions were plotted versus the ozone utilizations, it showed that equivalent coliform reductions are achievable, independent of contactor type and contact time. This suggested that perhaps log coliform reduction is a function of log ozone utilization. When all the data from all the experiments were transformed into $\log_{10}$ values and plotted, a highly significant log-log correlation between coliform reduction and ozone utilization was found.

TABLE 4 Mean Percent Ozone Utilization as a Function of Applied Ozone Dose (Increasing Y_1, constant Q_G/Q_L)*

Dose mg/dm^3	Mean percent ozone utilization (standard deviation)			Mean % U, all contactors
	Positive pressure injector	Packed column	Bubble diffuser	
1.5	84 (4)	79 (3)	89 (2)	84 (5)
4.5	68 (5)	68 (4)	80 (4)	72 (7)
7.5	62 (4)	62 (3)	77 (3)	67 (8)
Mean % U, All Doses	71 (10)	69 (8)	82 (6)	

*Q_G = carrier gas flow rate
Q_L = liquid flow rate
Y_1 = concentration of ozone in the carrier gas

TABLE 5 Mean Total Coliform Log Reduction (TCLR) as a Function of Applied Ozone Dose (Increasing Y_1, constant Q_G/Q_L)*

Dose mg/dm^3	Mean total coliform reduction (standard deviaition)			
	Positive pressure injector	Packed column	Bubble diffuser	Mean TCLR, all contactors
1.5	0.81 (0.68)	0.73 (0.59)	0.56 (0.40)	0.70 (0.54)
4.5	3.07 (0.43)	2.90 (0.39)	3.24 (0.49)	3.07 (0.44)
7.5	3.28 (0.40)	3.37 (0.30)	4.00 (0.49)	3.55 (0.50)
Mean TCLR, All doses	2.39 (1.25)	2.34 (1.25)	2.60 (1.58)	

*Q_G = carrier gas flow rate
Q_L = liquid flow rate
Y_1 = concentration of ozone in the carrier gas

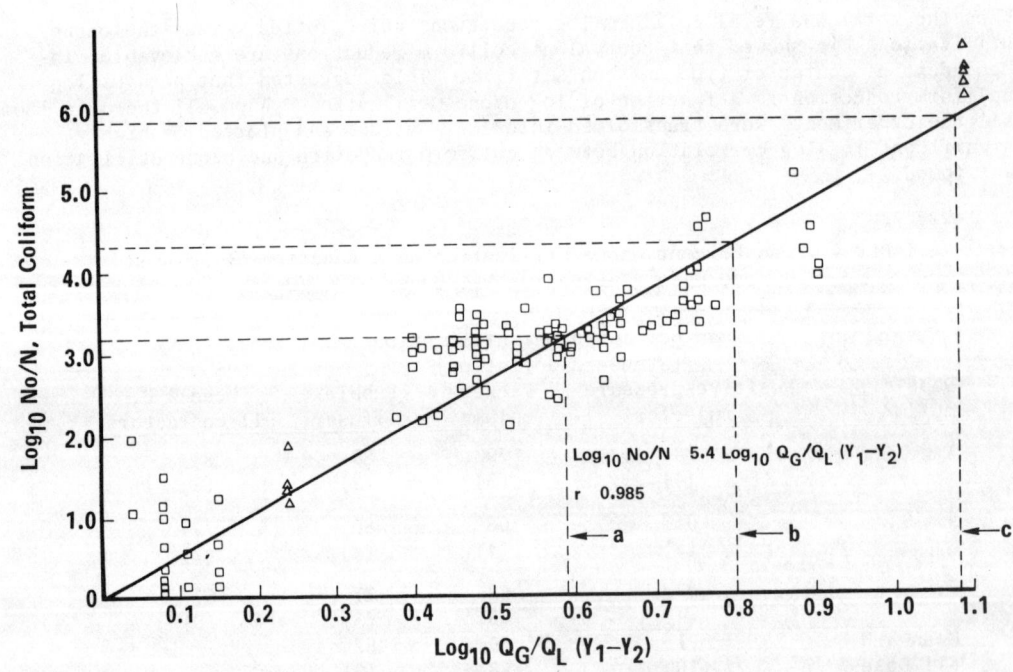

Fig. 4. Mean Log Total Coliform Reduction as a fuction of Log Ozone Utilization.

Fig. 4 includes 108 experimental data points shown as squares and 12 data points indicated by triangular symbols, that were obtained later for verification of the regression equations. It is possible other relationships may exist in the data. It was found that the linear log-log relationship is sufficient as a predictive tool. For example, to achieve a final effluent total coliform density of 1000/100 cm^3 starting with an initial number of 1.55 x 10^6 total coliform/100 cm^3, 4 mg/dm^3 ozone must be absorbed by the effluent. This is shown by dotted line labelled "a". To achieve a shellfish area standard of 70 total coliforms/100 cm^3 in final effluent, 6.5 mg/dm^3 ozone must be absorbed. This is shown by the dotted line labelled "b". To achieve the strict State of California standard of 2.2 total coliforms/100 cm^3, 12 mg/dm^3 ozone must be absorbed. This is shown by dotted line labelled "c". Only one of the three contactors studied would be capable of meeting this requirement with ozone generated from air. That contactor is the bubble diffuser.

CONCLUSION

Since log coliform reduction is a function of log ozone utilization independent of contact time, it follows that disinfection by ozone is mass transfer limited. The more ozone that can be transferred to the wastewater, the greater will be the bactericidal effect. When designing ozone contactors, this fact should be kept in mind.

Ozone appears to be a most promising alternative to chlorine. Its major drawback is the relative high cost of generation. Before ozone can gain widespread use for disinfection of wastewater, there will need to be improvements in: (1) economics of generation, (2) efficiency of gas-liquid contactors, (3) on-line analysis of residual ozone. Hopefully, the future use of ozone can be controlled by on-line measuring of the ozone residual, instead of relying on dosage.

NITROGEN FERTILIZER MANUFACTURING WATER EFFLUENTS AND CONTROL TECHNOLOGY

William J. Search

Monsanto Textiles Company, Guntersville, Alabama, U.S.A.

ABSTRACT

Water effluents in a nitrogen fertilizer plant originate from a variety of point and non-point sources. Often these sources are combined to take advantage of cost savings. This practice often causes problems, however in effluent segregation and treatment. The major components in the effluents are ammonia nitrogen (NH_3o and NH_4+), nitrate nitrogen, and organic nitrogen. Low concentrations of other constituents may also be present. This paper presents brief descriptions of the component nitrogen fertilizer production processes with emphasis on discharge point identification and characterization, quantification of potential pollution species, and available control technologies.

KEYWORDS

Nitrogen fertilizer manufacturing; water effluents; point sources; treatment methodology; pollutant species; ammonia; ammonium nitrate; urea; nitric acid.

PROCESS DESCRIPTION

The nitrogen fertilizer industry, for this paper, is defined as those primary fertilizers whose main nutrient contribution to the soil environment is nitrogen and from which all other nitrogen fertilizers are derived; i.e., synthetic ammonia, ammonium nitrate, and urea, with a consideration of nitric acid as an adjunct process for ammonium nitrate. A conceptual diagram of this industry is shown in Fig. 1.

---

[1] This paper presents portions of work done by the author while at Monsanto Research Corporation, Dayton, Ohio.

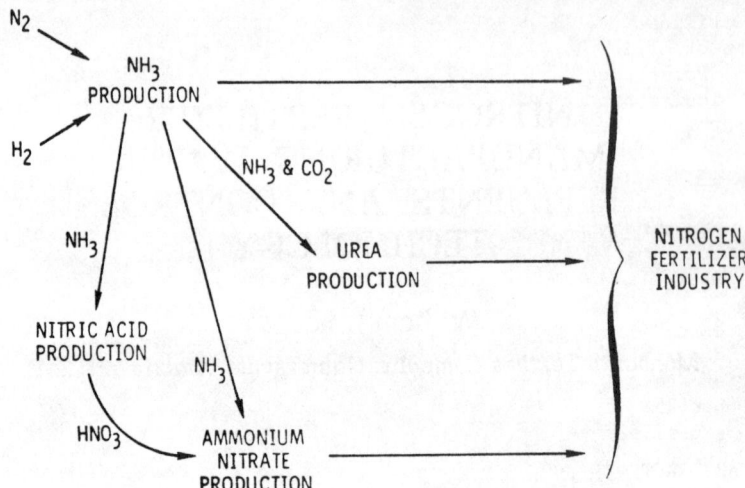

Fig. 1. Conceptual diagram of the nitrogen fertilizer industry.

Synthetic ammonia is commonly produced in the United States by the catalytic reforming of natural gas using a six-step process: 1) natural gas desulfurization, 2) catalytic steam reforming, 3) carbon monoxide shift, 4) carbon dioxide removal, 5) methanation, and 6) ammonia synthesis (Rawlings, 1977).

Ammonium nitrate solution is produced by an exothermic reaction of ammonia and nitric acid. The solution may be sold directly or blended with other fertilizer solutions (approximately 39% of the solution produced), or converted into solid form (approximately 61% of the solution produced). Solids are made by prilling, granulation, or graining. Prilling is the most common method of solidification, accounting for 92% of the solids produced in the United States (Search, 1977).

Urea is produced by a two-step reaction. In the first reaction ammonia and carbon dioxide react to form ammonium carbamate. In the second reaction the ammonium carbamate is dehydrated to form urea and water. Thirty-eight percent of the urea solution produced is sold as blending agent for other fertilizers or as a raw material in the synthesis of other chemicals. The remaining 62% is solidified by either granulation or prilling. Granulation is the most common method of solid urea production, accounting for 85% of the solid product (Search, 1977).

Nitric acid is produced in a three-step ammonia oxidation process: 1) ammonia is oxidized to nitrogen oxide or nitrogen in the presence of oxygen and a platinum catalyst, 2) nitrogen oxide is further oxidized to nitrogen dioxide, and 3) nitrogen dioxide reacts with water to form nitric acid and nitrogen oxide. The resulting solution stream is between 55% and 65% nitric acid (Search, 1977).

Modern, complex nitrogen fertilizer plants often incorporate a combination of these various component processes to take advantage of costs savings in raw material purchase and transportation and energy savings by interconnecting heat exchangers. While saving money in the production of the fertilizers, there are several ramifications of this communal arrangement in effluent segregation and treatment.

EFFLUENT ORIGIN AND CHARACTERIZATION

Origin identification of an effluent source is one of these ramifications. Two basic types of effluent origins exist in nitrogen fertilizer plants--point and non-point. Point sources are those which originate as a definite wastewater stream from a particular process. Non-point sources are those which originate from various leaks or from large areas within a plant.

Point Sources

In the ammonia manufacturing process, steam stripper exhaust condensate or spent regenerant used in the ion exchange purification of the recycled condensate can result in water effluent.

Point sources of wastewater in the ammonium nitrate manufacturing process may include condensate from neutralizer and evaporator exhausts and solutions from air pollution control equipment used on the cooler and/or dryer. It is estimated that less than 25% of the U.S. plants release process condensate to a receiving body or treatment plant. Also, effluent from air pollution control equipment may be blended into fertilizer solutions where markets exist.

Point sources in the manufacture of urea are condensate of the evaporator exhaust and filtrate from the concentration of urea solution when a crystallizer is used. Plants using a vacuum evaporator ($\approx$50%) condense the exhaust so that they can recover the nitrogen value in the stream as a dilute fertilizer (Search, 1977). When airswept evaporators are used the exhaust is scrubbed and vented to the atmosphere, and the scrubbing liquid may be recovered as a dilute fertilizer solution. Less than 25% of the plants use crystallizers (Search, 1977).

In general there are no point sources of water effluent from the actual nitric acid manufacturing lines (Train, 1974). There may be effluent from the cooling towers or spills and these are covered in the following sections.

A further point source common to many processes is cooling towers. Cooling tower blowdown or discharge, while non-contact streams, may contain ammonia or nitrate from absorption of ammonia from ambient air or from small leaks into the cooling system from process streams. Chemicals such as copper, chromium, and zinc compounds used as corrosion inhibitors will also appear in this stream.

Nonpoint Sources

In nitrogen fertilizer plants, there are various nonpoint sources that can contribute a major portion of the water effluent.

<u>Leaks and spills</u>. In any plant, a certain number of valve and pump leaks as well as random spills can be expected. In the nitrogen fertilizer plant, these leaks can become a major contributor due to the sometimes corrosive character of the material (Train, 1974). For example, the carbamate recycle is a highly corrosive slurry, and pump seals on such recycle lines must be replaced frequently. Spills also often occur at shipping facilities.

<u>Plant and equipment cleanup</u>. Spills are often not cleaned up until there is a general plant cleanup, unless they are large enough to require immediate attention. These spills generally occur around loading operations and are hosed down periodically.

Equipment cleanup generally involves the flushing of railroad cars or trucks. This normally occurs in a specified area with proper drainage designed specifically for this use.

Other cleanup sources may result when process vessels must be evacuated and flushed for maintenance and personnel entry.

<u>Additive filter cake backwash</u>. The use of additives in the manufacture of high density ammonium nitrate requires filtration to remove raw material insolubles. Periodic renewal of the filter cake results in backwashing entrapped nitrate solutions to the sewer.

<u>Runoff</u>. Rainfall runoff from a plant can be a significant contributor to a plant's total nitrogen loading because it collects ammonia and nitrate from the air and washes down the ground and buildings on which solid fertilizer material may have settled (Train, 1974).

Another ramification of current nitrogen fertilizer methodology is the characterization of an effluent discharge from a nitrogen fertilizer plant. The various combinations of processes, production capacities, process capabilities, level of recycle/reuse alternatives, collection systems, and treatment practices make it very difficult to characterize a representative effluent.

Three basic pollutant materials, however, have been shown to exist in most nitrogen fertilizer plant outfalls: ammonia nitrogen, nitrate nitrogen, and organic nitrogen. Ammonia nitrogen, when present in water, reaches a chemical equilibrium between unionized ammonia, ionized ammonia and hydroxide ions. At anticipated receiving water pH and temperature the unionized ammonia content at equilibrium was determined to range between 0% and 21% (Search, 1977). Other minor constituents as well as certain stream characterization parameters for several plants are shown in Table 1.

The following ranges of effluent concentrations have been presented for plants producing strictly urea (Search, 1977):

$$\text{ammonia nitrogen:} \quad 5\text{-}1{,}030 \text{ g/m}^3$$
$$\text{organic nitrogen:} \quad 1\text{-}3{,}640 \text{ g/m}^3$$

For plants producing strictly ammonium nitrate these effluent concentration ranges have been presented (Search, 1977):

$$\text{ammonia nitrogen:} \quad 0.1\text{-}9{,}111 \text{ g/m}^3$$
$$\text{nitrate nitrogen:} \quad .01\text{-}2{,}900 \text{ g/m}^3$$

Evaluation of these parameters should only be done with consideration given to plant capacity, discharge flow rate, receiving water flow rate, and treatment technology.

CONTROL TECHNOLOGY

Control technology as applied to the wastewater effluent from nitrogen fertilizer plants, focuses on the three basic pollutants previously discussed. Each plant may have a variation on the basic processes presented; however, the essential chemistry is the same.

TABLE 1 MINOR CONSTITUENTS PRESENT IN PLANT EFFLUENTS

Parameter	Ammonium nitrate plants					Urea plants		Plants producing both urea and ammonium nitrate						
	A	B	C	D	E	A	B	A	B	C	D	E	F	G
pH	7.2	7.7	6.3	7.6	6.3	7.1	7.5	8.2	7.2	8.4	7.4	8.3	7.6	8.0
Temperature, °C	13					13								
Total suspended solids, g/m^3			29	11	1.5			34	95	32	4	10	7	–
Total dissolved solids, g/m^3				18										
Chemical oxygen demand, g/m^3			67		29		100		72	74		12		
5-Day biochemical oxygen demand, g/m^3		8		<3	–									
Total hardness as CaCO$_3$, g/m^3	8					16								
Calcium hardness as CaCO$_3$, g/m^3	–					0								
Magnesium hardness as CaCO$_3$, g/m^3	10					16								
Total alkalinity as CaCO$_3$, g/m^3	2					–								
Sulfate (SO$_4$), g/m^3	4					0								
Chloride (Cl), g/m^3	4					3								
Sodium (Na), g/m^3	3					5								
Total iron (Fe), g/m^3	0.7					0.03								
Total phosphate (PO$_4$), g/m^3	0.1					0.7								
Manganese (Mn), g/m^3	0.12					0								
Copper (Cu), g/m^3	0					0.12		0.3						
Chromate (Cr), g/m^3	–					–							0.52	–
Silica (SiO$_2$), g/m^3						–								
Oil and grease, g/m^3	–												1.1	
Nickel (Ni), g/m^3	0					0.02								
Zinc (Zn), g/m^3	0.12											0.7		
Cyanide (Cn$^-$), g/m^3	3.6													
Residual Cl$_2$, g/m^3														

NOTE.—Blanks indicate data not available, and dashes indicate reported plant intake concentration greater than reported effluent concentration.

Containment

Containment is a basic technology for the control of low flow non-point sources such as pump seal leaks, spills and localized plant washdowns. It consists simply of a low dike (usually less than 0.15 m. high), surrounding the area. Evaporation controls the level and no discharge occurs. In some applications the collection area may be drained to a central collection basin. In these cases, another treatment technology is used before ultimate discharge.

Steam Stripping

Steam stripping is used primarily to remove ammonia, carbon dioxide and methanol from the process stream. The condensate stripper in its simplest form passes low pressure steam countercurrent to the condensate in a packed or tray tower. The stripped condensate is either discharged into a receiving stream or used as cooling water makeup or as boiler feedwater.

Under a grant awarded through the U.S. Environmental Protection Agency, the Louisiana Chemical Association participated in the study and development of a reflux steam stripper. The pilot unit achieved 98% and 99.8% removal of ammonia and methanol, respectively. Maximum overhead ammonia concentration was 6% (60,000 g/m^3) with a minimum bottoms concentration of 20 g/m^3 (Romero, 1976). A diagram of the pilot steam stripper is shown in Fig. 2.

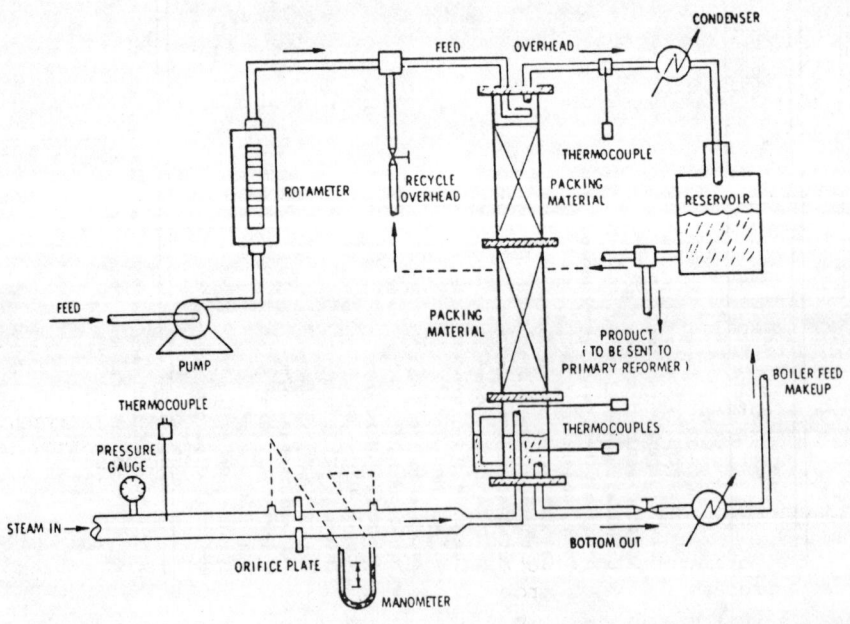

Fig. 2. Model of pilot steam stripper (Romero, 1976).

Air Stripping

An ammonia removal methodology previously used almost exclusively in the municipal wastewater treatment area is air stripping. Nitrogen fertilizer facilities are beginning to utilize this methodology especially in ammonia plants, because of its direct applicability without major modifications.

Testing has been done on air stripping ammonium nitrate process condensate (Train, 1974). Critical parameters are bed depth, transfer medium, surface loading rate, and proper pH. Disadvantages to air stripping are decreased efficiency due to cold weather and the deposition of calcium carbonate scale from the water on the column packing or internals resulting in plugging (Romero, 1976).

Urea Hydrolysis

The basic chemical principle of this process is to convert urea through a series of intermediate products back to ammonia and carbon dioxide which can then be driven off with steam. Hydrolysis can be achieved biologically or thermally. In the biological process, urease enzyme produced by Bacillus Pasteuri is supported on a fixed carbon bed. Oxygen and nutrients needed by the bacteria are added, because they are not supplied in the process condensate feed. This additional air stream must be scrubbed of ammonia. The biological process is also very sensitive to fluctuations in the feed stream and requires much time to achieve optimum column performance. As a result of these complicating factors, no biological hydrolysis systems are known to be currently in use in the U.S.

In the thermal urea hydrolysis process, heated urea process condensate is fed directly to the stripper tower where it hydrolyzes to ammonium carbamate which is then decomposed to gaseous ammonia and carbon dioxide. This process is carried out at temperatures above $100°C$ and under pressures up to 1.8 MPa (Train, 1974).

A modified hydrolysis stripping unit has been developed by CF Industries (Van Moorsel, 1975). This process uses a low pressure boiler, stripping column and deaerator to produce streams which can be recycled to the process (stripper overheads) or used as feedwater for a high pressure boiler.

Biological Treatment

A possible treatment mechanism for the removal of ammonia and ammonium nitrate for process condensate involves nitrification and denitrification. This process has been used for many years in municipal waste. The first step is biological oxidation of the ammonia to nitrate aerobically. The nitrates are then anaerobically denitrified in the presence of carbon to elemental nitrogen.

Biological treatment is highly dependent on 1) an adequate supply of oxygen, 2) warm temperatures, 3) sensitive pH control, and 4) availability of carbon. Because of these factors, biological treatment is not widely used.

Ion Exchange

Figure 3 is a schematic diagram of the Chemical Separations Corporation continuous ion exchange process. In this process, ammonium nitrate bearing fertilizer wastes flow to a strongly acidic cation exchange unit where the ammonium ion combines with the resin while the H^+ ion combines with the nitrate ion to form nitric acid.

The waste water then flows to an anion unit where the nitrate ion combines with the resin and water is formed. The following equations apply (Train, 1974):

$$\text{Cation:} \quad NH_4NO_3 + R_2H \rightarrow R_2NH_4 + HNO_3 \quad (1)$$

$$\text{Anion:} \quad HNO_3 + R_2OH \rightarrow R_2NO_3 + H_2O \quad (2)$$

The resulting water stream can be discharged to a receiving body or used as boiler feed water or cooling tower makeup. A representative feed and treated water effluent analysis is shown in Table 2 (Bingham, 1971).

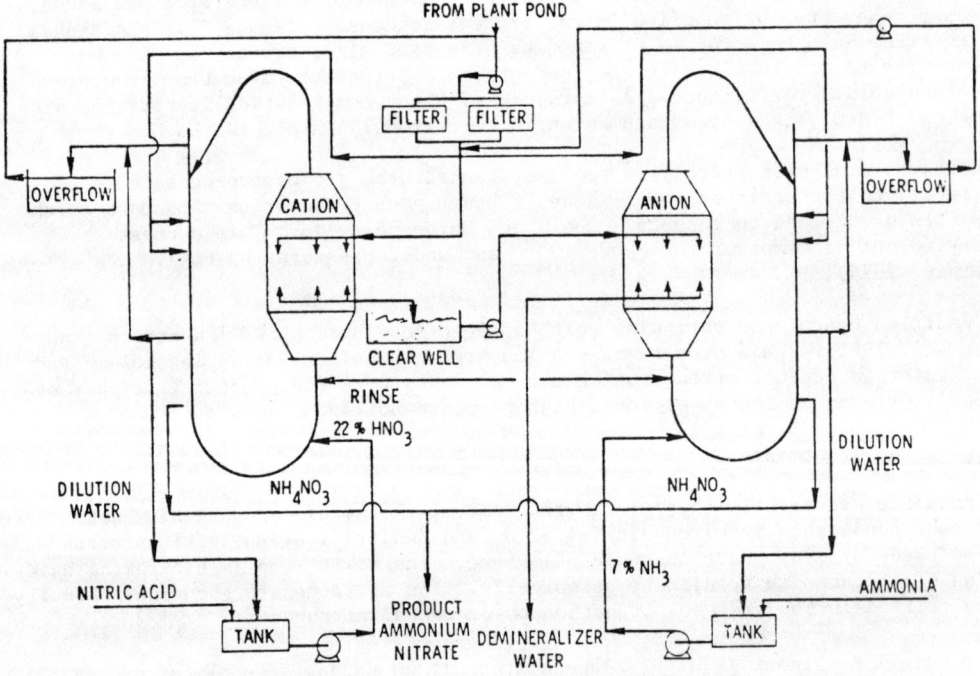

Fig. 3. Continuous ion exchange process (Train, 1974).

TABLE 2 REPRESENTATIVE WASTE AND TREATED WATER ANALYSIS FROM ION EXCHANGE (BINGHAM, 1976)

Component	Influent, g/m^3	Effluent, g/m^3
Ammonia (NH_3)	340	2 to 3
Magnesium (Mg^{++})	4.8	-
Calcium (Ca^{++})	60	-
Sodium (Na^+)	0	-
Nitrate (NO_3^-)	1,240	7 to 11
Chloride (Cl^-)	53	-
Sulfate (SO_4^-)	72	-
pH	5 to 9	5.9 to 6.4
Silica (SiO_2)	15	15
Ammonium nitrate removal is 99.4%		

The CHEM-SEPS® ion exchange system is operated in a continuous manner. Ion exchange resins, pulsed hydraulically through a closed loop, are divided into three sections for simultaneous demineralization of wastewater, washing of resin, and regeneration. To regenerate the cation and anion resin beds, 22% nitric acid (HNO_3) and 7% NH_3 solutions are used, respectively. The resulting streams will contain an ammonium nitrate concentration of 10% to 20%. The application of the CHEM-SEPS® process is limited to those manufacturers who have an outlet for this recovered ammonium nitrate solution. Many manufacturers feel that there are certain materials in the recovered solution that would prohibit evaporating this recovered solution to dryness because of their sensitizing qualities (i.e., they render the solid product more susceptible to explosions). This is of particular concern to those manufacturers who use some or all of their ammonium nitrate in the manufacture of explosives.

Ion exchange appears to be the growing technique by which ammonium nitrate plant effluents are being treated. As of January 1976, 10 major ammonium nitrate plants had decided to utilize the CHEM-SEPS® process (Table 3) (Brennan, 1976). Since that time several other plants have started using the CHEM-SEPS R process or some other modified ion exchange process.

TABLE 3 PLANT LOCATIONS USING ION EXCHANGE (BRENNAN, 1976)

PLANT	LOCATION
Hercules, Inc.	Louisiana, MO
Mississippi Chemical Co.	Yazoo City, MS
American Cynamid Co.	Hannibal, MO
Indiana Army Ammunition Plant (ICI America)	Charleston, IN
Standard Oil Co. of Ohio Vistron Division	Lima, OH
W.R. Grace & Co.	Wilmington, NC
Joliet Army Ammunition	Joliet, IL
Illinois Nitrogen Corp.	Marseilles, IL
CF Industries, Inc.	Chattanooga, TN
CF Industries, Inc.	Tunis, NC

Condensate Reuse

A potential control technology involves the direct recycle of neutralizer condensate as a feed stream. The effluent stream from the condensed neutralizer exhaust at an ammonium nitrate plant, for example, could potentially be used as a feed to the absorption column in the adjacent nitric acid plant. The value of the nitrate content in the wastewater would then be realized. Figure 4 is an example of ammonium nitrate effluent utilization.

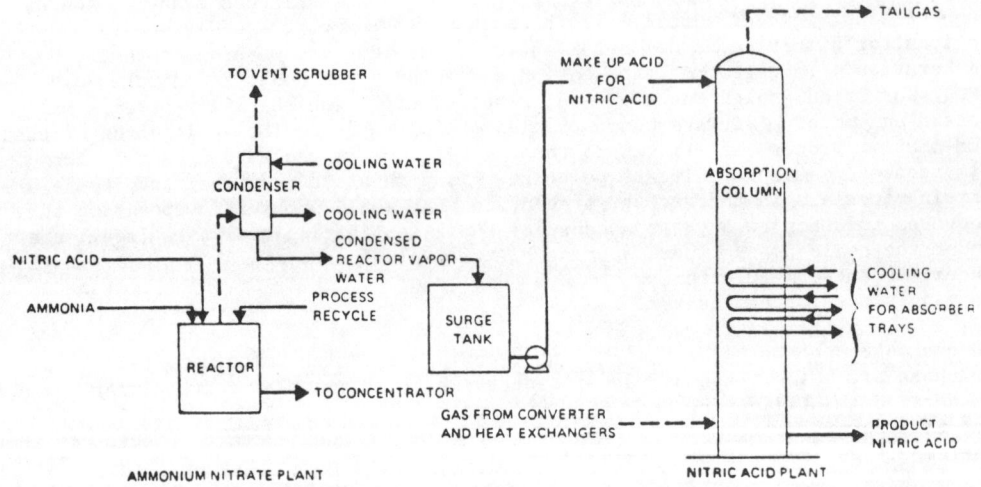

Fig. 4. Ammonium nitrate effluent utilization (Quartulli, 1976).

ACKNOWLEDGEMENT

The study of nitrogen fertilizer plants was a project performed under the contract "Source Assessment (Contract No. 68-02-1874)" with the Industrial Environmental Research Laboratory of the United States Environmental Protection Agency, Dr. Dale A. Denny - EPA Project Officer. Dr. R.A. Venezia served as EPA Task Officer for the study of the nitrogen fertilizer industry. Complete reports from this project can be found in the references.

REFERENCES

Bingham, C.U., and F.H. Yocum (1976). A closed cycle water system for ammonium nitrate products. Brochure, Chemical Separations Corporation, Oak Ridge, Tennessee, 11 pp.

Brennan, J.R. (1976). The CHEM-SEPS^R nitrogen recovery process; a pollution solution that works. In: Proceedings of the Fertilizer Institute Environmental Symposium (New Oreleans, Louisiana, January 13-16, 1976), The Fertilizer Institute, Washington, D.C. pp. 217-249.

Quartulli, O.J. (1976). Review of methods for handling ammonia plant process condensate. In: Proceedings of the Fertilizer Institute Environmental Symposium (New Oreleans, Louisiana, January 13-16, 1976), The Fertilizer Institute, Washington, D.C. pp. 25-44.

Rawlings, G.D., and R.B. Reznik (1977). Sources assessment: synthetic ammonia production. EPA-600/2-77-107m, U.S. Environmental Protection Agency, Research Triangle Park, North Carolina, 83 pp.

Romero, C.U., and F.H. Yocum (1976). Treatment of ammonia plant process condensate. In: Proceedings of the Fertilizer Institute Environmental Symposium (New Oreleans, Louisiana, January 13-16, 1976), The Fertilizer Institute, Washington, D.C. pp. 45-89.

Search, W.J., J.R. Klieve, G.D. Rawlings, and J.M. Nyers (1977). Source assessment: nitrogen fertilizer industry water effluents. EPA-600/2-78-004, U.S. Environmental Protection Agency, Research Triangle Park, North Carolina, 91pp.

Search, W.J., and R.B. Reznik (1977). Source assessment: ammonium nitrate production. EPA-600/2-77-107i, U.S. Environmental Protection Agency, Research Triangle Park, North Carolina, 78 pp.

Search, W.J., and R.B. Reznik (1977). Source assessment: urea manufacture. EPA-600/2-77-107l, U.S. Environmental Protection Agency, Research Triangle Park, North Carolina, 94 pp.

Train, R.E., A. Cywin, E.E. Martin and R. Strelow (1974). Development document for effluent limitations guidelines and new source performance standards for the basic fertilizer chemicals segment of the fertilizer manufacturing point source category. EPA-440/1-74-001-a, U. S. Environmental Protection Agency, Washington, D.C. 168 pp.

Van Moorsel, W.H. (1975). Method for treatment of urea crystallizer condensate. U.S. Patent 3,922,222 (to CF Industries, Inc.).

DEVELOPMENT OF ANION EXCHANGERS FOR SILICA REMOVAL: HISTORY AND PRESENT TRENDS

C. Calmon

*Water Purification Associates, Cambridge, Massachusetts 02142, U.S.A.
Consultant, Princeton, N.J. 08540, U.S.A.*

INTRODUCTION

With the increased demand for electricity after World War II, power utilities began to install boilers operating at high temperatures and pressures to increase the steam output for driving the turbines and increase productivity. It was already shown in the 1930's that as the temperature of the water in the boiler was increased, the solubility of salts in the steam also increases and the salts will deposit out on the blades of the turbines where the temperatures are lower (Splittgerber, 1939). Figure 1 shows the distribution of solutes between water and steam as a function of the ratio of the liquid density to the vapor density (Jonas, 1977). Silica deposits proved the most difficult to remove, forming on the turbine blades and resulting in loss of efficiency and increased outage time for cleaning. Therefore, as the boiler pressures increased, the silica limits in the boiler feedwater were lowered. The complete removal of salts, especially silica, became a major drive of the water treatment companies when boiler pressures began to reach in excess of 1500 psig.

STATUS OF SILICA REMOVAL TO 1946

The removal of strongly dissociated salts such as $CaSO_4$, $CaCl_2$, $Ca(HCO_3)_2$, $MgSO_4$, $MgCl_2$, NaCl, etc. could easily be achieved through the use of ion exchangers already commercially available. These included strongly acidic cation exchangers, R_cH, (e.g. carbonaceous, phenolic and styrene-divinylbenzene sulfonated products) and the weakly basic anion exchangers R_aN containing primary, secondary and/or tertiary amino groups which react readily in the basic form with the strong acids formed on the passage of the raw water through the cation exchanger in the hydrogen form. But the anion exchangers could not remove the weak acids such as silicic (H_2SiO_3) or carbonic (H_2CO_3) acids. The ion exchange reactions are represented by Equations (1) and (2).

$$R_cH + M^+X^- + M^+HSiO_3^- + M^+HCO_3^- \rightarrow R_cM + H^+X^- + H_2SiO_3 + H_2CO_3 \quad (1)$$
$$\hspace{7cm} \downarrow$$
$$\hspace{7cm} H_2O + CO_2$$

$$R_aN + H^+X^- + H_2SiO_3 + H_2CO_3 \rightarrow R_aNHX + H_2SiO_3 + H_2CO_3 \quad (2)$$
$$\hspace{7cm} \downarrow$$
$$\hspace{7cm} H_2O + CO_2$$

A two bed (i.e. R_cH & R_aN) demineralizing or deionizing system could reduce the strong salts to less than 3ppm. Carbonic acid was removed with a degasifier but silica remained in the effluent. Silica removal was accomplished by either precipitation of magnesium hydroxide or by absorption with the activated magnesia (Tiger, 1942). The quantity of magnesium required depended on the silica concentration in the raw water and on the reaction temperature; the higher the temperature, the lower the dosage. The very low silica content in the feedwater required for operating boiler pressures above 800 psig could not be economically attained by absorption of silica by magnesium compounds. The race by the manufacturers of ion exchangers for developing anion exchangers capable of removing silica speeded up.

FIRST ANION EXCHANGE PROCESS FOR SILICA REMOVAL

It was indicated above that the weakly basic anion exchangers could remove strong acids but not silicic acid having pK values of 9.8 and 11.8. These values indicate the acid to be very poorly dissociated. To utilize weak base anion exchanger for silica removal, it was essential to convert the silica to a strongly dissociated electrolyte. Lourier and Kliachko (1945) reacted silicic acid with hydrofluoric acid (HF) to form fluorosilicic acid, H_2SiF_6, which is a strongly dissociated electrolyte (i.e. $H_2SiF_6 \rightleftarrows 2H^+ + SiF_6^=$). $SiF_6^=$ could then be readily removed by the weak base anion exchangers.

As hydrofluoric acid is not a desirable product to be handled by boiler room personnel, the HF was formed by adding NaF to the water prior to passage through the cation exchanger in the H form.

The reactions involved in the process are given by Equations (3), (4) and (5).

$$R_cH + NaF \rightarrow R_cNa + HF \quad (3)$$

$$6HF + H_2SiO_3 \rightarrow H_2SiF_6 + 3H_2O \quad (4)$$

$$2R_aN + H_2SiF_6 \rightarrow \begin{array}{c} R_aNH \\ R_aNH \end{array} \Big\rangle SiF_6 \qquad (5)$$

The dosage of NaF depended on the concentration of silica in the water (Bauman, Eichhorn and Wirth, 1947; Calise and Lane, 1949).

There were objections to the process:

a. HF formation was considered dangerous to untrained help.

b. $SiF_6^=$ emission as a waste was not wanted.

c. The cost was increased through the use of NaF and the increased electrolyte content which had to be removed by ion exchange. The higher the salt concentration, the more regenerant has to be used.

However, anion exchangers soon became available which could remove silicic acid.

FIRST ANION EXCHANGERS WITH QUATERNARY NITROGEN GROUPS

The first anion exchangers to have strong base groups capable of absorbing silica reversibly was prepared by Dudley and Lundberg (1949) and Kressman (1950). These resins were prepared by the condensation of epichlorhydrin and aliphatic polyamines and contained 10 to 15% strong base groups. They are referred to as medium basic anion exchangers because they contain both strong and weak basic groups (Gilwood, 1947). These resins are now manufactured by all the United States ion exchange manufacturers.

This resin made possible almost complete deionization through the following two systems:

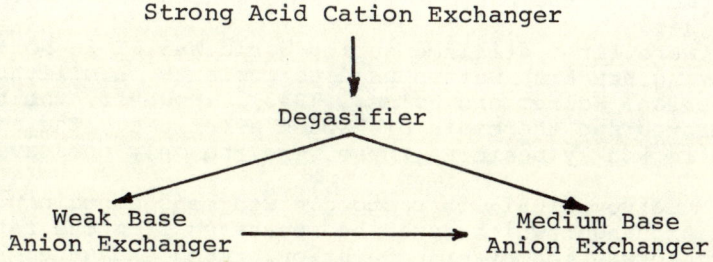

But, as the pressure of the boilers increased, the demand for better quality feedwater also increased. Similarly, at the same time the transistor and microcircuit manufacturers demanded purer water than required for boiler feedwaters operating at critical pressures.

ANION EXCHANGERS WITH ONLY QUATERNARY BASIC GROUPS

In 1944 the D'Alelio patents (1944) for preparing ion exchangers with styrene-divinylbenzene (DVB) matrices were issued. In 1947 Bauman and Eichhorn reported successful results with these cation exchangers in bead form. Within several years an array of both weak base and strong base anion exchangers with styrene-divinyl benzene as the matrix appeared. The strong base resins were of two types (Bauman, 1952; McBurney, 1952): Type I and Type II. Type I gave very low silica leakage, but had a lower regenerant efficiency and lower capacity than Type II resins, which had a higher capacity but a higher leakage. In addition the Type II resins lost some of its strong base capacity on repeated cycles (Gilwood, Calmon and Greer, 1952).

In spite of the availability of these resins, better quality water was required than could be produced by systems combining strong acid exchange units with strong base anion exchange units.

EXTREMELY HIGH PURITY WATERS OBTAINABLE DUE TO MIXED BED DEVELOPMENT

Since the ion exchange process is a reversible process and the beds in commercial operation are never fully regenerated between cycles, it is difficult to obtain complete desalination with a two-stage system, i.e. a cation exchanger followed by an anion exchanger, since the H^+ formed in the cation exchanger replaces some of the cations on the cation exchangers at the bottom of the bed. Since the released ions at the bottom of the bed form salts, the cations pass through the anion exchanger unchanged. The degree of leakage depends on the degree of regeneration of the cation exchanger and the concentration of salts in the raw water.

To improve the effluent quality of waters with relatively high salt concentrations, e.g. > 300 ppm, multistage systems were suggested (Richter, 1939) as early as 1939. In these systems water is passed through a series of units, each containing cation and anion exchangers. But with each added unit, the capital cost increased. However, the development of the use of mixed beds solved the problem.

Mixed beds were first utilized during World War II in North Africa for deionizing brackish waters used as coolants in military vehicles (Pemberton, Walker and Holmes, 1943). However, the beds were not regenerated and therefore discarded after use. The anion exchangers were weakly basic as these were the only ones available.

Kunin and McGarvey (1951) introduced mixed beds containing a strong base anion exchanger which could be separated from the cation exchanger by backwashing after exhaustion. Each exchanger was regenerated according to the effluent quality requirements. This innovation made it possible to obtain very high quality water at a reasonable price.

Table I gives the quality water, obtainable by two-bed, mixed-bed and two-bed plus mixed bed systems.

TABLE I

Residual Component	Two-Bed	Mixed-Bed	Two-Bed Plus Mixed-Bed
Electrolytes, in ppm	2.5 to 3.5	0.2 to 0.5	0.3 to 0.10
Silica, in ppm as SiO_2	0.95 to 0.1	0.01 to 0.1	0.05 to 0.02
Max Specific Resistance in megohms-cm	100,000	2,000,000	18,000,000

Constant improvement in techniques of operating mixed beds have improved the quality as well as length of runs between regenerations.

FOULING OF STRONGLY BASIC ANION EXCHANGERS

But the millennium did not arrive with the development of the strongly basic anion exchanger since it was soon discovered that these exchangers were subject to fouling by large organic anions usually found in surface waters (McGarvey, 1954; Bogers, 1954).

Many techniques were applied to avoid fouling. Among these were:

(1) Superchlorination and coagulation followed by filtration first through anthrafilt then through an activated carbon column before the water contacted the ion exchangers (Calmon, 1966; Applebaum, 1968).

(2) Use of lower crosslinked resins so that the greater porosity of the exchangers when wet permitted better diffusion for the large anions during regeneration.

(3) Use of strong base anion exchangers in the salt form which acted as scavengers for the organic anions before the process stream contacts the strong base anion exchangers in the OH form (Applebaum, 1964).

(4) Treatment of the strong base anion exchanger with a solution of salt and caustic on a regular schedule for the removal of accumulated organics (Applebaum and Crits, 1966).

(5) Use of a weak base anion exchanger after the cation exchanger but prior to a strong base anion exchanger and/or a mixed bed. In many waters the weak base anion exchanger proved not to be stable.

(6) Fouling in a mixed bed due to sloughing from the cation exchanger was eliminated through the use of a cation exchanger having a higher crosslinking than the standard 8.5% crosslinked resin (Calmon and Simon, 1970).

(7) Once fouling was observed, various chemical treatments were recommended: oxidation of the organic foulants with sodium hypochlorite; or treatment of the exchanger with salt and caustic (Kunin,

1972).

DEVELOPMENT OF MACROPOROUS EXCHANGERS

It was indicated above that low crosslinked gel type resins when wet are more porous than the standard gel resins. The gel resins are sulfonated or aminated copolymers of styrene and di-vinylbenzene. However, these low crosslinked resins, while they release the large organic anions more readily and permit more rapid diffusion of these anions on regeneration, are generally structurally weaker and are more susceptible to oxidation. Also, because of non-uniform polymerization, the low crosslinked resins can still contain some small pores which enables some fouling to occur.

During polymerization, Abrams (1956) introduced an insoluble uncrosslinked polymer, which upon introducing a polar group on the final polymer by either sulfonation or amination, became soluble, thus leaving pores in the structure. However, Abrams kept the standard crosslinking leaving the resin with the same stability as the standard gel resin.

Meitzner (1960) developed resins, coined macroreticular, which had pores even when dry and which also had a very high degree of crosslinking, thus achieving the goal of minimizing both fouling and susceptability to oxidation. This was accomplished by using a solvent present during the polymerization mixture which was soluble in the monomers but in which the polymer, once formed, was insoluble. The precipitated polymer formed the walls of pores filled with solvent which was later extracted (Kunin, 1962).

Since both low crosslinked resins and the macroreticular resins are macroporous when wet, the writer prefers to refer to the macroreticular resins as "fixed pore" resins, since pores exist in the gel type resins only when wet.

Both weakly and strongly basic resins, as well as strongly acidic resins of this type are available. Fouling, as well as susceptibility to oxidation because of their high crosslinking are greatly reduced in these resins. Amberlite IRA-93, a weak base resin, followed the cation exchanger and preceded the strong base anion exchanger or mixed bed.

Resins with uniform pores were prepared in England. These resins are referred to as isoporous resins (Millar, Smith and Kressman, 1965). They were prepared by preventing chain entanglements during polymerization. However, these resins do not have a high degree of the crosslinking of the fixed pore resins, so aside from the uniform pores, their stability properties are the same as the standard gel resins (Arden, 1968).

LAYERED OR STRATIFIED BEDS

With the development of fixed pore weak base anion exchangers, which proved to be less susceptible to organic fouling and fairly stable to oxidation, while at the same time having the basic property common to weakly basic anion exchangers of high regenerant utilization, the macroporous (or fixed pore) weakly basic anion exchangers became a major component in many ion exchange desalination systems. A number of these systems are given below:

Let CEXH = cation exchanger H

MWBAX = macroporous weak base anion exchanger

SBAXOH = strong base anion exchanger OH

MB = mixed bed

#1	#2	#3	#4	#5
CEXH	CEXH	CEXH	CEXH	CEXH
MWBAX	MWBAX	MWBAX	MWBAX	--
SBAXOH	SBAXOH	--	--	SBAXOH
--	MB	MB	--	MB

In both #1 and #2 systems, two separate units are used for the anion exchangers increasing capital costs. In place of two units layered or stratified beds, i.e., the use of both strong and weak base anion exchanger in the same unit could be used. Usually the weak base resin has a lower density than the strong base resin so the bed could be backwashed for separation prior to use or regeneration. Thus, with layered beds high capacity, silica removal, better regenerant efficiency, less fouling (water is in contact first with the macroporous weak base resin) could be obtained (Downing, 1967).

However, various difficulties occurred (Downing, 1967). Mixing of the resins takes place and silica precipitation in the weak base portion was observed. As the anion concentration ratio in the water varied, changes in the ratio of strong base anion exchangers to weak base exchangers have to be made. Since waters frequently vary in composition, the predetermined ratio was not correct for changes that took place. As a result, other anion exchangers were developed which overcame these problems.

ACRYLIC TYPE ANION EXCHANGER

Due to existing patents on chloromethylating of copolymers of styrene-DVB which on amination became anion exchangers, Ionac manufactured resins with ethyleneglycoldimethacrylate as the crosslinking agent in place of DVB (Kressman, 1953; Greer, 1957). These strong base resins were less susceptible to fouling.

However, it was Gregor of Columbia University who insisted for some years that the aromatic groups in the matrix of ion exchangers have an affinity for the organic constituents found in waters, making it difficult to remove the organics. He suggested the use of aliphatic matrices, which are hydrophilic, so the kinetics of elution of the organics would be better. Today strong base resins with acrylic (aliphatic) matrices are being evaluated both in Europe (Mansfield, 1976) and in the United States. While these resins may be less susceptible to fouling, the questions still remains: "Are they as stable as the aromatic based anion exchangers?"

EFFLUENT QUALITY IMPROVEMENT BY VARIOUS DESALINATION TECHNIQUES

(a) Mixed beds are used as polishers to desalination systems containing two stage ion exchange units. These units contain a cation exchanger followed by either a strong base anion exchanger or a weak base anion exchanger. However, it has been shown that the residual ions in many effluents may be due to the release of occluded NaOH ion the anion exchanger which has not been completely rinsed out and cationic amines from the anion exchanger itself. The introduction of a cation exchanger in the hydrogen form after the two units gives excellent quality waters (Jackson and Smith, 1977). In fact there are such systems which give such excellent quality that there is no need for mixed beds for polishing.

(b) Strong base anion exchangers Type II were found to give the same quality of effluent silica as Type I if operated countercurrently.

(c) In condensate demineralization longer runs between regeneration are made possible through either replacing the sodium in traces of cation exchanger particles found in the anion exchanger due to difficulties in completely separating the mixed bed, with either ammonium (Applebaum and Crits, 1947) or calcium (Michel, 1970; Sisson, 1970) or by special separation techniques of the mixed bed prior to regeneration of the individual components (Salem and Gall, 1972).

(d) A new development is the use of an inert resin with a density between the cation exchanger and the anion exchanger so there will be no mixing of the resins at the interface. Thus, better separation could be obtained.

(e) Particulate metallic entities, especially those of iron, found in condensates tend to settle in ion exchange beds and constitute another foulant. These can be removed magnetically before reaching the bed (Babcock and Wilcox). If iron has settled on the ion exchangers, it can be removed ultrasonically (Holloway, 1971).

MACROPOROUS ANION EXCHANGERS CONTAINING ON THE SAME POLYMERIC MATRIX BOTH STRONG AND WEAK BASE ANION EXCHANGE GROUPS

A recent new development (Akzo) in anion exchangers is a macroporous resin with both strong and weak base anion exchange groups on the polymeric matrix, thus, eliminating the need for two units or layered beds. The fixed pores of the resins reduce the fouling tendency of the resin, the strong groups remove the silica and the weak groups give a high regenerant efficiency. This resin can replace either the weak base or the strong base anion exchanger or both prior to the mixed bed or can be used by itself if ultrapure water is not needed. Because the resin removes silica, the load on the mixed bed in many systems is reduced so the mixed bed can be operated for longer periods between regenerations, resulting in savings in regenerant and rinse and reducing the quantity of wastes.

The total capacity of the exchanger runs from 28 to 34 Kgr. as $CaCO_3$ per cu ft (1300 to 1600 meq/liter) of which about 20 to 25% are strong base groups.

MISCELLANEOUS TYPES OF ANION EXCHANGERS

Colloidal Silica Removal Resin

In some waters silica exists in aggregated or colloidal form. This silica is not removed by usual strong base anion exchangers due to the size of the aggregates. Amberlite IRA-938 has extremely large pores so colloidal silica is reversibly absorbed (Kun and Kunin, 1966).

Absorbing Resins

Akzo Chemie of Holland, one of the earliest companies to apply ion exchangers to treating sugar solutions, developed low crosslinked, very porous anion exchangers designed for decolorizing. These exchangers are designated by Asmit. Similarly Diamond-Shamrock, originally working with the Dole Company for recovering sugars from pineapple wastes, developed several decolorizing resins which are still being used by quite a few plants marketing commercial syrups (Abrams, 1971).

Thermally Regenerable Anion Exchangers

CSIRO, the Australian scientific research organization, has developed ion exchange resins which utilize water (i.e. H^+ and OH^- of H_2O dissociation) at elevated temperatures ($\approx 90°C$) as the regenerant for converting cation exchangers to the H^+ form and anion exchangers to the base form. At normal temperatures the resins have an affinity for being in the salt form. Desalting plants utilizing these resins are under construction (Bolto, 1970).

Metallo-organic Anion Exchangers

A great deal of work is going on in the preparation of metallo-organics as catalysts. A metallo-organic polymer anion exchanger with better stability at higher temperatures has been reported in the literature (Ito and Kenjo, 1968).

Inorganic Anion Exchangers

Natural calcium hydroxy apatites have been utilized in fluoride ion removal. Aluminum oxide and freshly precipitated iron oxide have been used for the removal of specific anions with limited success. Similarly the synthetic hydrous oxides of zirconium, tin and titanium show anion exchange properties (Merz, 1959).

Open Structured Anion Exchangers

In biochemical work the scientist frequently has to deal with large anions, e.g. proteins, nucleic acids, hormones, etc. It is, therefore, necessary for him to have ion exchangers with open structures so the particular ions can be readily admitted to the active sites and then readily diffused out on elution. Wood celluloses, crosslinked and then reacted with various amines to form weakly, medium and strongly basic anion exchangers, are available for biochemical work (Peterson and Sober, 1956).

Another type of open structure ion exchange resin is made from crosslinked polysaccharide dextran which have been aminated. They are sold under the trade name of Sephadex. They are applied to sensitive and high molecular weight biogenics (Porath, 1957).

Ion exchangers made with wood chips as the matrix have been applied to organic waste treatment (Rowe, 1979).

Specific Anion Exchangers

The field of utilizing specific groups on matrices for reacting with specific ions is being investigated in many laboratories. Some products have been made by analytical chemists (Calmon and Gold, 1976).

AUXILIARY PROCESSES FOR SILICA REDUCTION

Reverse Osmosis

In the reverse osmosis process most of the ions remain in the reject stream. The traces (1 to 2%) of the ions in the permeate can be removed by ion exchange. Since the reverse osmosis membranes are subject to fouling, pretreatment is essential. Reverse osmosis is being applied to desalination of sea and brackish waters and to normal waters where high purity is desired, as in the production of microcircuits and makeup water for boiler feed in electric power plants (Southworth and Applegate, 1956).

Ultrafiltration

It has been noted that ultrafiltration used prior to ion exchange protects the strong base anion exchanger from fouling and also produces better quality waters. However, colloidal silica remains in the reject stream (Connelly, 1976).

Immobilized Enzymes and Bacteria

Although these are still in the early stages of development, they may have potential applications in the pretreatment of water, e.g., the removal of silica (some bacteria utilize silica in their metabolism) and the destruction of organics in surface and wastewaters to prevent fouling (Burnett, 1979).

Conversion of Micro Ions to Macro Ions

Ion exchangers are used for obtaining ultra-pure water, for household use, as a scavenger after reverse osmosis, and for desalination of waters less than about 400 ppm in concentration. However, the wastes produced by ion exchange processes, when the regenerant effluents cannot be utilized, may become an obstacle to further utilization of ion exchangers. As a result, further work will be directed towards utilizing more membrane processes and specific ion exchangers.

One method the author feels is promising is to form a complex by reacting ions with some large molecule and then removing the complex by ultrafiltration. This process could also be made for removing specific ions without removing inocuous ions, e.g., the removal of low concentrations of iron and manganese from drinking waters where calcium and magnesium are within permissible limits.

CONCLUSION

The writer began working with ion exchangers, their synthesis, characterization and application a few years after Adams and Holmes, in 1935, introduced the concept of attaching polar groups on polymeric matrices, thus, converting ordinary resins into ion exchangers (Adams and Holmes, 1935). The first anion exchanger he worked with was made from condensed emeraldine dyes (Austerweil and Fiedler, 1937) which had the tendency to leak chlorides and color. He still has the vivid memory of pink cellulose acetate produced at two large plants after the white cellulose was washed with effluents from the newly installed deionization plants containing emeraldine as the anion exchanger.

In the forty years of work in the field, he has seen all the developments described in this paper. He still believes that the ultimate has not been reached with polymers containing active sites whether they be in the form of beads or membranes. In addition, other energy forms than chemical, e.g., magnetics, thermal will be utilized in conjunction with ion exchange, ion recovery or ion separation. In other words, the field is still dynamic and evolving (Weiss, 1970).

REFERENCES

Abrams, I. M. (1956). Ind. Eng. Chem., 48, 1469.
Abrams, I. M. (1971). Sugar. Y. Azucar, 66 (#5), 31-34.
Adams, B. A. and Holmes, E. L. (1935). J. Soc. Chem. Ind., 54, 1T.
Commercial bulletins of manufacturers. (e.g. Akzo Chemie Nederland).
Applebaum, S. B. (1968). "Demineralization by Ion Exchange", Chapter 4, Academic Press, N.Y.
Applebaum, S. B. and Crits, G. J. (1947). U. S. Patent 3,336,747.
Applebaum, S. B. and Crits, G. J. (Sept. 1964). Ind. Water Eng.
Applebaum, S. B. and Crits, G. J. (1966). Proc. Intern. Water Conf. Eng. Soc. West Pa.
Arden, T. V. (1968). "Water Purification by Ion Exchange", pp 99-108, Penum Press, N. Y.
Austerweil, G. V. and Fiedler, A. (1937). Compt. rend., 205, 1234.
Babcock & Wilcox Bulletin, "Electromagnetic Filters".
Bauman, W. C. and Eichhorn, J. (1947). J. Am. Chem. Soc. 69, 2830.
Bauman, W. C., Eichhorn, J. and Wirth, L. C. (1947). Ind. Eng. Chem. 39.
Bauman, W. C. and McKellar, R. (1952). U. S. Patent 2,614,099.

Bogers, P. (1954). Water (Holland) 38, 299.

Bolto, B. A., et al (1970). "Thermal Regeneration of Weak-electrolyte Resins" in Ion Exchange in the Process Industries, p. 270, Soc. Chem. Ind., London.

Burnett, F. (1979). "Immobilized Enzymes", Chapter 16, Vol. II in "Ion Exchange Processes for Pollution Control", Calmon, C. and Gold, H., editors, CRC Press.

Calise, V. J. and Lane, M. (1949). Inc. Eng. Chem. 41, 2554.

Calmon, C. and Kingsbury, A. W. (1966). "Preparation of Ultrapure Water" in "Principles of Desalination", K. S. Spiegler, ed. Academic Press, N. Y.

Calmon, C. and Gold, H. (1976). Env. Science & Technology, 10, 980.

Calmon, C. and Simon, G. P. (1970). "Behavior of Ion Exchangers in Ultra-Pure Water Systems" in "Ion Exchange in the Process Industries", Soc. Chem. Ind., London.

Connelly, E. J. (1976). Proc. American Power Conf., 38, 950.

D'Alelio, G. F. (1944). U. S. Patents 2, 340,110-11 and 2,366,007-08.

Downing, G. D. et al (1967). Intn. Water Conf., Pittsburgh, Pa., Vol. 28.

Dudley, J. R. and Lundberg, L. A. (1949). U. S. Patents 2,469,683-4.

Gilwood, M. E. (1947). Proc. Ann. Water Conference, Eng. Soc. West. Pa., 7, 94.

Gilwood, M. E., Calmon, W. and Greer, A. H. (1952). J. Am. Water Works Assoc., 44, 1057.

Greer, A. H. (1957). U. S. Patent 2,788,331.

Gregor, H. P., Lectures given on ion exchange at various occasions attended by writer.

Holloway, J. and Hollsfield, J. P. (1971). Proc. American Power Conf., Vol. 33.

Ito, T. and Kenjo, T. (1968). Bull. Chem. Soc. Japan, 41, 614.

Jackson, E. W. and Smith, J. H. (1977). Proc. Intn. Water Conf., Pittsburgh, Pa., Vol. 38.

Jonas, O. (March, 1977). Combustion, p 2.

Kressman, T. R. E. (1950). British Patent 643,543.

Kressman, T. R. E. and Aueroyd, E. I. (1953). British Patent 649,778.

Kun, K. A. and Kunin, R. (1966). Proc. Inten. Water Conf., Vol. 27.

Kunin, R. (1972). "Ion Exchange Resins", pp 368-374. Robert E. Krieger Pub. Co. Huntington, N. Y.

Kunin, R. and McGarvey, F. X. (1959). Ind. Eng. Chem., 43, 734 and U. S. Patent 2,578,837.

Kunin, R., Meitzner, E. and Bortman, N. (1962). J. Am. Chem. Soc. 84, 305.

Lourier, Y. Y. and Kliachko, V. A. (1945). Compt. Rend. Acad. Sc., USSR 49, 40.

Lundberg, L. A. and Dudley, J. R. (1949). U. S. Patent 2,469,692.

Mansfield, G. H. (1976). Chapter 25 in "Theory and Practice of Ion Exchange", Soc. Chem. Ind., London.

McBurney, C. M. (1952). U. S. Patent 2,591,573.

McGarvey, F. X. and Reents, A. C. (1954). Chem. Eng. 61(9), 205.

Meitzner, E. F. and Olive, J. A. (1960). Fr. Patent 1,237,343.

Merz, E. Z. (1959). Electrochem, 63, 288.

Michel, R., Kingsbury, A. W. and Calmon, C. (1970). Proc. American Power Conf., Vol. 32.

Millar, J. R., Smith, D. G. and Kressman, T. R. E., J. Chem. Soc., p 304.

Pemberton, R. T., Walker, A. J. R. and Holmes, E. L. (1943). British Patent 553,233.

Peterson, E. A. and Sober, H. A. (1956). J. Am. Chem. Soc., 78, 751 & 756.
Porath, J. (1957). Arkiv. Kemi, 11, 259.
Richter, A. (1939). Angew. Chem., 52, 679.
Rowe, M. C. (1979). "Cellulose Micro Ion Exchange Resins", Chapter 23, Vol. II in "Ion Exchange Processes for Pollution Control", CRC Press.
Salem, E. and Gall, G. P. (1972). Proc. Intn. Water Conf., Pittsburgh, Pa., Vol. 33.
Sisson, A. B., et al. ibid.
Southworth, F. C. and Applegate, L. E. (1956). Proc. American Power Conf. 38, 957.
Splittgerber, A. (1939). Mitt. Ver. Grosskesselbesitzer, 73, 206.
Tiger, H. L. (1942). Trans ASME 54, 49.
Weiss, D. E. (1970). "Some Ion Exchange Processes of Energy Transduction" in "Ion Exchange in the Process Industries", p 294, Soc. Chem. Ind. London.

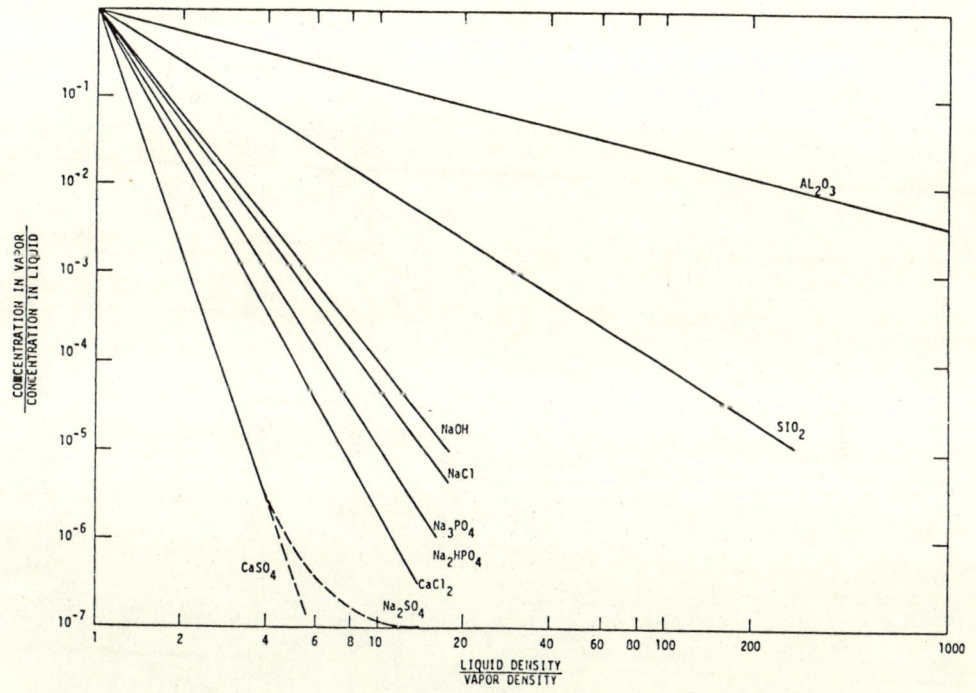

Figure 1 K_D, distribution of solutes between water and steam.

DEMINERALIZATION OF BRACKISH WATERS BY AN ION-EXCHANGE/ ELECTRODIALYSIS PROCESS

G. Boari and G. Tiravanti

Istituto di Ricerca sulle Acque, Consiglio Nazionale delle Ricerche, Via De Blasio, Bari, Italy

ABSTRACT

The growing shortage of low salinity water sources requires brackish water demineralization which is a very important process for many industries. Ion-exchange is generally used in most cases, but it needs expensive chemicals for regeneration and gives rise to disposal problems of high salinity wastes.

This work describes a demineralization process where the anion exchange section has been substituted by an Electrodialysis unit. Acid water eluating from the cation exchangers is fed in an ED stack and a product concentration lower than 10^{-3} N is reached. Complete demineralization is then achieved by a polishing mixed-bed ion exchanger.

The main advantages of this coupled process are:

1- elimination of large amounts of NaOH for the regeneration of the anion exchangers. NaOH cost represents at least 50% of the total demineralization cost by the Ion-Exchange process;
2- recovery of consistent amounts of HCl from the ED concentrate streams, which is reused for the regeneration of the cation exchangers.

Experimental results showed that, for a 10^{-2} N feed water, the electric power consumption of a commercial ED stack, with 0.6 - 0.7 coulomb efficiency, was of the order of 0.5 kWh/m^3.

An economic evaluation of this process shows higher investment costs but much lower operating costs, with respect to the Ion-Exchange process alone; IE/ED coupling is still favourable if energy requirements, which take into account energy for chemical manufacturing, are compared.

KEYWORDS

Demineralization; ion-exchange and electrodialysis; recovery of acid regenerant; process economic evaluation.

INTRODUCTION

The electrodialysis (ED) process is a non destructive unit process able to separate and concentrate ionic species of a low molecular weight in solution. Its main applications are desalination of brackish and sea waters, purification and neutralization of liquid food, enrichment of radioactive isothopes, and other utilizations in the biological field (Wilson, 1960; Spiegler, 1966; Winnicki and co-workers, 1975; Mc Rae and Leitz, 1976; Iaconelli, 1977). This process was also applied (Katz, 1971) to produce ultra pure water for boiler feeds and rinsing water in electronic industry. However it is not economic and unreliable with respect to the Ion-Exchange (IE) process in this case, due to the high electric resistance resulting from high dilution and to the presence of scaling ions.

The IE generally uses strongly acid cation exchangers in acid cycle to eliminate cations, and weakly and strongly anion exchangers in basic cycle, to remove anions including silica. Finally the last traces of dissolved salts are removed by a mixed-bed ion-exchanger (Helfferich, 1962; Applebaum, 1968).

The ion-exchange regeneration needs acids and bases with quantities proportional to the feed water salinity. During the service, the acid effluent from the cation-exchange resins is neutralized by the OH^- ions released by the anion-exchange resins.

The disadvantage of ion-exchange process is waste waters from regeneration. Their salt content is more than twofold that of the feed water, and they often require neutralization. This may cause some difficulty in reaching the concentration limits fixed by the present laws, if enough process waters are not available for their dilution as, for example, in the case of a thermo-electric power plant.

These reasons suggested the idea of a new process based on the coupling of ion-exchange and ED; the main advantages are regenerating chemicals recycle and reduced waste waters discharge, which actually means economic and energetic savings.

This paper aims to verify the possibility to apply this process for the production of demineralized water from brackish waters. This method is also compared with the traditional ion-exchange process.

DESCRIPTION OF THE PROCESS

It is well known that ED is usually used for desalination of brackish waters with salt contents between 1000 and 10000 mg/dm^3 TDS.
The treated water salinity is between $250 \div 500$ mg/dm^3 TDS (drinking water).
Lower salinity values could be economically obtained with feed solutions constituted of electrolytes with higher electric conductivities than those normally found in waters to be desalted.

Fig. 1 presents HCl and NaCl conductivity as a function of their concentration.

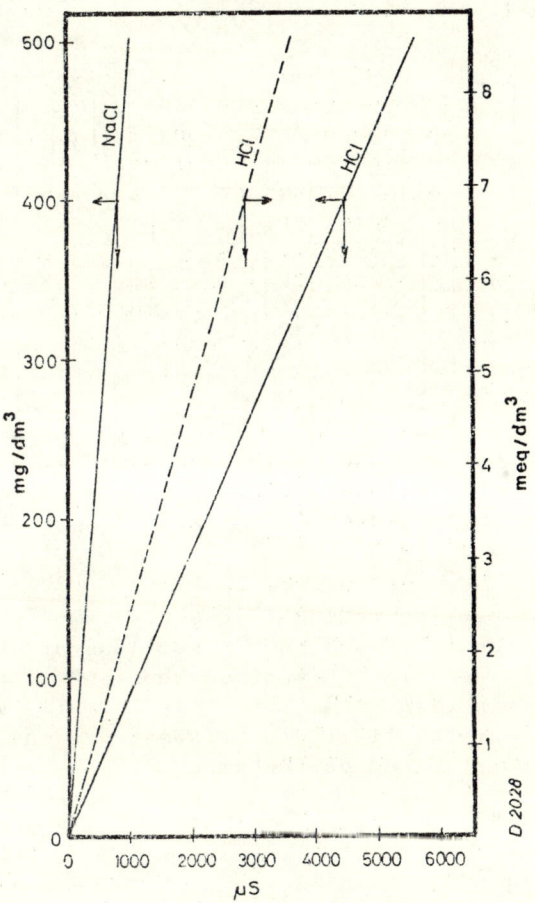

Fig. 1. Concentration of HCl and NACl VS conductivity.

This Fig. 1 shows, as an example, that a 250 mg/dm^3 NaCl solution has the same conductivity of a 45 mg/dm^3 HCl solution. Starting from these considerations, a coupled process IE/ED was developed.
The treatment of a feed water is performed in four different sections, as shown in Fig. 2. In the first section, by means of strong cation resins (SCE), a nearly complete removal of the cations is obtained and the transformation of HCO_3^- ions in CO_2 is achieved. The effluent of this section contains mainly HCl, enters the second section, made up by ED unities, in which most of the HCl is removed from the water solution, reaching product levels of the order of 1 meq/dm^3. It is also possible to get a concentrated solution of the same acid, which can then be reused for the regeneration of the cation-exchange resins. In the third section strong anion exchange resins (SAE) remove silica

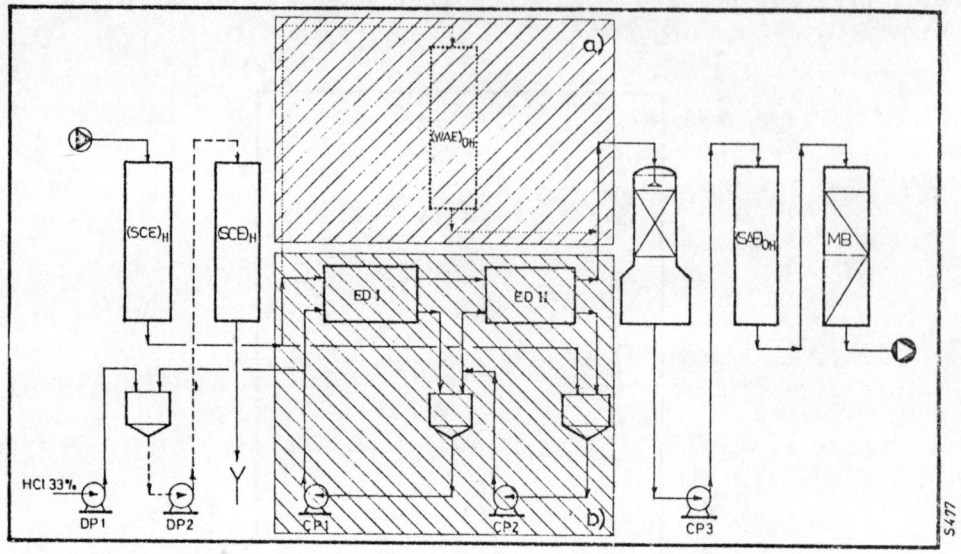

Fig. 2. Overall layout of the IE/ED demineralization process.

while in the last section residual ions are removed in a mixed-bed (MB). The first, the third and fourth sections are similar to the traditional IE process. The ED section then substitutes the weack anion-exchange resin (Fig. 2 a), which is normally used in the IE process when the water salinity to be treated is higher than 200 mg/dm^3 TDS, in order to reduce the amount of bases used for regeneration.

EXPERIMENTAL

Experiments were performed on commercial cation and anion-exchange membranes Ionics (Nepton 61-CZL-183 and 111-EZL-219) and Asahi Chemical (Aciplex K-101 and A-101).
Samples of cation and anion-exchange membranes were previously treated with 2 N solution HCl, then rinsed and stored in distilled water. The membrane potentials (emf) were measured with the Dawson and Meares (1970) method in flow conditions. Within the cell, the extremities of a strip, 5 cm thick, of the studied membrane were in contact with the HCl solutions at different concentrations. In this way, as Kwak (1972) showed, the transport through the membrane was negligible and interfering concentration polarization phenomena at the membrane/solution interfaces, caused by the electrolyte diffusion and by the direct osmosis, were eliminated.
Measurements were carried out within an air thermostat at 25°C $\pm$ 0.1°C in static (no flow) conditions, using saturated calomel reversible electrodes. Membranes were previously conditioned in the diluted solu-

tions. The potentials were measured by a 610 C Keithley electrometer with an overall accuracy of $\pm$ 0.1 mV; the membrane potential equilibrium was reached after about one hour.
The laboratory cell used in the ED experiments has been already described by Passino and co-workers (1976, 1978) and is similar to the one used by Rajan and co-workers (1966).
Ag/AgCl electrolytic reversible electrodes, made according to Jves and Janz (1961) completely insulated except in the working area, were inserted between 5 cell pairs of the stack making possible the measurements of the applied voltage without feed electrode overpotential interferences. The feed flow velocity of the brine and diluted HCl concentrations was 14 cm/s.
The voltage and electric current measurements were carried out using a HP data acquisition system, already described by Passino and co-workers (1976). All the components of the experimental set up were made of non metallic inert material, to avoid corrosion due to HCl solutions and contamination due to other ions.

RESULTS AND DISCUSSION

The suggested process works only with ion-exchange membranes keeping high permselectivity when high HCl solution ratios are reached.
The membrane potential measurements are a best help in determining the suitability of the ion-exchange membranes for this process. Literature data concerning membrane potentials for HCl solutions are generally referred to concentration ratios in the range between 2 and 10 (Arnold and Swift, 1967). It was then necessary to measure membrane potentials on the studied membranes for concentration ratios up to 100, fixing the diluted concentrations respectively at 10^{-2} and 10^{-3} N.
The Figure 3 shows that the experimental values for cation membranes are close to the theoretical ones, calculated by the Nernst equation, using for HCl activity coefficients in solution the data of Robinson and Stokes (1959). This means that the membrane H^+ transport numbers are about one, even for the high concentration ratios used.
On the contrary, the results obtained on the anion-exchange membranes show that their membrane potentials are very different from the theoretical values, when both the brine concentrations and the concentration ratios increase.
Furthermore, the Asahi membrane shows a selectivity higher than the Ionics membrane, due to the higher concentration of the membrane fixed charges (6.42 VS 2.57 meq/g solvent for Asahi and Ionics respectively) and thus to the higher electrolyte Donnan exclusion (Helfferich, 1962). These results show that, in the ED process, in order to obtain good coulomb efficiency, it would be necessary to operate with concentration ratios lower than 80, because over this value, sharp decreases of anion permselectivity could occur, as shown in the Figure 3.

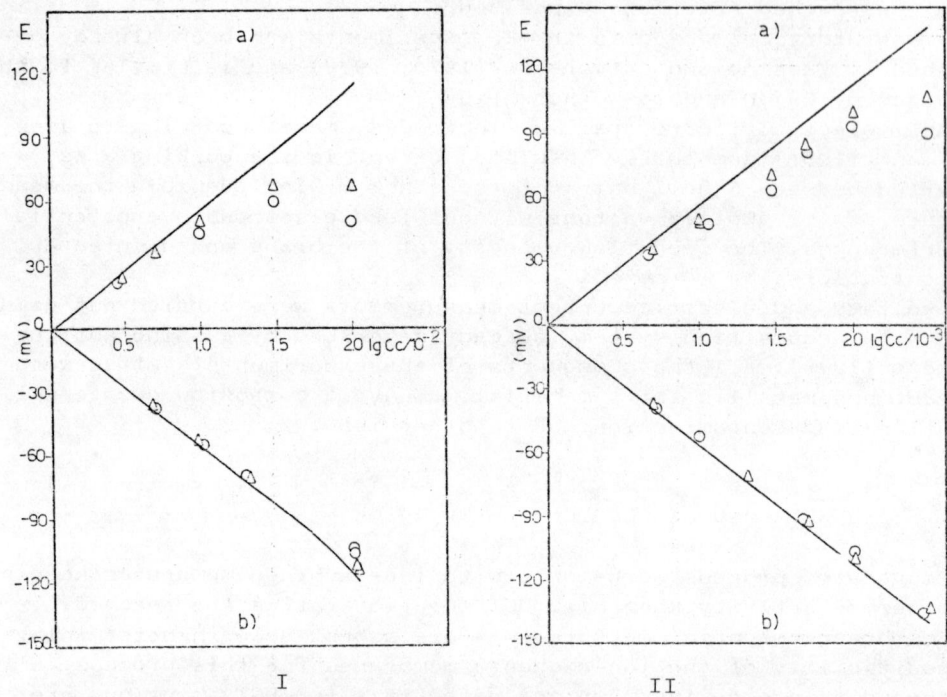

Fig. 3 Electric potential arising across the tested membranes in HCl solutions as a function of $\log C_c/10^{-2}$ (I) and of $\log C_c/10^{-3}$ (II). The continuous lines represent the maximum theoretical Nernst potentials.
a) anion membranes; b) cation membranes.
O Ionics membranes; Δ Asahi membranes.

A mass balance was made, taking into account the process parameters of Table 1.
Fig.4 shows that the above concentration ratio is not reached when two ED stages are used in series, and when the brine of the second stage is used as feed of the brine circuit of the first one. ED blow down concentration is 235 meq HCl/dm^3. The concentration of the regenerant solution, after the fill up with the necessary concentrated HCl, is 2.8%, as the ED blow down is completely utilized. Its amount is sufficient (Duolite data bulletin, 1971) for the regeneration of the strong cation-exchange resins, with a regenerative level of 210%; it follows that, according to water composition reported in Table 1, a saving of 28.6% of the acid regenerant is possible.
The dose of acid for regeneration can be decreased to 120% of the stoichiometric value when weak cation-exchangers are placed before the strong cation-exchanger section, as the former are regenerated by the exhaust acid solutions coming out from the strong cation-exchange

TABLE 1 Operative Parameters of the Coupled IE/ED Process, for the Complete Recycle of HCl Recovered by ED Concentrate Stream

WATER COMPOSITION

Cations	meq/dm^3	Anions	meq/dm^3
Na$^+$ + K$^+$	10.0	Strong anions (Cl$^-$ + SO$_4^{2-}$)	10.0
Total hardness	5.0	Bicarbonates	5.0
Total	15.0	Total	15.0

STRONG CATION EXCHANGE SECTION

Ion exchange capacity	55.0 g CaCO$_3$/dm$_r^3$	Regenerant HCl concentration	2.5-3%
Regenerative level	210%	Filled up HCl concentration	33%

ELECTRODIALYSIS SECTION

Number of stages ED in series	2	Feed HCl concentration 1st stage	10.0 meq/dm^3
Maximum ration C_c/C_d for each stage	80	Feed HCl concentration 2nd stage	3.0 meq/dm^3
		Product HCl concentration 2nd stage	1.0 meq/dm^3

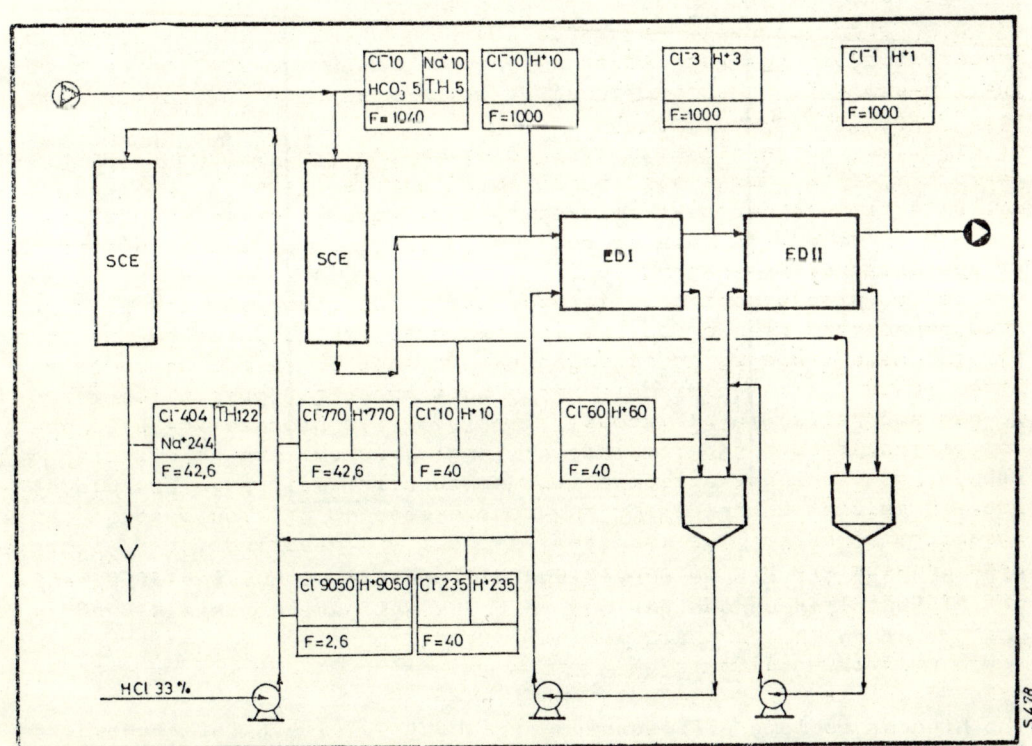

Fig. 4. Ion and mass balance, for the first (SCE) and the second (ED) sections of the process. Conc. = meq/dm^3; Flow rate F= dm^3/h.

resins. In this case, with the suggested process, a saving up to 50% of the regenerant solution is possible. Its concentration (1.85% HCl) is still high enough to regenerate the strong cation-exchange resins (Duolite data bulletin, 1971) since they are exhausted mainly by monovalent cations.

The brine and dilute HCl concentrations calculated from the mass balance of Fig. 4 were used in the experiments performed on an ED laboratory cell, utilizing the more selective Asahi membranes.

Figures 5 and 6 show the current/voltage polarization curves and the coulomb efficiencies, at different HCl concentrations, corresponding respectively to the desalination range of the first and second stage ED.

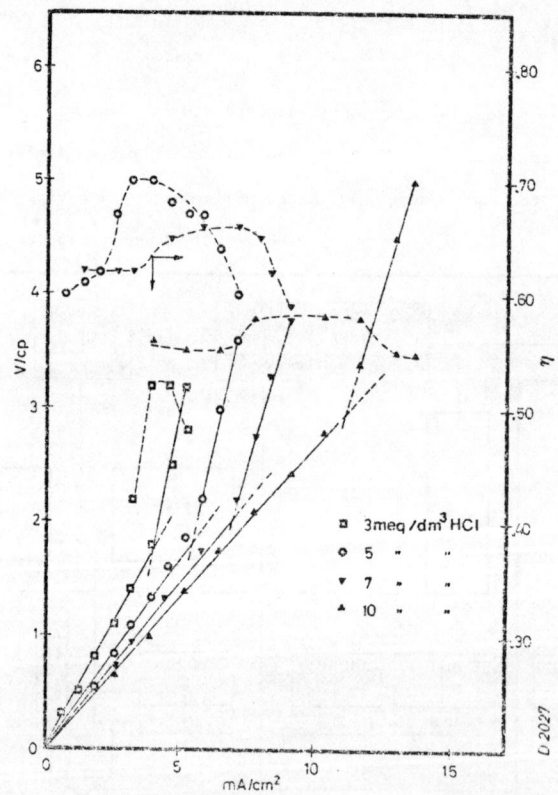

Fig. 5. Current/voltage curves and coulomb efficiency for different HCl feed concentration, in the desalination range of the first stage ED. $C_C = 240$ meq/dm^3.

The highest coulomb efficiencies are about 0.7 - 0.8 for the two stages, in a good agreement with what foreseen from the membrane potential measurements.

The data of the Figures 5 and 6 were used to find the best operating

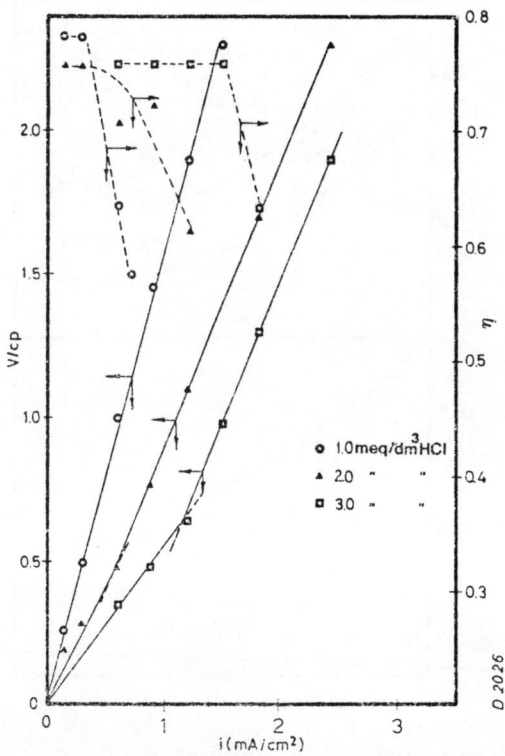

Fig. 6. Current/voltage curves and coulomb efficiency for different HCl feed concentrations, in the desalination range of the second stage ED. $C_c = 60$ meq/dm^3.

conditions in the batch experiments, simulating the first and second ED stages behaviours.

Fig. 7 shows, as an example, the obtained experimental data, for the first (Fig. 7 a) and second stage (Fig. 7 b), in terms of product/feed concentration ratio VS the recirculation time, and the corresponding theoretical curves (dashed lines) calculated assuming ideally selective membranes.

In the same Fig. 7 the trends of the observed current densities, for an applied voltage of 1.2 V/cell pair in both stages are reported. This voltage was selected utilizing the data of the Figures 5 and 6, in order to reach a good compromise between operating current densities and power consumption. In this way it was possible to minimize the decrease of coulomb efficiency, due to phenomena such as concentration polarization, electrolyte back diffusion and water transport, especially on the anion membranes.

Table 2 shows the best operational ED parameters obtained. The coulomb efficiencies were 0.74 and 0.63 for the first and second stage

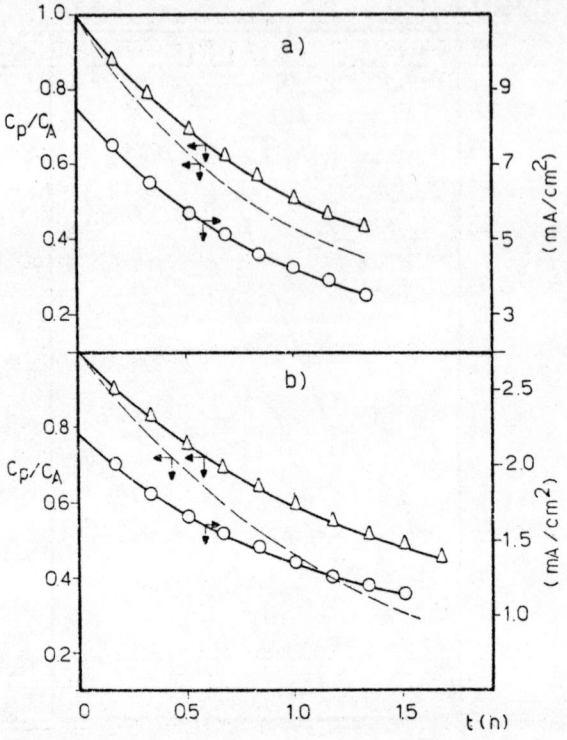

Fig. 7. Product/feed concentration ratio, C_p/C_a, and current densities VS recirculation time for the first stage (a) and second stage (b) ED. The dashed lines represent the theoretical curves.

TABLE 2 Operating Conditions and Power Consumption on ED Batch Experiments with HCl

Stage	I	II
Number of cell pairs	5	5
Membrane area (cm^2)	38.7	38.7
HCl feed concentration (meq/dm^3)	9.94	2.88
HCl brine concentration (meq/dm^3)	240.0	60.0
Solution volume recirculated (dm^3)	7.0	7.0
Constant voltage applied (V/cp)	1.2	1.3
Mean current density (mA/cm^2)	5.5	1.5
Mean coulomb efficiency	0.74	0.63
Electric power consumption (kWh/m^3)	0.37	0.10
Electric power consumption (kWh/eq)	0.053	0.050

respectively, with a total electric power consumption, related to the desalination alone, of 0.052 kWh/equivalent of removed HCl.

ECONOMIC CONSIDERATIONS

The above results show that the treatment of HCl solutions, for demineralization purposes, by ED equipped with commercial ion-exchange membranes, is technically possible.
As far as economical feasibility is concerned, it is well known that the NaOH used for the regeneration of the anion-exchange resins is produced by electrolysis of concentrated NaCl solutions.
The electric energy required to produce one equivalent of NaOH is about 0.12 kWh (Milazzo, 1963). Taking into account that a 20% more than the theoretical is required to regenerate the weak base anion-exchange resins, the specific energy becomes 0.144 kWh/eq of removed strong anions.
Table 3 reports a comparison between the specific energy consumptions and the related costs, for the removal of strong anions by using weak anion-exchangers or ED.

TABLE 3 Comparison Between Specific Energy Consumption and Related Costs, for Strong Anions Removal by Weak Anion-Exchangers or by Electrodialysis

PROCESS	SPECIFIC ENERGY REQUIRED (kWh/eq)	SPECIFIC COST (Italian Lire/eq)
Anion Exchange	0.144	12.0
ED (1st + 2nd stage)	0.052	1.56
Ratio IE/ED	2.8	7.7

It is then clear that the ED process appears to be the best from an energetic point of view; moreover, the comparison is still more favourable to ED, if the corresponding costs, in Italy, of the NaOH and of the electrical power are taken into account. This is probably due to the higher incidence of the manufacturing and transport costs for the soda.
The ED process involves some plant complications and its investment costs are nearly the double of those of the anion-exchangers. The corresponding higher amortization costs are widely compensated by a saving of about 30 - 50% of the HCl required for the regeneration of

the cation-exchange resins. However, amortization costs are of the order of 5 - 10% of the anion-exchangers operating costs; their incidence to the above economical considerations is low.

CONCLUSIONS

The studies of perm-selectivity of commercial membranes in HCl solutions and the laboratory experiments in electric current flow conditions proved the feasibility of the coupled IE/ED process for the demineralization of brackish waters.
A feature of this process is, at the moment, the relatively low coulomb efficiency obtained in the ED, due to the poor exclusion of H^+ ions by the anion-exchange membranes, at high HCl concentration values. The development and use of anion membranes with higher perm-selectivity as shown by Kawahara and co-workers (1977), could improve the process, saving energy and reaching a higher HCl blow-down concentrations. This improvement would be even more necessary when the ratio between bicarbonates and strong anions, in the water to be treated, is lower than considered in this study.
However, the economic and energetic advantages are considerable still now, using the commercially available membranes, when the water salinity to be demineralized is more than 10 meq/dm^3.
Further research, to be carried out on pilot plants, is needed to confirm the feasibility of this process on industrial scale.

ACKNOWLEDGEMENT

The authors would like to express their gratitude to dr. A. Brunetti, Mr. C. Marra and Mr. N. Palmisano, for assistance in the laboratory studies.

REFERENCES

Applebaum, S.B. (1968). Demineralization by Ion-Exchange, Academic Press, New York.
Arnold, R. and D.A. Swift (1967). Australian J. Chem., 20, 2575.
Dawson, D.G. and P.Meares (1970). J. Colloid. Sci., 33, 117.
Duolite data Bulletin D.T.S. 0.006 A (October 1971).
Helfferich, F. (1962). Ion exchange, Mc Graw - Hill Book Co., New York.
Iaconelli, W.B. (1977). Paper presented to the Spring Meeting of the Electrochemical Soc., Philadelphia 8-13 May, paper N. 252.
Jves, D.J.D. and G.J. Janz (Ed.) (1961). Reference Electrodes, Academic Press, New York.
Katz, W.E. (1971). Proc. Am. Power Conf., 33, 830.
Kawahara, T., H. Shibata, T. Ishida (1977). Paper N. 246 presented to the Spring Meeting of the Electrochemical Soc., Philadelphia, May 8-13.

Kwak, J.C.T. (1972). Desalination, 11, 61.
McRae, W.A. and F.B. Leitz (1976). In Norman L. Li (Ed.), Recent Development in Separation Science, CRC Press, Cleveland, p.157-170.
Milazzo, G. (1963). Elettrochimica, Ed. Studium, Roma, p. 479.
Passino, R., A. Rozzi, G. Tiravanti (1976). In R. Passino (Ed.), Study Week on Biol. and Artificial Membranes, Pontificia Academia Scientiarum, Rome, 14-19 April 1975, Elsevier Sc. Publ. Co., Amsterdam.
Passino, R., G. Boari, A. Rozzi, G. Tiravanti (1978). Desalination, 24, 55-82.
Rajan, K.S. (1966). For U.S. Office of Saline Water, Res. and Dev. Progress Report N. 222.
Robinson, R.A. and R.H. Stokes (1959). Electrolyte Solutions, 2nd Edition, Butterworths, London.
Spiegler, K.S. (Ed.) (1966). In Principles of Desalination, Academic Press, New York.
Wilson, J.R. (Ed.) (1960). Demineralization by Electrodialysis, Butterworth Sci. Publ. London.
Winnicki, T., M.A. Gostomczyk, M. Manczak and A. Poranek (1975). Env. Protection Eng., 1, 37.

ULTRA-PURE WATER AT THE SERVICE OF TECHNOLOGY

Martin P. Steiner

Th. Christ Ltd., Water Treatment Division, CH-4147 Aesch, Switzerland

ABSTRACT

"Water treatment" stands for a multitude of separation processes which remove dissolved and undissolved particles with diameters between 10^{-1} and 10^{-10} m. Ion exchange is one of the most efficient techniques among all and leads to pure and ultra-pure water by complete desalination where water of almost theoretical purity namely with a specific electrical resistance around 20 MOhm cm and a germ count below 100 per 100 ml, can be obtained.
Two standard systems for the production of pure and ultra-pure water are presented: One for the pharmaceutical industry which yields water according to most pharmacopoiae of the world, the other for the micro-electronics industry. Both systems are based on the MAXISTIL mixed bed desalination unit, a sophisticated circulation system and the OSMOSTIL reverse osmosis unit. The latter improves the relative capacity of the mixed beds by ten times and holds back organics and micro-organisms leading to extremely long operation time of the sterile filters. Production costs are below 1 % as compared to distillation, especially when all possibilities of recirculation are used, and make-up water quantities are below 10 % of the total required ultra-pure water volume.

KEYWORDS

Ion exchange; reverse osmosis; electronical industry; pharmaceutical industry; ultra-pure water; ultra-pure water circuits; sterile ultra-pure water; process water for pharmaceutics; rinsing water for electronics.

INTRODUCTION

"Water treatment" essentially stands for a multitude of separation processes applied to a specific field and which are of similar importance in the chemical and food industries. As a matter of fact all these processes under consideration are somehow filtration processes from merely mechanical filtration (like in a sand filter) to ion filtration (semipermeable membranes), without a very well definable transition from one to the other, especially where the two classes of filtration are combined in one and the same process (Fig. 1).

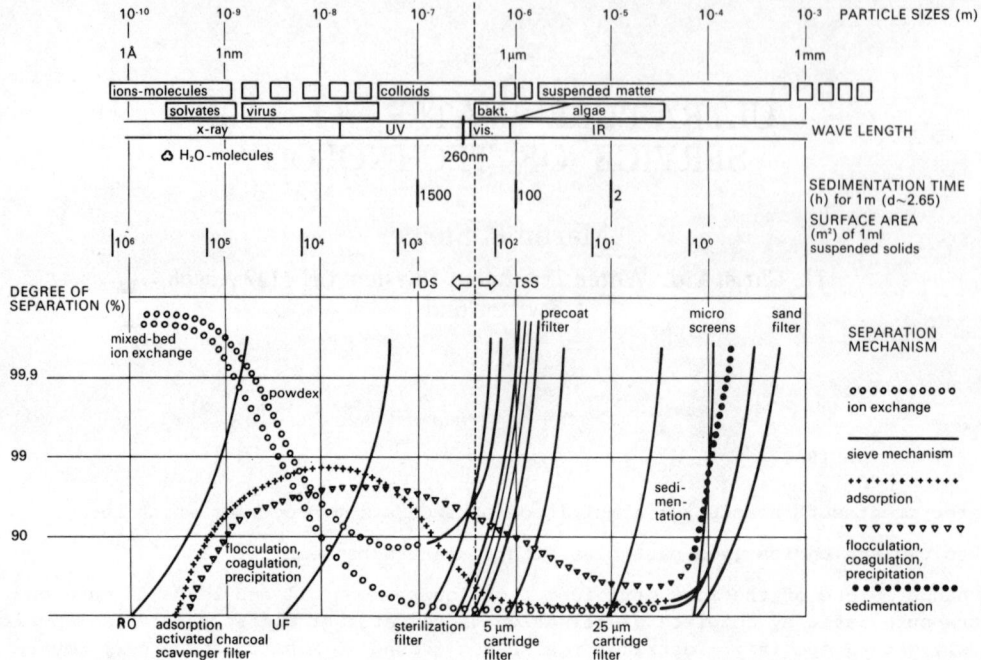

Fig. 1. Water treatment and various particle sizes

The totality of these processes is shown to cover every theoretically possible particle size according to the specific retention capability of the chosen process, for dissolved or undissolved particles as well as for polar and non-polar substances. Therefore <u>every particle size</u> from a few millimeters down to a few "Ångström Units", i.e. from 0.01 down to 0.000'000'000'1 m finds an optimal retention in one of the filtration processes as shown in the figure.

<u>Ion exchange techniques</u> as part of the water treatment technology are based on selective adsorption of water soluble undesirable ions (loading period) in exchange with desirable ones. Thus ion exchange resins alter the water quality in such a way that the ionic composition of the treated stream is shifted into a desired direction. This shift is the result of quite a number of dynamical equilibria determined by:
- hydration energy
- swelling or shrinking of the resin beads
- porosity and cross-linking of the resin material
- osmotic properties of the resin
- electrodynamic potential drops within the system
- velocity of the flow

and other lay-out parameters.
We are merely mentioning a few of those components and do not want to go as deep into the physico-chemical aspects of ion exchange as to enhance such important conceptions as the regeneration level or the "Donnan Potential" a.s.f. But we want to point out that sophisticated water treatment technology- and especially ion exchange-has become an comparative index for industrial development as such.

Among the outstanding phenomena of the ion exchange techniques as compared to all other techniques which form the water treatment technology we should mention the following:
- ion exchange technique provides most selective processes for the absorption of specific ions out of an ion mix.
- ion exchange processes can be reproduced under identical working conditions, and therefore
- ion exchange processes provide to a far extent predictable results under different working conditions (Fig. 2).

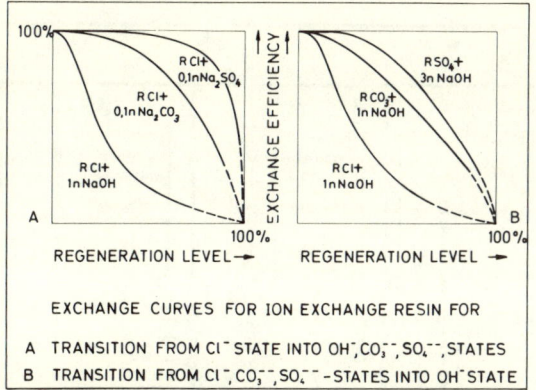

Fig. 2. Predictability of ion exchange processes

The mentioned properties of ion exchange technique represent considerable advantages from the point of view of plant reliability and safety as well as from the economical point of view. There should be mentioned a few disadvantages too, inasmuch as the above outstanding properties of this technique limit the flexibility of an ion exchange plant once built under specific premises: a certain resin volume will suffer what is called "fouling" and other aging processes which alltogether are almost independent of the intensity of exploitation but depend rather on the way the users "treat" their system. (There is e.g. a CHRIST plant near Warsaw where the only change during a 15 years working period has been a single renewal of the exchange resins despite the fact that the two lines composing the plant once built for alternating operation are since long run simultaneously because of increasing quantity requirements, which among other things means twice as many regenerations). But as this plant has been serviced regularly, no disadvantages have been reported until today.

From the point of view of hydraulics as well, the lay-out of an ion exchange plant is not very flexible and may be adapted to new premises (as changing raw water quality or increasing capacity requirements) only within certain limits.

One of the different possible applications of ion exchange techniques is complete desalination where all cations dissolved in the water are to be exchanged against protons from the cation resin and all anions from the water against hydroxyl ions from an anion resin so that at the plant outlet all dissolved solids should be taken out of the treated stream.

Special attention has to be paid to the complete desalination of very dilute solutions since here especially thermodynamics and kinetics become even more controlling for the water quality to be expected after treatment. This is of special importance in nuclear power generation systems, where in addition to dissolved salts undissolved and radioactive particles and ions respectively have to be removed from a treated stream (as e.g. in the condensate polishing system) and where only such advanced processes as e.g. the POWDEX powdered resin pre-coat technique can match the high demand in quality and reliability. They lead to ultra-pure water and to almost quantitative recirculation of the water volume once brought into the system. They are also very economical in exploitation (Fig. 3).

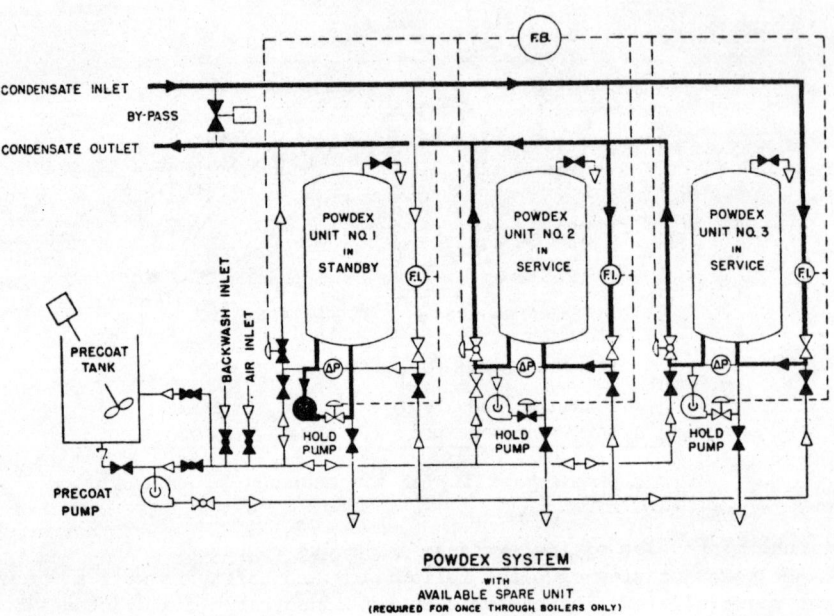

Fig. 3. Ultra-pure water by POWDEX condensate treatment

ULTRA-PURE WATER - TODAY'S STATE OF THE ART

Ultra-pure water today is far from being a privilege of nuclear power stations. Deionized water in general has become a more and more common intermediate in the industrialized countries: people involved with the economics of an enterprise as well as engineers have become aware of the fact that investment in water treatment plants are not productive as such but may lead to rationalization and to considerable improvements in quality of the production to which they are linked - and especially to export products such as food, liquors, electrical energy - and electronic products.

When low power "MOS Integrated Systems" as they are contained in watches, TV sets and other products of today's electronics industry were commercialized, people soon began to realize that one of the most important production steps granting for proper operation and long-term stability of these circuits is careful rinsing of the crystalline surfaces. The aim is now to rinse such components to the highest possible extent. And it has come as no surprise to chemists that the most effective rinsing agent for this purpose turned out to be ultra-pure water. Ultra-pure water is chemically almost inert, a neutral solvent with outstanding properties and especially high dissolving capacity. It's low costs as a raw material and a low production cost increase when treated, together with high working safety and plant reliability made it soon indispensable in the field of electronic industry.

Now what does "pure" and "ultra-pure" really mean ? What is the difference between conventional desalination and such a plant as an MRS ultra-pure water system ?

Industrial Pure Water Quality

Let us go back to the two-column desalination system - a precursor of the ultra-pure water system. Here the stream of treated water - after a suitable mechanical filtration and may-be de-ironing step - flows first through a column filled with cation resin and from there directly through an anion filter. In the cation filter an equilibrium is built up between the water flowing through the system and the cation exchange resin mass which leads to an enrichment of the resin with cations from the water and to a corresponding concentration of protons in the treated stream. Cations from the water are absorbed by the resin, with preference for smaller ions and those with a higher positive charge, so that at the outlet of the cation filter no cations - or most likely a very small concentration of sodium cations will be found in this stream. This is the so-called sodium leakage of the cation filter. In the anion filter the corresponding reaction occurs with the anions dissolved in the treated stream. The anion resin will absorb practically all anions which are dissolved in the treated stream. Small anions and those from the dissociation of strong acids, like chloride (Cl-) or sulphate (SO_4-2), disfavour ions like bicarbonate (HCO_3-) or silicate (SiO_3-2) with lower charge density. If all anions are absorbed, the "purified" water will contain a small amount of sodium hydroxide, the hydroxyl ions not having got a proton as a partner to build a water molecule. If there is a certain leakage of bicarbonate or silicate, the treated water will contain what corresponds to a small concentration of $NaHCO_3$ or Na_2CO_3 - or the analogous sodium salts with silicate. The criterion for purity is conductivity of course and a "raw water" of drinking water quality, with a specific conductivity of may-be 300 to 500 μS/cm will thus lead to a "pure" water of around 10 to 20 μS/cm normally. This seems to be a relatively high specific conductivity for a "completely demineralized water" as compared to the theoretical value of chemically pure water which is at 0,042 μS/cm at 20º C. But the given 10 μS/cm are obtained with such a minute impurity as 2 ppm Na OH corresponding to less than 0.05 val/m3 (Fig. 4).

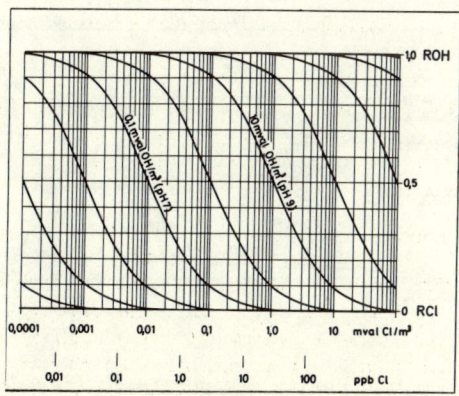

Fig. 4. Chloride concentration, pH and conductivity

If we wish to obtain a lower conductivity value under the same premises we may use a mixed bed system.

The mixed bed system, where cation resin and anion resin are contained in the same vessel and mixed thoroughly before the loading period begins, represents what might be called an infinite multitude of two-column systems formed by the vicinity of cation and anion resin beads:
As soon as the absorption of cations leads to a free proton concentration in the stream this is neutralized by the hydroxyl ions desorbed in exchange with the anions from the stream, so that an equilibrium is established and practically a new raw water composition gets into touch with the rest of the resin. The mixed bed system consequently may be compared to an extraction system in the chemical industry where over a multitude of separation steps ultra-pure solutions can be isolated from mixtures, and therefore the water resistivity after a mixed bed desalination step will normally be up to 400 times higher than after a two column desalination system, i.e. a specific conductivity value below 1 μS/cm corresponding to a resistivity of 1 MΩcm may be obtained. But for ultra-pure water this is not enough. Today the standard of quality of rinsing water in the electronics industry is about 18 MΩcm at 20° C. How can such a high quality be obtained ? And what is the difference between pure water and ultra-pure water ?

Ultra-Pure Water: How to do it ?

(Fig. 5)   Let us start from good drinking water. This water may look pure to us: it is transparent, without bad odour or taste, it contains dissolved solids at limited concentrations and no health hazardous substances at all. If we want to purifiy this water further on, we will have to remove:
- dissolved solids (ions, molecules, gases)
- undissolved particles (colloids)
- organics (microorganisms, germs, degradation products)

Our picture shows quality requirements for a few applications and especially pharmaceutical requirements and those related to micro-electronics. This graph may lead to the simple definition that ultra-pure water contains none of the above named impurities. In practice zero concentrations are unrealistic. But mechanical filtration combined with ion exchange processes and decontamination systems present a good solution to the problem. The lot of these methods can be effectively backed by reverse osmosis.

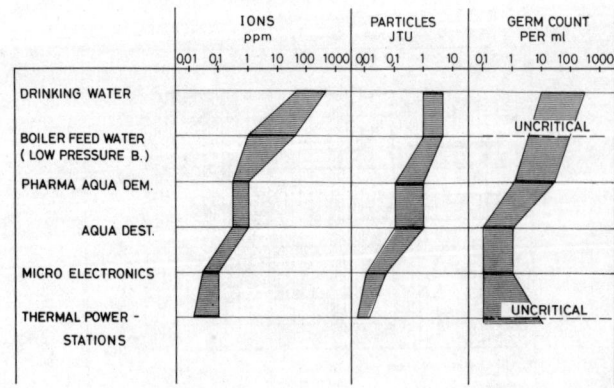

Fig. 5. Varying purity demands for different purposes

MRS - Standard for Pharmaceutics

The MRS system (Fig. 6) has been developed especially for the pharmaceutical industry. The raw water (drinking water quality) passes alternatively one of two mixed bed filters and a sterilization device before flowing to the end-user. The pump in-between allows even at maximum consumption a good portion to circulate back to the mixed bed units so that normally the raw water portion fed to the mixed beds is small as compared to the other portion, in order to keep the output at constant quality. An automatic resin desinfection device may back up the system.

Microbiotic contamination depends heavily on regeneration intervals: after a certain running time the number of germs will increase considerably. Every regeneration represents an important disturbance for the growth of germs - chemically and mechanically. However too short regeneration intervals will favour the appearance of resin particles. Therefore the system has to be balanced by optimum regeneration intervals for the sake of low germ growth and low resin attrition. When after a number of regenerations the number of germs cannot be brought down to a given value, a desinfection step may be indicated.

The recirculation system as well forms an important part of the MRS system: it has to be laid out in such a way that a constant flow may be maintained in every corner of the pipeline and that the end-users get the purified water as quickly as possible and at short distance from the main line. In addition it must be possible to desinfect this part too - chemically or with steam.

The characteristics of the MRS system are:
- high availability
- high operation security
- fully automatic operation and regeneration
- minimum servicing
- extremely small waste water volume
- extremely low microbiotic contamination

The following purified water parameters are typical:
- conductivity below $1 \mu S$/cm at 20° C
- less than 30 germs per ml

This corresponds to most pharmacological prescriptions of the world (Pharmacopoiae) and leads to a quality suitable for almost all pharmaceutical purposes.

The mixed bed units in the system are today's result from a development over 30 years when CHRIST launched the first mixed bed systems ever produced in Europe. They are protected by patents in Switzerland and abroad. Licensing is possible.

The MRS system is composed of standard units and laid out for maximum capacities between 3 and 24 m3/h and daily output quantities between 3 and 100 m3.

The relative capacity between regenerations can be multiplied by 10 if a reverse osmosis system preceeds the MRS system. After the reverse osmosis (RO) not only dissolved solids will be removed to a great extent but organics and suspended matter as well. This will improve the system capacity and also favour storage of the demineralized water.

An MRS analogous system can be built for minimum make-up water quantities down to the range of a few liters per hour. Here the mixed bed desalination units are replaced by desalination cartridges from the MINISTIL cartridge programme. Here too conductivity will be below 1 μS/cm together with low microbiotic contamination of the purified water.

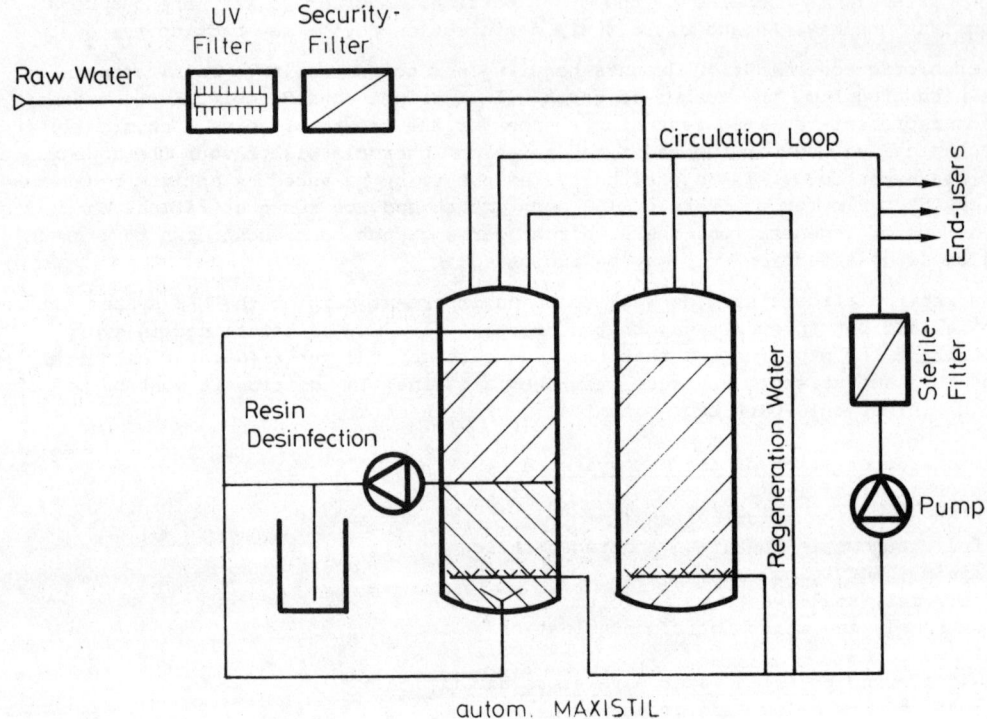

Fig. 6. The "MRS" pure water system

<u>MGR - Standard for Electronics</u>

Let us now proceed to the ultra-pure water system for micro-electronics - the MGR system. In this field of industry water quality must satisfy even higher demands: the MGR ultra-pure water system leads to the following <u>parameters</u>:
- plugging index below 1
- conductivity below 0.55 μS/cm (25° C)
- particle count below 100 per ml
- particel size below 0.5 μm
- germs less than 1 per ml

- KMnO$_4$ consumption below 1 mg per liter

- electrical resistance beyond 18 MΩcm
- output minimum 3 m3/h
- output maximum: no practical limits

The MGR ultra-pure water system represents a combination of almost all water treatment techniques: pre-coat filtration, reverse osmosis, sterilization, ion exchange (Fig. 7). The MGR ultra-pure water system is the full answer to the extreme demands by the micro-electronics industry. Softening as a pre-treatment may be necessary. The FILTAC ® pre-coat filtration may hold back suspended matter down to the micrometer range. It also will take care of oxidants from drinking water chlorination and a few other substances which are hazardous for the RO membranes. The reverse osmosis part consisting of one or several standard blocks will considerably reduce the salt content and the organic contaminants. An intermediate basin may give the possibility of having the RO running 24 h a day even with considerable fluctuation at the consumers side and collect the recirculated rinsing water as well as the purified water from the circulation line.

A de-acidification filter (ion exchange) will hold back impurities from the rinsing processes and three mixed bed units working in a merry-go-round mode will provide the highly desalinated water as the actual MGR part.

UV treatment proceeding the MGR part and sterile filters after the same will hold down the germ count of the produced ultra-pure water. Here too the importance of an appropriate lay-out of the circulation pump and the distribution network springs to the eye. A corresponding line of "Mini-systems" with the same quality parameters is given by a package unit combining an OSMONETTE RO unit with MINISTIL cartridges, leading to maximum make-up quantities of about 20 liters per hour up to a few m3/h. In these very small systems the recirculation of contamined water after use is omitted by obvious reasons.

It is certainly worth mentioning that the sterilization devices in the MGR systems are required because of the contaminants brought into the system by the recirculated rinsing water after use rather than because of the make-up. Microbiotic decontamination may therefore optionally be backed by dosing a desinfectant into this recirculation line. The described MGR system needs a make-up volume of only 10 % of the total water volume required. Sterile filters are reported to have had a running time of over 2 years in such a system in Switzerland, and the availability of the system has been influenced by the electronic manufacturing processes rather than by the system itself. Two systems have been mentioned with which ultra-pure water of a very defined quality can be produced with a maximum availability and at 1 % of the costs for distillation for the same purpose. We might as well consider photochemistry, haemodialysis, high performance cooling systems, thermal power generation and others. Wherever highly developed industry and technology are found, high strung quality demands for water are a natural consequence.

This is a big challenge to us who are specialized in water treatment: a challenge for innovational effort, for sophisticated engineering, for still better systems in order to assist our partners in the productive field of today and tomorrow with appropriate solutions and clear answers.

„MGR" ULTRA-PURE WATER SYSTEM

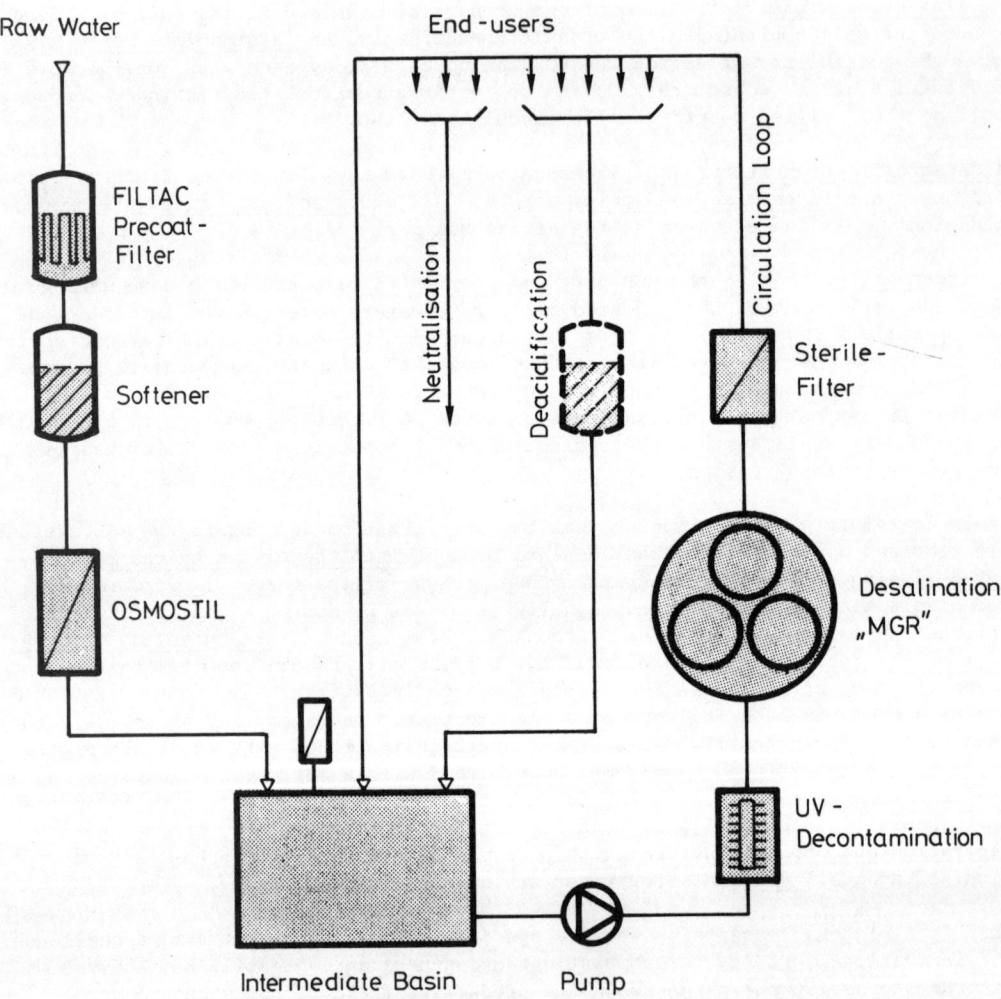

Fig. 7. The "MGR" ultra-pure water system

ELECTROCHEMICAL METHOD FOR PHOSPHORUS PRECIPITATION

P. Ennet, M. Hannus and H. Mölder

Tallinn Polytechnic Institute, U.S.S.R.

ABSTRACT

The electrochemical method for phosphorus precipitation in activated sludge process has been studied. Iron electrodes were placed in aeration basin.

The aim of the laboratory experiments was:
1. to determine the metal dissolution rate depending on current intensity;
2. to determine removal of phosphorus depending on the rate of dissolved metal.

Experiments were also carried out in field conditions. Formulas for calculation of the parameters of phosphorus electrochemical precipitation process are given.

KEYWORDS

Wastewater treatment; biological-electrochemical method; removal of phosphorus; laboratory experiments; pilot plant tests.

LABORATORY EXPERIMENTS

The electrochemical method for phosphorus precipitation in activated sludge process is based on anodous dissolution of an electrode metal. The electrodes are usually made of aluminum or iron, the latter being cheaper.

Two sets of laboratory experiments were performed. The first series of static tests was run in 3-dm^3 glass cylinders. In order to imitate the conditions of simultaneous precipitation, the solution used in the tests was activated sludge.

The aim of static tests was:
1. to determine the metal dissolution rate in the conditions of simultaneous precipitation;

2. to determine the necessary metal quantity depending on initial phosphorus and dissolved oxygen concentrations.

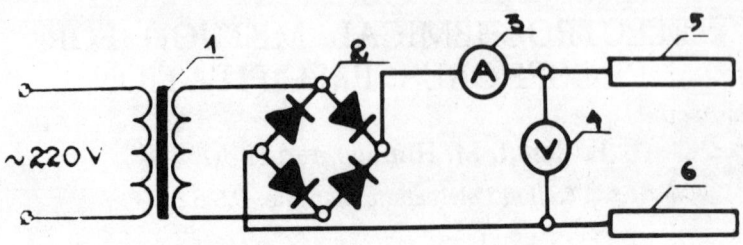

Fig.1. Electrical scheme of static tests.
1 - transformator; 2 - rectifier;
3 - ampermeter; 4 - voltmeter;
5 - anode; 6 - cathode.

Interelectrode spacing was 10 mm, each electrode was made of an iron plate 50 x 50 x 4 mm. The scheme of the laboratory set is shown in Fig.1.

The quantity of dissolved metal was determined at potentials of 2.0 V, 5.0 V and 10.0 V, with the current intensity 21 to 134 mA and the current density of 8.4 to 53.6 A/m². The total iron concentration in the solution was determined at 1 hr intervals. One series of experiments lasted for 7 hours. The coefficient of dissolved metal can be calculated according to the formula:

$$\eta = \frac{Me}{Me_n} \qquad (1)$$

where
η - metal dissolubility coefficient;
Me - actual quantity of dissolved metal;
Me_n - theoretical quantity of dissolved metal.

At the voltage of 2.0 V, the coefficient η ranged from 0.87 to 1.04, the average being 0.95; at 5.0 V potential it was 0.93 to 1.10, average 1.0; and at 10.0 V potential 0.91 to 0.98, average 0.94. These results show that in case the tests are run in an aeration chamber the coefficient η remains in the same range as by conventional electrochemical precipitation of phosphorus (Onstott and others, 1973 Sadek, 1970).

The second series of static tests was carried out to determine the quantity of dissolved metal necessary for precipitation of phosphorus. The tests were run at different initial phosphorus concentrations (5.6, 8.6 and 10.8 mg P-PO$_4$/dm³). Further studies were conducted in order to determine the effect of dissolved oxygen concentration upon phosphorus removal:
a) with saturated dissolved oxygen concentration ($\square$, $\triangle$, O in Fig.2)

b) with dissolved oxygen concentration of 1.0 mg O_2/dm^3 (■ , ▲ , • in Fig. 2).

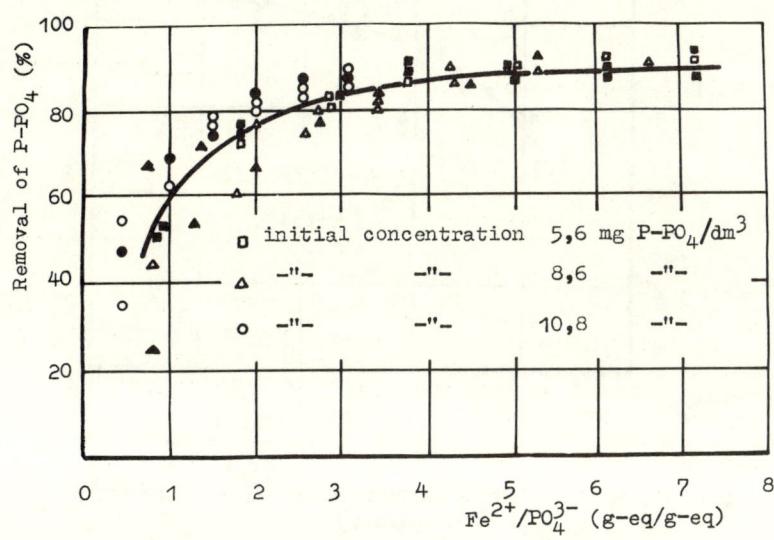

Fig. 2. Phosphorus removal efficiency (static tests).

Figure 2 shows that the fluctuation in the dissolved oxygen concentration in the activated sludge solution has no effect on precipitation of phosphorus. The results show fluctuation of 40 to 70% in the removal of phosphorus in case $Fe^{2+}/PO_4^{3-} = 1$. 80% removal of phosphorus in the static tests requires 2.5 g-eq. of Fe^{2+} per 1g-eq. of $P-PO_4$ removal.

PILOT PLANT EXPERIMENTS

In order to obtain data on the electrochemical precipitation of phosphorus in the course of the biochemical wastewater treatment, the 2nd series of tests was run in a pilot plant BIO-100 (300 dm^3/day). Iron sheet electrodes measuring 25 x 25 cm with 20 mm spacing were used. The electrical scheme is identical to that of the static tests. The metal dissolubility coefficient was assumed to be 1.0. The results given in Fig.3 coincide with the static tests. The initial phosphorus concentrations were 10 to 12 mg P/dm^3. Earlier investigations (Ennet, Molder, 1978) show that the biochemical purification process is not inhibited by simultaneous electrochemical precipitation. Table 1 presents wastewater purification parameters depending on electrolysis intensity.

At electrolysis intensity up to 500 C/dm^3 biochemical process has high efficiency (BOD_5 removal higher than 95%). The activated sludge settleability is good (Mohlman index ~70 cm^3/g), suspended solids concentration in the effluent ~10 mg/dm^3.

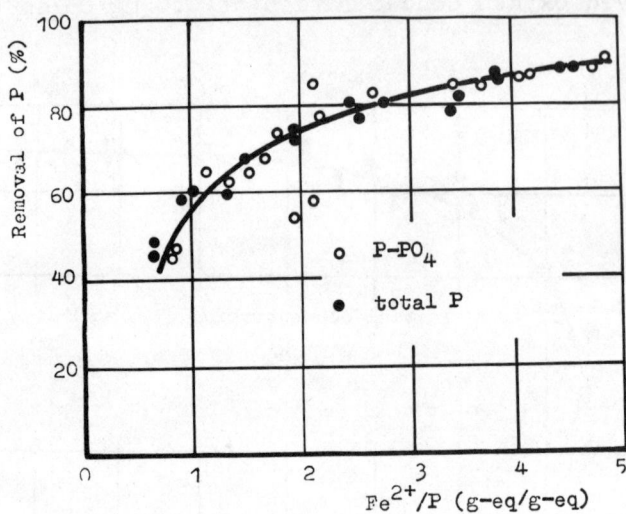

Fig. 3. Phosphorus removal efficiency (pilot plant)

TABLE 1 Effect of Electrolysis on Wastewater Treatment Rate (C/dm^3)

Electrolysis intensity C/dm^3	Sludge index (Mohlman index) cm^3/g	Treated sewage Suspended solids mg/dm^3	BOD_5 mg/dm^3	BOD_5 removal %
-	96	16	14	96
1 - 100	72 - 88	12 - 16	2.3 - 4.0	97 - 98
101 - 200	66 - 83	5 - 12	3.8 - 5.2	96 - 97
201 - 300	68 - 75	8 - 12	4.4 - 8.0	94 - 97
301 - 400	64 - 69	4 - 6	4.8 - 5.2	96 - 97
401 - 500	70 - 76	7 - 9	5.0 - 6.4	96 - 97

FIELD TESTS

Field tests were carried out in a packaged aeration plant at extended aeration. The wastewater flow ranged from 110 to 150 m^3/day (population of the settlemant being 700).

The electrical scheme was identical to that of laboratory tests. Electrodes measuring 30 x 80 cm with spacing of 15 mm and 5 mm thick were used. The intensity of the process was regulated by variations in voltage (6.5 to 17 V) and by the number of electrodes. The results are listed in Fig. 4. To protect against corrosion the steel aeration tank is connected with the cathode. The electrodes may be made of metal waste.

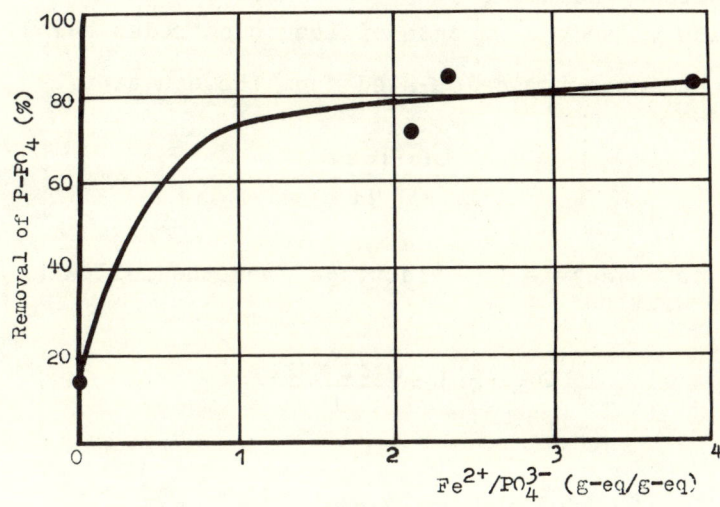

Fig. 4. Phosphorus removal efficiency (field tests)

Taking into account the easiness of exploitation, the electrochemical precipitation of phosphorus in small treatment plants may be considered highly effective.

CALCULATION FORMULAS

The parameters of phosphorus electrochemical precipitation process may be calculated as follows (Ennet, 1978):
Quantity of dissolved metal

$$Me = \beta m M \frac{AJt\, n_1}{Fz} \qquad (2)$$

where
- Me - dissolved metal in kg/day;
- β - relation between the actual activated ion quantity and the metal activated ion quantity, calculated from the stoichometric equation (in case of 85% phosphorus removal $\beta = 2.5$);
- m - relation between the activated ion equivalent weight of the metal and phosphorus equivalent weight (with iron electrodes $m = 2.7$, with aluminum electrodes $m = 0.87$);
- M - phosphorus quantity in sewage (kg/day);
- A - atomic weight of aluminum electrodes $A = 27.0$, iron electrodes atomic weight $A = 55.8$;
- J - current intensity, A;
- t - time, seconds ($t = 86400$ sec.);
- n_1 - metal dissolubility coefficient $n_1 = 0.8 - 1.0$;

F - Faraday number ($F = 96500$ C/g-eq);
z - ion valency (in case of iron electrodes $z=2$).

The current intensity, calculated from the equation 2:

$$J = \frac{\beta m M F z}{A t \eta_1} \qquad (3)$$

In order to calculate the electrode area, the following equation should be written

$$J_a = \eta_2 J_c = \frac{u \varkappa S_a \eta_2}{l} \qquad (4)$$

where
J_a - actual current intensity, A;
J_c - current intensity, calculated by Ohm's law, A;
η_2 - coefficient, taking into account the influence of electrode passivation (depends on the electrolysis regime, i.e. with the voltage of 2V $\eta_2 = 0.2$, at periodical electrode polarity reversal the coefficient increases, the maximum value being $\eta_2 = 1.0$);
u - voltage between the electrodes, V;
$\varkappa$ - specific conductivity, Ω^{-1} ;
S_a - anode surface area, m^2;
l - interelectrode spacing, m;

The total anode surface area is calculated from the equation 4:

$$S_a = \frac{J_a l}{u \varkappa \eta_2} \qquad (5)$$

Electrodes should be introduced into the aeration tank as compact blocs, each anodous electrode inside the bloc working with both sides. In order to ensure a good circulation in the interelectrode space, the electrode width ought to be less than 30 cm, and spacing more than 1 cm.

Total electrode number:

$$K = \frac{S_a}{S} + n \qquad (6)$$

where
K - electrode number ;
S - area of one surface of an electrode, m^2;

n - number of electrode blocs;
S_a - anode surface area (from equation 5);

To avoid electrode passivation and to ensure uniform electrode usage their polarity is periodically reversed (reversal period being not more than 1 day).

Electrode working time:

$$T = \frac{S_a b \rho n_3}{Me} \qquad (7)$$

where
 T - working time, days;
 b - electrode thickness, m;
 ρ - specific weight of electrode material, kg/m^3;
 n_3 - electrode usage coefficient $n_3 = 0.8$

CONCLUSIONS

The technique for removal of phosphorus in extended aeration process is based upon anodous dissolution of an electrode in the aeration basin. Iron electrodes made of metal waste should be preferred because of their cheapness. 80% removal of phosphorus requires 2.5 g-eq. of Fe^{2+} per 1 g-eq. of phosphorus.

The parameters of the process (calculated with the initial phosphorus content in the wastewater 10 mg/dm^3 and retention time of 24hrs) are as follows:
- voltage between the electrodes 3 - 6 V;
- expenditure of electric current 2.4·10^5 C/m^3;
- current density 2.8 A/m^3;
- relative electrode density - 0.15 m^2 of anode surface per 1 m^3 of aerotank volume.

To avoid electrode passivation and to ensure uniform electrode usage, their polarity is periodically reversed.

The electrochemical method for removal of phosphorus also affords corrosion protection of the packaged steel aerotanks.

REFERENCES

Ennet, P. O., A. H. Molder (1978). Electrochemical method for phosphorus precipitation. *Publications of Tallinn Polytechnic Institute*, 445, 13-18.
Ennet, P. O. (1978). Calculations by designing the simultaneous phosphorus precipitation in the extended aeration process. *Publications of Tallinn Polytechnic Institute*, 445, 19-23.
Onstott, E. J., W. S. Gregory, E. F. Thode and K. L. Holman (1973). Electroprecipitation of phosphates. *Environ. Eng. Div. Proc. Amer. Soc. Civ. Eng.*, 99, 897-907.
Sadek, S. E. (1970). *An electrochemical method for removal of phosphates from waste waters*, Washington.

AMMONIUM PHOSPHATE RECOVERY FROM URBAN SEWAGES BY SELECTIVE ION EXCHANGE

L. Liberti, N. Limoni, R. Passino* and D. Petruzzelli

Istituto di Ricerca Sulle Acque, C.N.R., 5 via De Blasio,
70123 Bari, Italy
*Istituto di Ricerca Sulle Acque, C.N.R., 1 via Reno,
00198 Roma, Italy

ABSTRACT

Some practical aspects of the process for nutrient recovery from secondary effluents by ion exchange are investigated in detail. These aspects comprise ion speciation, $MgNH_4PO_4$ production, resin regeneration and resin fouling. Fouling by bio-resistant organic pollutants appears to affect the anion resin performances in the absence of a properly designed scavenging section.

KEYWORDS

Ion exchange; ammonium and phosphate removal; eutrophication; wastewater tertiary treatment; clinoptilolite; resin fouling; fertilizer recovery.

INTRODUCTION

Recycle and non-waste technology is the key for future approaches to environmental problems if pollution is to be coped with regard to paucity of non-regenerable sources and to inflation waves.
Along these lines, recovery techniques should be preferred to chemical precipitation and biological nitrification-denitrification processes to abate phosphates and ammonium from treated urban sewages, in order to prevent man-made eutrophication. Selective ion exchange appears a promising candidate to this aim, as recently shown, among others, by Higgins (1976), Gregory (1976), Yoshikawa (1978) and Kataoka (1978). After extensive investigations conducted at IRSA since 1976 (Liberti and co-workers, 1976, 1977a, 1977b, 1979), a tertiary treatment process, where NH_4^+ and $HPO_4^=$ are selectively removed by a natural zeolite (cli-

noptilolite) and a weak base anion exchange resin respectively,both
regenerated with NaCl,was set up.Recovered NH_3 is stripped out from
the cationic eluate and transferred to the spent anionic regenerant,
where the premium quality,slow release fertilizer $MgNH_4PO_4$ is quantitatively precipitated by addition of a Mg salt.Regenerant solutions
can then be recycled for further use.
The process pilot plant tests showed also the need for a better evaluation of the following aspects:ion speciation,$MgNH_4PO_4$ production,
optimization of regeneration and resin life.
This paper presents the experimental conclusions reached on the above
points.

EXPERIMENTAL

The pilot plant used has been already described in detail elsewhere
(Liberti,1979).It comprises a sand column (e.s. 0.052 cm,u.c. 1.48,
d 24 cm,h 100 cm),followed by an activated carbon section (3 columns
with Filtrasorb 300 from Calgon Corp.,Pittsburg,Pa.,with a geometry
similar to filtration) for the pretreatment of a secondary effluent.
The anionic and cationic columns (d 5.6 cm,h 162 cm) were loaded with
the weak base anion exchange resin Kastel A 102 from Montedison,Milan
Italy,and with Hector,Calif.,M-4 clinoptilolite from Baroid,Houston,
Texas respectively.Wastewater purified with ion exchangers was a secondary effluent taken after the chlorination stage from the sewage
treatment works of the town of Bisceglie,near Bari,and stored in a
refrigerated reservoir at IRSA laboratories.
The composition of wastewaters used in the investigation is shown in
Fig. 1.Details of the laboratory apparatus for $MgNH_4PO_4$ precipitation
are given in Fig. 2.
Laboratory investigations on regeneration and on resin life were made
using Plexiglass columns (d 2cm,h 120 cm).Here,before each regeneration,performed upflow,the resins were loaded with 80 v/v_r of a synthetic solution reproducing Bisceglie's average secondary effluent,
at a downflow rate of 22.5 v/v_r h.
In tests on anion exchange resin fouling,about 20 ppm of Aerosol OT
(a linear alkyl-sulphonate,$C_{20}H_{37}NaO_7S$),or of commercial humic acid,
or of a 1:1 mixture of both were added to the synthetic feed.
All reagents were analytical grade.Analytical procedures and other
details have already been described (Liberti,1979).

RESULTS AND DISCUSSION

Ion speciation

Pilot plant evaluation of the process over a period of about 6 months
confirmed the possibility to reach on wastewater the concentrations
of P and N required by the present regulations (10 ppm P and 15 ppm
NH_4 respectively).This is clearly seen in the A section of Fig. 1,

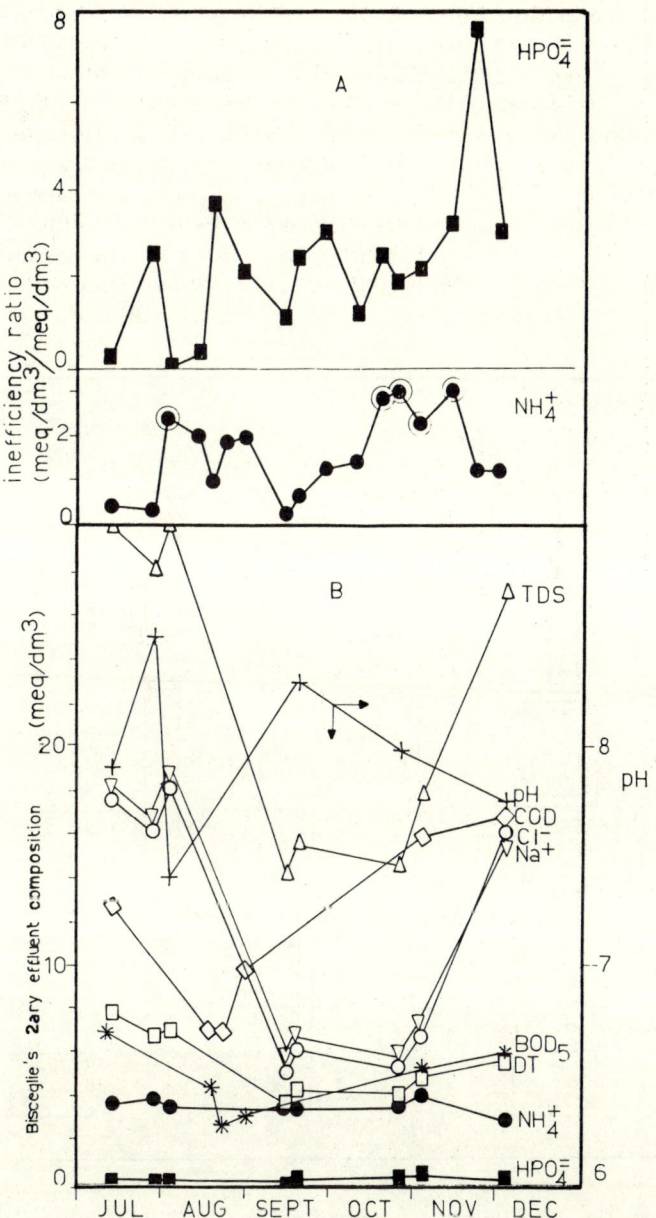

Fig.1.A) Inefficiency ratio(= nutrient concentration in the treated effluent/total nutrient fed to the resins per cycle) during the pilot plant tests.(⊙ concentration exceeding the MAC, maximum allowable concentration).
B) Bisceglie's secondary effluent composition during the investigation (COD and BOD_5 x 0.1).

where acceptable plant performances, expressed by the inefficiency ratio, are recorded with only the NH_4^+ MAC occasionally exceeded. Higher leakages of nutrients, however, are always connected to peak concentrations of Na^+ and Cl^- ions in the feed (see B section of Fig. 1) As already found in previous investigations, mass balances between exhaustion and regeneration of the resins apparently showed a progressive build-up of non regenerated ions on the resin phase (i.e., about 3.5 eq.NH_4/dm_r^3 accumulated in 17 cycles!). Accurate checks through each section of the pilot plant during the exhaustion, however, demonstrated that each of the investigated species can be retained by any of the process stages, as shown in Fig. 3.

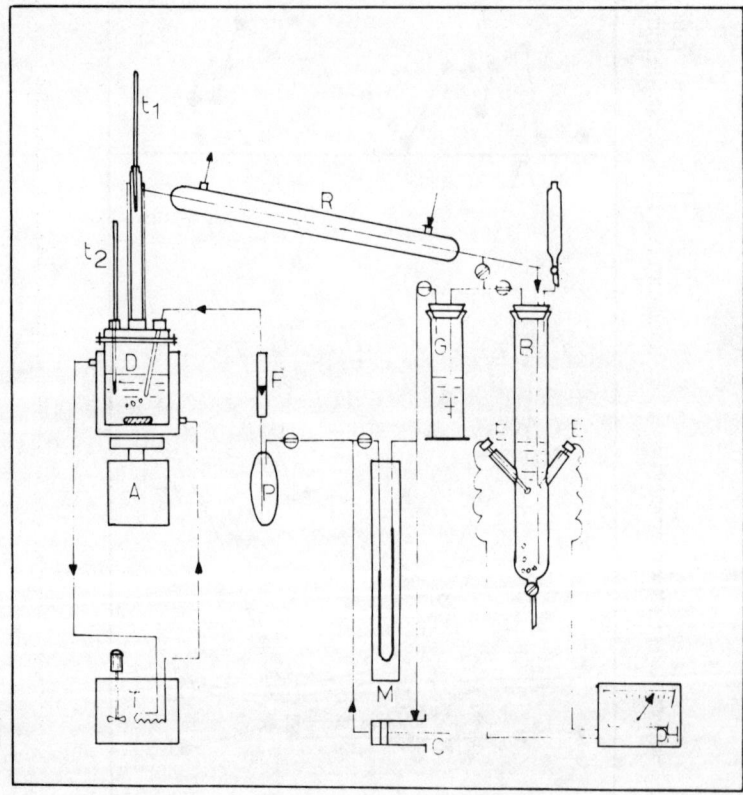

Fig.2 Laboratory apparatus for $MgNH_4PO_4$ precipitation. (A=magnetic stirrer, B=anionic absorber, C=compressor, D=cationic stripper, E=pH electrodes, F=gas flowmeter, G=HNO_3 or H_2SO_4 trap, L=$MgCl_2$ solution, M=manometer, P=pneumatic valve, R=refrigerator, T=thermostat)

Clearly, the existence of complex forms (chelates, complexes, etc.) of single ions makes the application of the usual distinction of "cation" and "anion" to wastewaters somewhat questionable, confirming the necessity of carefully identifying ion speciation for mass balances to be rigorously met in these systems.

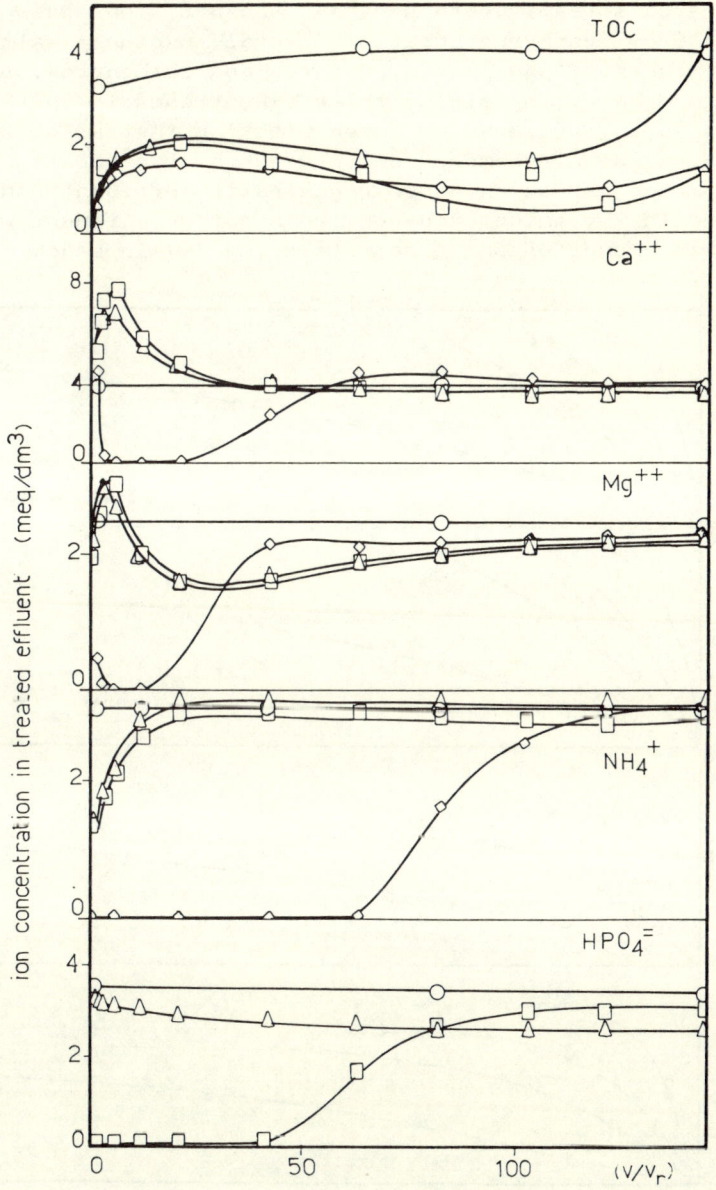

Fig.3. Breakthrough curves for various species after the different sections of the pilot plant in a typical run.
(concentrations after: o filtration, △ adsorption, □ anionic and ◇ cationic sections).

$MgNH_4PO_4$ production

Two concentrated streams, containing NH_4^+ and $HPO_4^=$ ions respectively, are obtained from resin regeneration. Several insoluble salts, namely calcium and magnesium phosphates, sulfates and carbonates, would coprecipitate with $MgNH_4PO_4$ by mixing these two streams. In the absence of insoluble Na salts, pure $MgNH_4PO_4$ precipitation instead occurs if, by the aid of a vacuum air compressor between the two solutions, NH_3 is stripped from the cation exchange regeneration effluent and continuously absorbed in the anion exchange regeneration effluent, where a stoichiometric amount of an Mg salt (i.e., $MgCl_2$) is added.

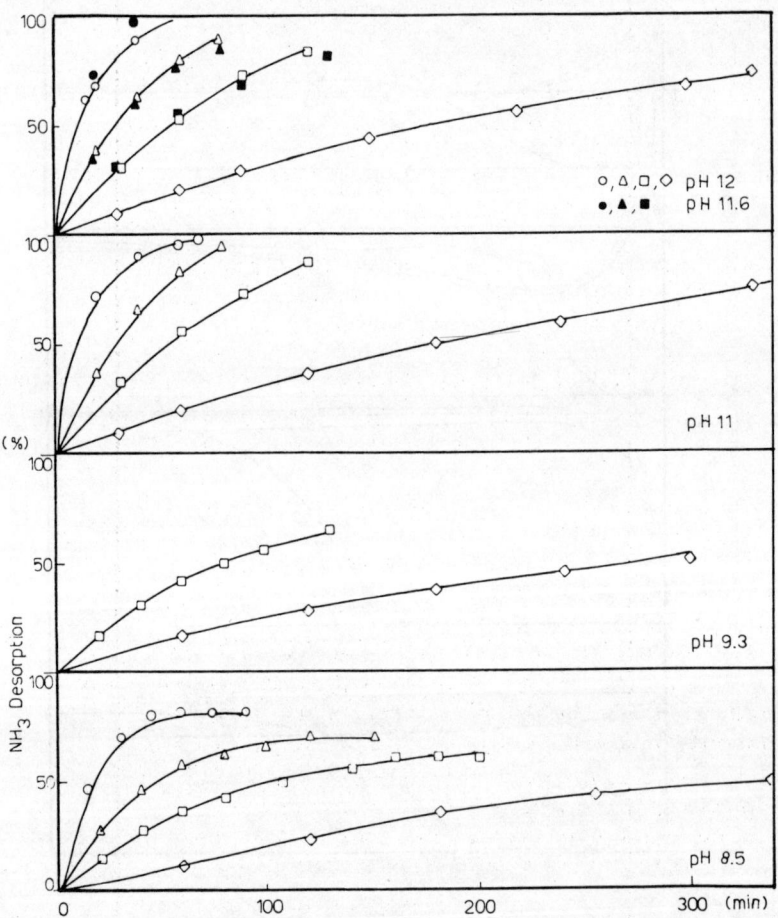

Fig. 4. NH_3 stripping curves from the cationic eluate as a function of pH and temperature. (o 80/65, △ 60/45, □ 45/30 and ◇ 20°C, referred to thermometers t_2 and t_1 in Fig. 2).

The laboratory apparatus described in Fig. 2 was used to determine suitable conditions for such operation.The pH and temperature of the cation exchange regeneration effluent exert an influence on NH_3 stripping (Fig. 4),whilst phosphate precipitation is controlled by the pH of the anion exchange regeneration effluent.
As described in detail elsewhere (Liberti,1977b),a pH = 11 and t = 35°C for NH_3 stripping and pH 8-9 for quantitative phosphate precipitation provide the optimum conditions for an almost pure,readily setteable and filterable salt,with the chemical formula $MgNH_4PO_4 \cdot 6 H_2O$, to be obtained in a few minutes.$Mg(OH)_2$,precipitated during lime correction to pH = 11 of the cation regeneration effluent,and NH_4NO_3 or $(NH_4)_2SO_4$,formed by recovery of the excess NH_3 in the corresponding acids,were also obtained as useful process by-products.

Optimization of regeneration

Caustic brines are usually proposed in the literature for clinoptilolite regeneration(Mercer and Ames,1970;Koon and Kaufman,1975).Accordingly,NaCl 2M added with lime up to pH 11.5 and with HCl to pH 3 was employed in the referred pilot plant runs as cation exchange and anion exchange regenerants respectively.A preliminary economic evaluation of this process (Liberti,1979),however,indicating an over-all (capital plus operating) cost of about 0.03 U.S.$/m^3 of treated wastewater,showed that 50% of it should be accounted for by chemicals. An evaluation of resin performances under different regeneration conditions was then undertaken to reduce NaCl requirements and,if possible,to substitute it with sea water (which could be approximated to a 0.6M NaCl solution).
Regeneration parameters varied as follows:
- NaCl concentration,from 0.1 to 2M
- regeneration level,from 0.2 to 50 eq $NaCl/dm_r^3$
- pH,from 3 to 12.5
- flow rate,from 10 to 40 v/v_r h.

The results for the clinoptilolite section,reported in Fig. 5,confirm that little benefits,if any,are associated to NaCl concentrations higher than 0.6M,while pH higher than 11 in these conditions accelerates resin consumption,as already anticipated by Mercer and others(1970). Furthermore,regeneration flow rates up to 40 v/v_r h are shown by Fig.6 to be compatible with efficient regeneration.
The following conditions can thus be adopted for optimal regeneration of clinoptilolite:20 v/v_r of neutral (pH 7) 0.6M NaCl at a flow rate of 30 v/v_r h.
Somewhat similar results were obtained for the anion exchanger section,typical regeneration histories of which,at different NaCl concentrations,are reported in Fig.7.Here,4 v/v_r of acidic (pH 3) 0.6M NaCl at a flow rate of 6 v/v_r h ensure a satisfactory regeneration. Appreciable economies in chemical requirements are clearly introduced by these results.Further benefits may be expected by efficient use of sea water and/or recycle of the spent regenerants after the

$MgNH_4PO_4$ precipitation, as will be tested in the course of further developments of this research.

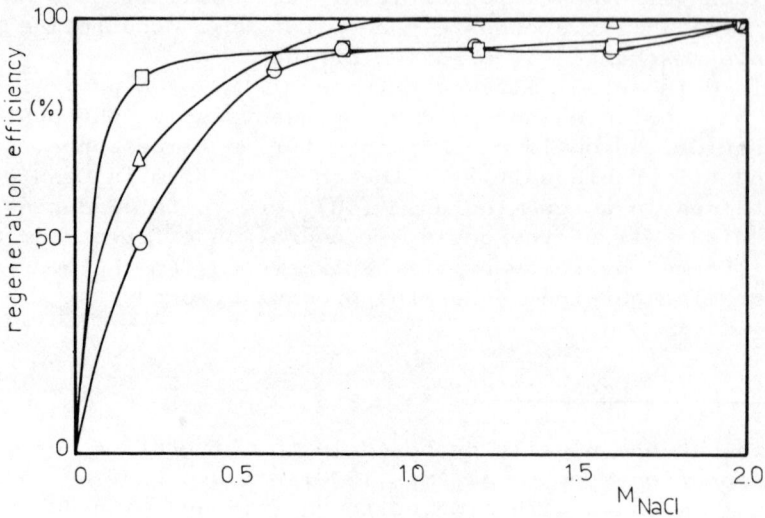

Fig. 5. Regeneration efficiency for clinoptilolite at various NaCl concentrations.
(pH = □ 12, △ 11.5, o 11)

Resin life

In the referred economic analysis, a clinoptilolite yearly consumption as high as 30% was prudentially assumed, to account for wear and attrition. A greater indecision concerned the anion exchange resin life, the weak basicity and the gel-type matrix of which make it particularly vulnerable to fouling by bio-resistant organics still present in a secondary effluent. Furthermore, previously laboratory tests (Liberti, 1979) indicated that the presence of bicarbonates in the feed causes the disactivation of the weakly basic anion exchanger groups of the resin through the hydrolysis reaction

(1) $\quad RNH_3Cl + HCO_3^- \rightleftharpoons RNH_3OH + CO_2 + Cl^-$.

Automatization of the pilot plant is presently under completion to allow for accelerated testing of resin resistance to fouling.
Meanwhile, the anion exchange resin resistance to some typical organic pollutants has been evaluated, using a synthetic solution containing a humic acid, a linear alkylsulfonate or a mixture of both. According to the copious, although often contradictory, literature on resin fouling (Abrams, 1962; Wolff, 1971; Rowe, 1975; Kim and co-workers, 1976; Bolto and others, 1978), the humic acid is representative of irreversibly adsorbed organics, while the ionic character of the alkylsulfonate should make it easier to be released from the resin.

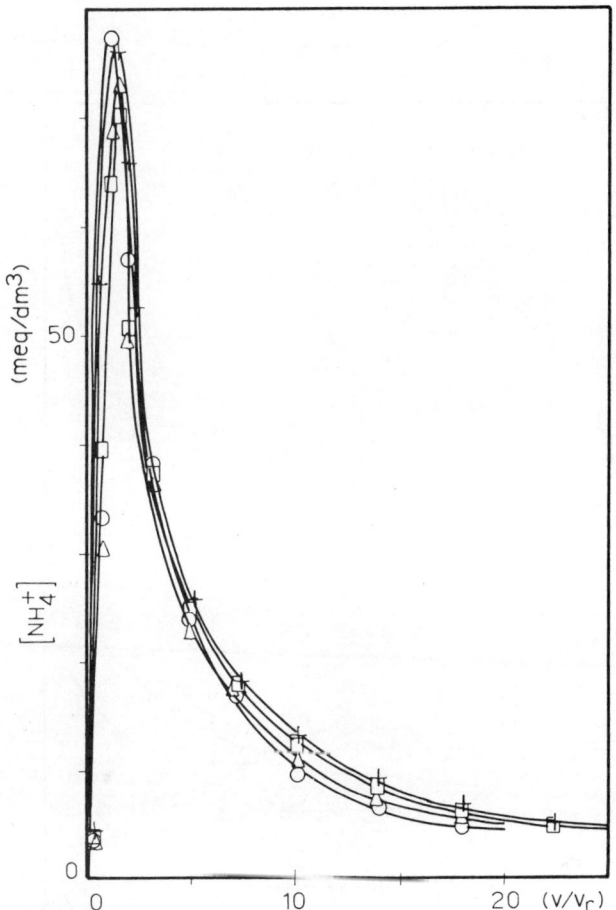

Fig. 6. NH_4^+ regeneration histories from clinoptilolite as a function of regenerant flow rate. Regenerant 0.6M NaCl fed upwardly at pH 7. (+ 38, ◻ 30, △ 20 and o 10 $v/v_r h$).

As shown by Fig. 8, these assumptions have been confirmed as more than 80% of Aerosol OT (Fig. 8 A) against about only 40% of humic acid (Fig. 8 B) was desorbed. Furthermore, the fouling capacity of the latter prevented even the Aerosol OT to be eluted from the resin when this was loaded with the mixture of both chemicals (Fig. 8 C).

It should be noted that only inorganic regeneration treatments have been tested, combining the ion exchange action of various electrolytes (NaCl, HCl and NaOH) with the osmotic shocks of water. Better results can reasonably be expected by added use of solvent extraction and similar organic treatments.

The tendency of the employed weak anion exchange resin, however, to be fouled makes it worth considering to substitute it by other adsorbent

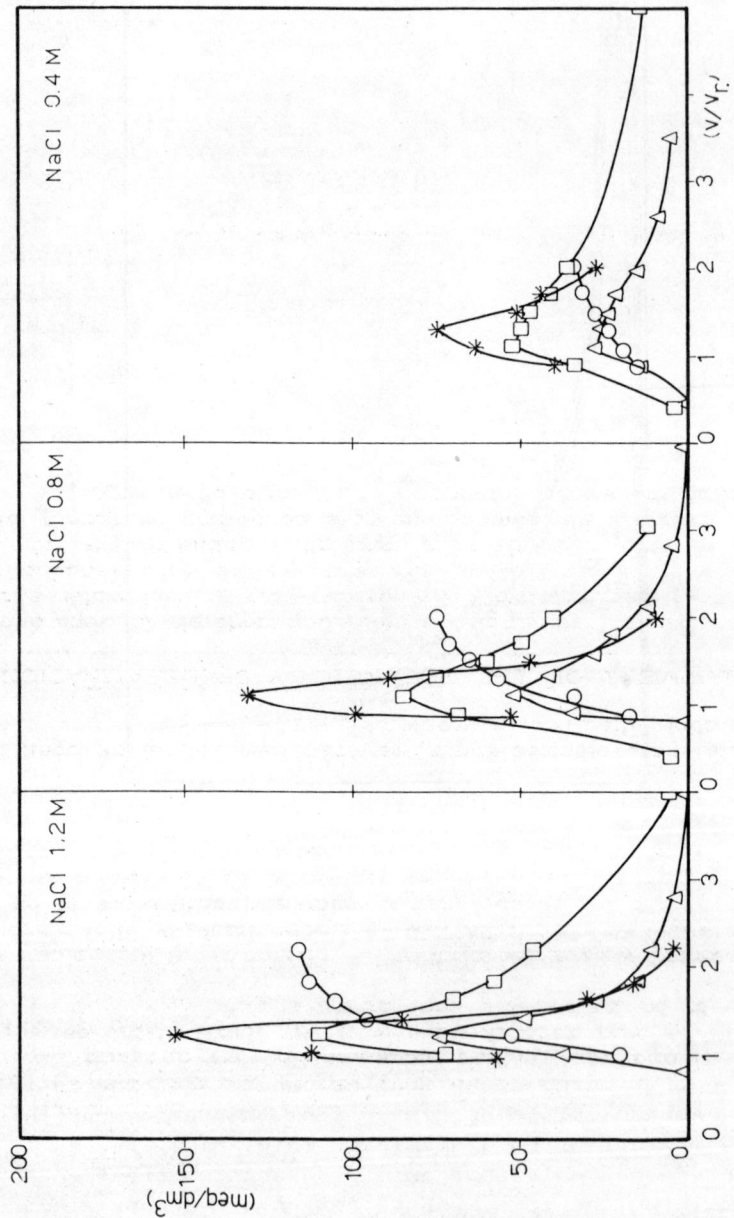

FiG. 7. Regeneration histories from anion column at different NaCl concentrations. (* $SO_4^=$, □ HCO_3^-, △ $HPO_4^=$, ○ Cl^- x 0.1).

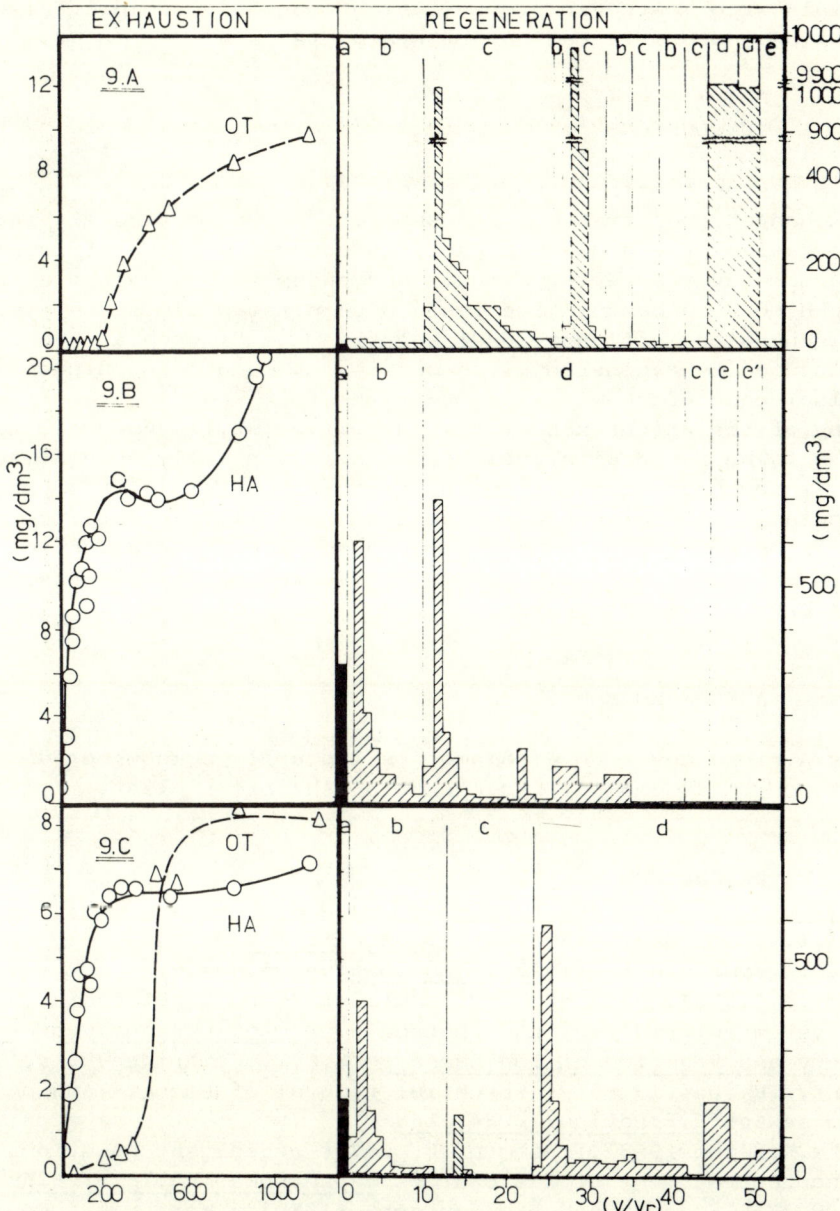

Fig. 8. Fouling tests for the weak base anion resin.
(concentration histories during the exhaustion
and regeneration for --- Aerosol OT and ——
humic acid, alone (A or B) or in mixture (C)).
Regeneration treatments: a, backwash; b, 2M NaCl;
c, distilled water; d, 0.5M NaOH; d', 0.25M NaOH;
e, 0.6M HCl; e', 0.3M HCl).

materials with scavenger properties,and possibly avoiding also the adsorption and/or filtration pretreatments.

CONCLUSIONS

Experiments on tertiary treatment of urban wastewater by selective ion exchange demonstrated the possibility to significantly reduce the concentration of P and N inorganic compounds in these effluents. Furthermore,by recovering ammonia and phosphates,a valuable fertilizer,$MgNH_4PO_4$,can be quantitatively precipitated almost pure from resin regeneration effluents.NaCl solutions as diluted as sea water have been used as regenerant,with consistent reduction in quantity and cost incidence of chemical consumptions.
Fouling of the anion exchange resin,however,still appears a serious problem to be faced with.Prolonged testing,on fully automatized equipment,is now under evaluation for further analysis of the process.
Particular attention will be paid to resin life and to the utilization of the resin beds in triple purpose (i.e.,filtration,adsorption and ion exchange),by substitution of the weak anion resin with a strong base,scavenger resin.

ACKNOWLEDGMENTS

Messrs.A.Pinto and L.De Girolamo are sincerely acknowledged for the assistance in assembling and operating the pilot plant.

REFERENCES

Abrams,I.M.,and S.M.Lewan (1962).J.A.W.W.A.54,537-43.
Bolto,B.A.,and others (1978).Desalination 25,45-59.
Gregory,J.(1976).Anion exchange equilibria and kinetics in the ternary system:chloride-sulfate-phosphate.Proceedings Int.Conference on Theory and Practice of Ion Exchange,S.C.I.,Cambridge,U.K.,10.1.
Higgins,I.R.(Oct.1976).Preferential removal of ammonia and phosphates from sewage.U.S.Pat.no.3,984,313.
Kataoka,K.(June 1978).Treatment of waste regenerant solutions in ion exchange treatment of wastewaters.Japan.Kokai no.78 67,675.
Kim,B.R.,V.L.Snoeynk and F.M.Saunders (1976).J.Wat.Pollut.Contr.Fed.48, 120-33
Koon,J.H.,and W.J.Kaufman (1975).J.Wat.Pollut.Contr.Fed.47,448-65.
Liberti,L.,G.Boari and R.Passino (1976).Water Res.10,421-28.
Liberti,L.,G.Boari and R.Passino (1977a).Water Res.11,517-23.
Liberti,L.,D.Petruzzelli and M.Polemio (1977b).Ann.Fac.Agr.,Univ.Bari XXIX,701-16.
Liberti,L.,G.Boari and R.Passino (1979).Water Res.13,65-73.
Mercer,B.W.,and L.L.Ames (Oct.1970).Mobile pilot plants for the removal of ammonia and phosphates from wastewater.Int.Wat.Conf.Eng.Soc.

West.Penn.,Pittsburg,Penn.
Mercer,B.W.,L.L.Ames,C.J.Touhill,W.J.Van Slyke and R.B.Dean (1970). J.Wat.Pollut.Contr.Fed.42,R 95-107.
Rowe,M.C. (Oct.1975).Effl.Wat.Treat.J.,519-24.
Wolff,J.J. (1971).Trib.Cebedau 336,487-90.
Yoshikawa,T.,and M.Hirai (May 1978).Treatment of wastewaters containing phosphates.Japan.Kokai no. 78 58,154.

UTILIZATION AND DESALINATION OF SALINE MINE WATERS

Jozef Kępiński

Technical University, 70-310 Szczecin, Poland

ABSTRACT

The development of large seale desalting of sea water and other saline waters is exposed. Distillation plants both LTV and MSF are characterized along with scale prevention and materials of construction. Membrane methods: electrodialysis and reverse osmosis are described including membrane properties and configuration. The examples of large desalination plants are cited. Huge quantities of mine waters associated with black coal mining pose quite a serious problem in Poland. The concentration of salt reaches as much as 200 g per litre. Both main rivers in Poland, the Vistula and the Odra, are handicapped by these saline mine waters.

The most concentrated coal mine brines with above 70 g/l TDS are utilized according to the technology of the Central Mining Institute. The brines are evaporated in a MSF demonstration plant running since 1975. Gypsum and table salt are getting crystallized out, and vapors are condensed to yield water for industrial and common use purposes. In view of future expansion of coal mining in the next decade, more than ten plants of the same range of capacity have to be built in the coal mining region.

For less concentrated waters the application of membrane technology namely reverse osmosis - and/or electrodialysis is under study. A pilot plant for the reverse osmosis is designed. Potable water and brines are to be the products of this process. The concentrates should be processed further by evaporation as pointed out above.

KEYWORDS

Desalination, distillation, electrodialysis, reverse osmosis, black coal, mine water, salt production, distillate, corrosion.

DEVELOPMENT OF THE TECHNOLOGY OF DESALINATION

From the physicochemical standpoint desalination consists in separation of a solution - saline water into two constituents - the solvent water and the solutes, predominantly salts. The current sense of the word desalination is recovery of usable water from seawater and other saline waters. The remaining part of the saline water, the concentrate containing practically all the salts is returned to the sea or disposed of in some other way.

Distillation

Distillation is the dominant method of desalination, accounting for 85 per cent of the worls desalinating capacity by the end of 1974 and until that time - the only practical method of desalting seawater. Two types of distillation plants are in use: The multiple-effect and the multistage-flash plants. The multiple-effect evaporators were in extended use for a very long time in the chemical process industries. After a rather short period of installing the typical submerged type evaporators the long tube vertical evaporators (LTV) were adapted to this end and suitably developed. Instaead of 3 to 4 effects as usual in the chemical process industries, 10 or even more effects are in use. The size of the distillation trains in the to day desalination plants is several times larger than of these encountered elsewhere. The size of the first generation plants as installed in the early sixties reached 4000 t distillate a day. For example a LTV plant of 1 Mgal capacity was successfully demonstrated in the US in the early sixties. Now with the second generation the range between 10 and 20 thousand t distillate a day is the most frequently encountered. The LTV technology was developed for many years in the Soviet Union. Fig. 1 demonstrates an evaporator as installed in a plant at Shevchenko at the Caspian Sea.

The other concept of distillation used to a very large extent is the multistage-flash distillation. In the flash evaporators, sea water is first heated to 90 - 120°C and then made to evaporate in several (up to 50) chambers in which the pressure is lowered consecutively. Since the vapour simply flashes off the warm liquid the resulting precipitates form primarily in the liquid. Fig. 2 demonstrates the principle of MSF distillation. The multistage-flash plants accounted for two-thirds of the cumulated capacity of the desalination plants in 1974. The largest simple distillation train is a MSF train of 36 000 t daily capacity in Porto Torres, Sardegna, Italy and also the largest plant composed of six MSF trains in Hong-Kong with an output of more than 180.000 t distillate daily. The possibility is being studied in the USA, Japan and Israel of building the MSF plants of third generation with daily output of 100 - 200 thousand t distillate coupled with nuclear reactors as source of heat. To date only one plant of second generation was built in Shevchenko in the Soviet Union coupled with a nuclear reactor. Shevchenko is the largest center of desalination in the USSR with cumulated capacity of about 130 000 t distillate a day by 1980. The coupling with thermal power plants is rather common this bringing some advantage and some limitations as well. The other concepts - vapour compression distillation and solar distillation are limited to small capacity.

The prevention of scale and corrosion are of outstanding importance in distillation processes. The formation of alkaline scale is prevented by decarbonization with mineral acids or by application of suitable inhibitor, like Hagevap or new synthetic liquid Belgard EV. No practicable chemical method exists of preventing the formation of calcium sulphate scale. In most sea water distillation plants the scaling is prevented by keeping the heating temperature below 120°C. In some cases the seeding principle is being applied. The components of scale are precipitated on the surface of small crystals circulated in the brine. Calcium sulphate, barium sulphate or calcium carbonate are used as seeding crystals.

The choice of suitable materials is a very difficult and responsible task. For chambers and shells of heat exchangers carbon steel and for piping of heat exchangers brass and cupro-nickel are used. In the Soviet Union a rather extended use is made of stainless steel and carbon steel clad with stainless steel. Titanium alloys passed several pilot plant trials and proved suitable

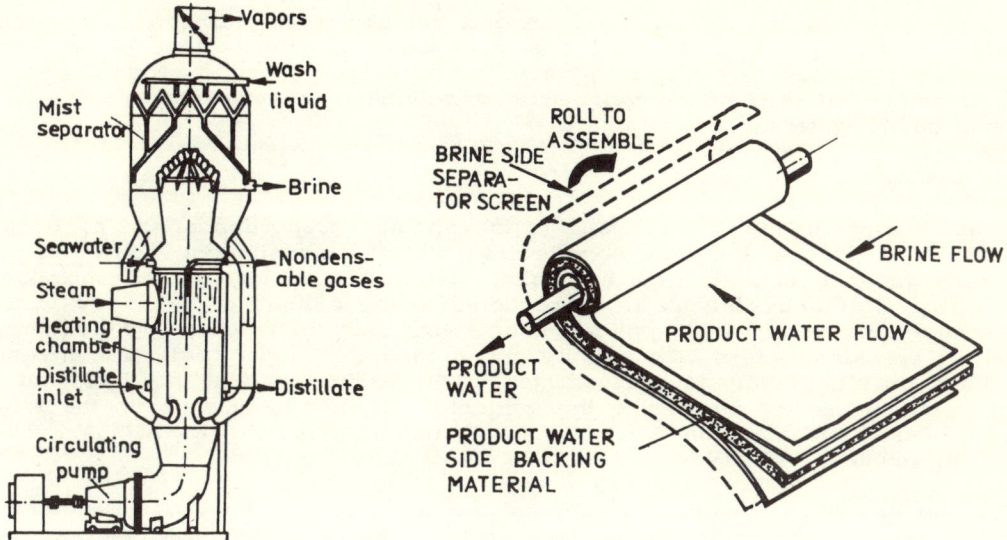

Fig. 1. The Shevchenko LTV evaporator.

Fig. 3. The spiral wound module.

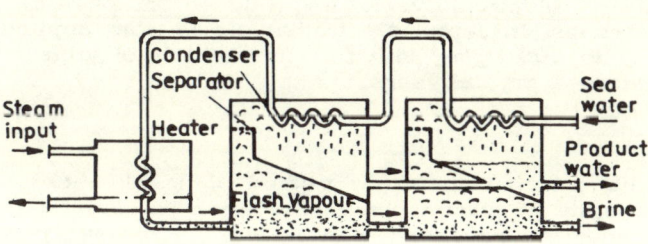

Fig. 2. The principle of MSF distillation.

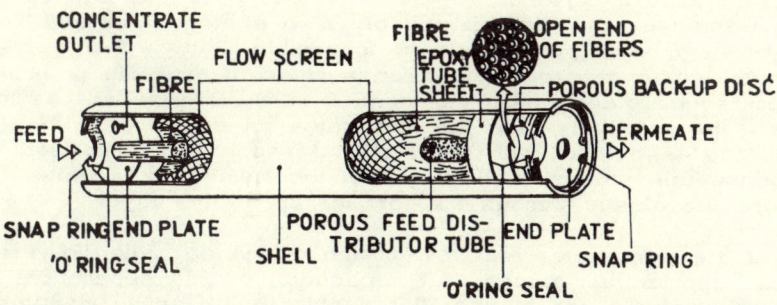

Fig. 4. The hollow fiber module – Du Pont „Permasep" permeator.

in form of thin pipes for heat exchangers. All aluminium plants passed sucessfully the pilot plant stage. MSF chambers made of concrete are under investigation. One may say that distillation, the well known unit operation, was successfully adapted and developed in desalination of sea water and some other saline waters.

Electrodialysis

In electrodialysis the use is made of the specific properties of ion exchange membranes. This is the only commercial method of desalination based on removing the ions of salts from the saline water. Accordingly the consumption of energy in electrodialysis is proportional to the salinity of the inflow water. Therefore the economical application of electrodialysis is limited to desalination of brackish waters with salinity not exceeding 5 kg/m^3 salt. The investigations on electrodialysis were started in the Netherlands at the TNO Institute with the aim of recovering the valuable components of whey in the forties. Besides desalination electrodialysis found some applications in the food and agricultural industries and also in wastewater treatment.

The ion exchange membranes, anionic and cationic, are the most important components of electrodialytic equipment; the spacers serving as conduits and liquid distributors in the narrow compartments of the electrodialyser, and electrodes are the other components of the electrodialytic stack. Only a few firms manufacturing membranes remain on the market with Japanese companies leading in the field. The costs of the membranes are relatively high and membrane replacement is a major item in the cost of desalted water. In the Soviet Union a line of electrodialysers is produced for several years using home made membranes. In Japan electrodialysis is also applied to preconcentrate seawater up to 200 kg/m^3 salt for manufacture of table salt in multiple-effect evaporators - crystallysers.

Reverse osmosis

Reverse osmosis (hyperfiltration) is a rather new method of desalination (idea of E.Reid, 1953). It consists in forcing the saline water against a semipermeable membrane under the pressure higher than the osmotic pressure of the saline water (about 25 atm for seawater). In practice the working pressure must be substantially higher in order to achieve reasonable filtration rates: 30 - 40 atm for brackish water, 60 - 80 atm for sea water. The manufacturing of suitable membranes is the crux of reverse osmosis. The membranes are formed mainly by the method established by Loeb and Sourirajan in the early sixties. Modified cellulose diacetate was for several years the most widely used material in reverse osmosis research and commercial equipment. Cellulose diacetate is dissolved in acetone and modified by water solution of magnesium perchlorate or formamide. Many other polymers were also tried, but to date only membranes made of aromatic poliamide reached the commercial stage of development. The membranes are asymmetric in structure - the relatively high filtration rates as compared with other polymer films are associated with the fact that separation is caused by an extremely dense thin skin while the rest of the membrane is quite permeable and offers the elastic porous support for the active layer.

The configuration of the membranes and the design of the so called reverse osmosis modules is important for successful application of this separation method. Several configurations are in use: flat sheets in the spiral modules (Fig. 3.), tube membranes about 20 mm in diameter, spaghetti membranes and modules - about 3 mm in diameter. All these configurations necessitate mechanical support elements, made of metal or plastics. The capillary membranes (about 0,5 mm outside diameter) are selfsupported (Fig. 4). The commercial Du Pont hollow fiber modules B-9 for desalination of brackisch water are fairly compact devices capable to supply 9,5 to 30 m^3 per day of permeate (drinking water) when fed with inflow water of up to 6 g per litre salinity. The external size of these modules is 119 mm dia. and 1190 mm long and 300 mm dia. and 1200 mm long respectively.

The installations for electrodialysis and reverse osmosis accounted for only 15 per cent of the cumulated world desalting capacity by the end of 1974, i.e. about 300 000 m^3 per day. Nevertheless in the following five years the reverse osmosis expanded very rapidly, although some revival of interest was also noted in electrodialysis. Several very large reverse osmosis plants were erected in different parts of the world; in Saudi Arabia, five plants of about 200 000 m^3 capacity were installed 1976-1978 in order to desalinate brackish well water to supply drinking water for the population of the capital city Riyad.

A very important step in the development of reverse osmosis was the commercialization of reverse osmosis hollow fiber modules for desalination of sea water (Du Pont B-10 modules). Working under 56 atm pressure these modules are capable of one - pass desalting of sea water, demonstrating practical salt rejection in plus of 98,5%, while flux is diminished to the half of the performance of B-9 modules. In Poland several reverse osmosis plants were installed in the electronic industry, applying Du Pont B-9 modules. **Reverse osmosis is incorporated in the preparation scheme of pure water** for rinsing the electronic components along with ion exchange. The largest plant installed in 1973 is of 1400 m^3/d capacity. The inflow water is municipal water with about 500 g/l TDS. Salt rejection of about 90% and recovery of 75% were achieved in the long run. After some period of initial troubles the plant showed to be very reliable. The installation of one more plant with two trains of 1080 m^3/day each started in 1978. The preliminary results of running the first of these trains are quite satisfactory.

Other Methods of Desalination

Several pilot plants were built in course of development of the freezing concept. The ice crystallized out of saline water is composed of pure solid water. The freezing may be accomplished at more elevated temperatures - about 10°C - if hydrates of hydrocarbons or freons are crystallysed out. Upon melting of the ice crystals fresh water is obtained. When hydrates are decomposed the mixture of water and organic liquid is obtained.

The desalination by pressure dialysis or piezodialysis is in the early research stage. In piezodialysis specific ion exchange membranes, so called mosaic membranes, similar to those used in electrodialysis are combined with pressure as moving force, therefore the process principles and those developed for equipment are expected to be similar to reverse osmosis.

Energy Consumption and Cost of Desalination

Desalination is an energy consuming process. In theory the heat of evaporation is entirely recovered during condensation. Nevertheless, because specific heat and evaporation heat of water are very large, huge quantities of heat are transported and the associated losses are considerable. The distillation trains are characterized by the evaporation ratio of about 10, indicating that 10 tons of distillate is produced per ton of low pressure heating steam. The costs of distillate in the American distillation plants of

1 Mgal capacity were about ₿ 0,25 per ton, and were not substantially reduced in more recent larger second generation distillation plants. The consumption of energy is several times lower when desalting brackish water using membrane methods. The cost of desalted water is also considerably lower. In the last years the costs of desalted water increased substantally following inflation and increase of fuel and energy cost. This is true for distillation, where the cost of energy amounted to 20-40% of the reached 60-80%.

Despite of raising desalination cost the cumulated world desalination capacity is expected to double in the years 1975-80 owing to erection of new plants in the Middle East, America and Soviet Union.

The literature on desalination is now rather very voluminous. The newcomer in the field will be best served by the short monograph by Spiegler, 1977. A systematic review is presented in a quite recent Gmelin volume (Delyannis, 1974). For those working in the field the journals: Desalination and Membrane Science, published by Elsevier, are of first rank importance.

The above review indicates that there exist several technical possibilities of recovery of water from sea water and other saline waters. The limitations are rather of economical nature.

DESALINATION OF SALINE MINE WATERS

Desalination of Saline Mine Waters Abroad

The occurrence of saline mine waters is a very common phenomenon, but very few examples of management or utilization of these waters are to be found in the literature.

A large electrodialysis plant was installed in order to utilize brackish mine waters in a South-African mine at the early stage of development of this desalination method. The plant of 11 000 m^3/d capacity operated at initial salinity of 3,1 g/L. After a few years of rather successful operation the plant was dismantled when the salinity of the mine waters diminished substantially. The venture was desribed in detail in a monograph (Wilson, 1960).

The Southeast part of the United States - the State of Pennsylvania and parts of other states, are plagued by acid mine waters. Drainage from coal mines is a significant source of pollution of rivers and streams. Oxygen from the air reacts with the pyritic sulphur that is present with the coal, and sulphuric acid is formed. The acid is then dissolved by water; this solution also dissolves significant quantities of iron, calcium and magnesium. Iron hydroxide is precipitated in the stream beds upon aeration. This precipitate presents serious hazards to fish and wildlife as well as to municipal water supplies. The drainage is acid with pH 2,5 to 3,5 and may contain a few grams of TDS per litre. The reverse osmosis process was tried with success at two abandoned coal mines in Pennsylvania (Schultz and Newley, 1966).

The Donbass region in the Soviet Union is a densely populated and industrialized area with important coal mining. The rivers of this area are rather small with variable flow, some of them drying out completely in summer. About 500 000 m^3 daily of saline mine waters were drained about 1970. The project of building a pipeline to the Sivash Gulf for draining mine waters and industrial waste waters was discussed at that time, coupled with LTV evaporation of some brines, and electrodialysis for desalination of brackish waters (Sergeenko, 1967).

Characterization of Coal Mine Waters in Poland

In Poland, the principal sources of salinity handicapping the rivers and streams are mining of coal, copper and sulphur as well as the chemical manufacturing, with soda industry in the first place. The principal source of saline waters is the black coal mining. Other operations, like mining of sulphur and copper are less important, and to date present no immediate menace.

There are three main black coal mining regions in Poland. Two of them, the Wałbrzych and Lublin regions, do not present serious problems now and during the next decade. In the Upper-Silesian Coalfield huge quantities of saline mine waters are pumped, with salinity ranking from potable water to the brines of 200 g salt per litre. The main constituent of those waters is sodium chloride, reaching up to 90 per cent of TDS (less than 80 per cent in seawater). Generally, the concentration of salt in mine waters is increasing with deepening of coal mines and mining of the most valuable coking coal.

The area of the Upper Silesian Coalfield, almost at the sources of the Odra and Vistula rivers, creates very unfavourable conditions for the drainage of saline mine waters. The situation is worsened by the considerable concentration of other industries and the high density of population. There is a general shortage of usable water in this region.

According to the recent inventory more than 900 thousand m^3 of mine waters are pumped each day. The less concentrated waters are used for municipal and industrial use. There remain about 250 thousand m^3 daily of more saline waters with daily load of 3300 tons of salts. The load of salts in the Odra tributary prevails with 2100 t/d of salts, thus raising the level of the sum of chloride and sulphate ion at the average low water level by 800 mg per litre. It was found that six mines are contributing in an excessive manner to the above indicated salt load. The salt load contained in 15,4 thousand m^3 of brines with salinity of more than 70 g per litre equals to 1.220 tons per day.

By utilization of the rather small volume of these brines, the burden of salts released into Odra river may be halved (Gisman and Szczypa, 1979).

Utilization of the Coal Mine Brines by Evaporation

The investigation centered on solving this problem started in the early sixties, in the Central Mining Institute in Katowice. MSF distillation method was adapted to evaporation of the brines and crystalization of sodium chloride. The salient feature of this technology is equalization of the concentration of Ca^{++} and SO_4^{--} ions followed by crystallization of calcium sulphate dihydrate on seeds of appropriate size of the same substance circulated in the brine. Therefore, besides distillate and table salt, the gypsum is the one more product leaving the plant. It is dumped along with some quantity of the magnesium and potassium salts purged as the concentrated brines upon the crystallization of salt (sodium chloride.).

Following the exploitation of a pilot plant, a demonstration plant was put in operation, in 1975, for utilization of 2400 m^3 daily of brines with salinity of about 100 g per litre (Fig. 5). The plant is composed of ten flash evaporators connected in series. Each evaporator is fitted with condenser acting as a regenerative heater of the incoming brine except for the three

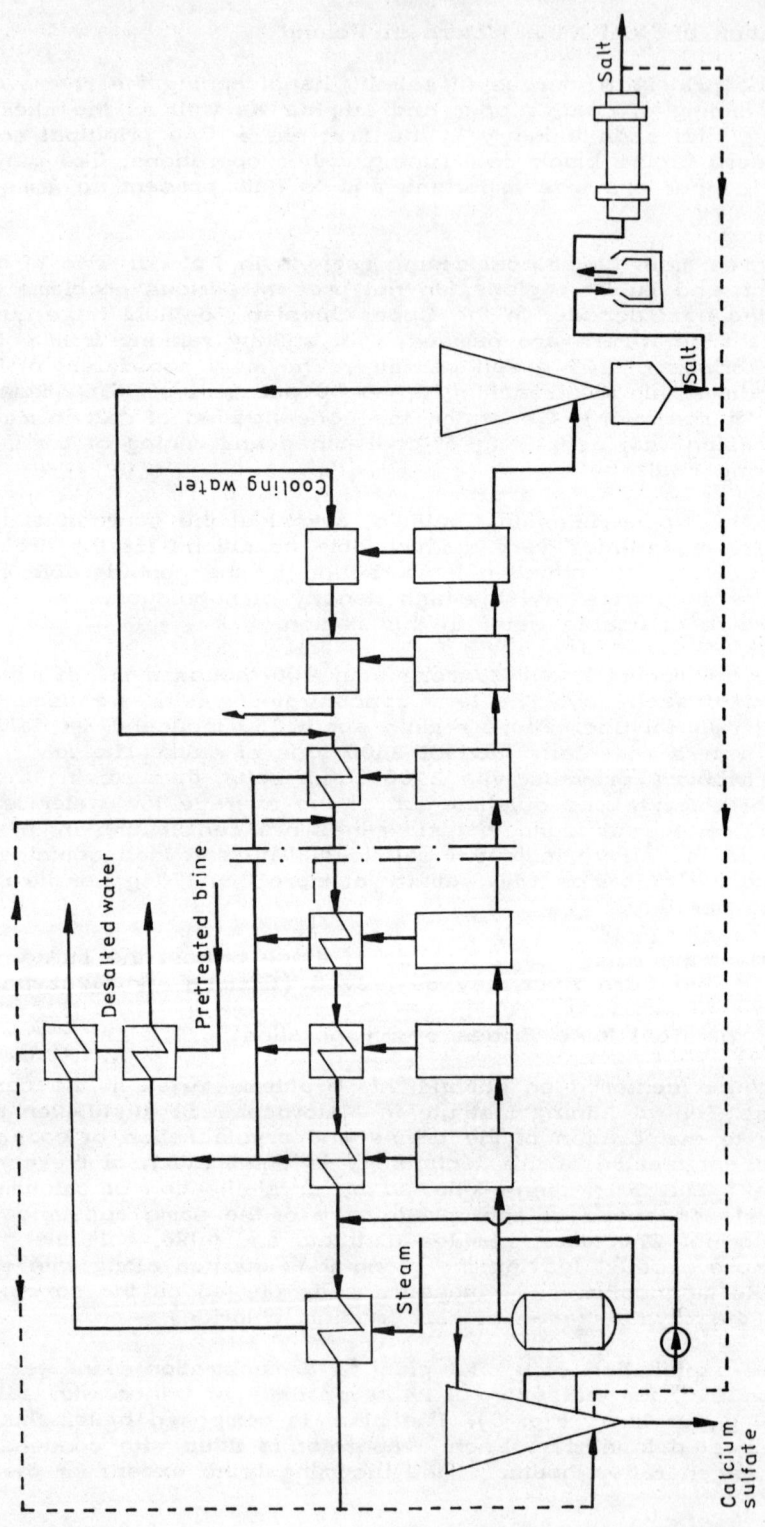

Fig. 5. Flow-sheet of the Central Mining Institute plant for utilization of coal mine brines.

last evaporators cooled with rejected water. Before entering the first flash evaporator the brine is heated to 120°C in the steam heater and passed through the gypsum crystallizer where calcium sulphate dihydrate is precipitated in the form of a loose suspension with particle size less than 20 μm. The gypsum crystals are separated in sand filters. Sodium chloride is crystallized in the last stages, and upon separation in the salt separator is further processed. No mesh separators are used in the flash evaporators this resulting in about 100 mg salt per litre of distillate. The salt contains 98,6 - 98,8 per cent sodium chloride (for flow-sheet-see Fig. 5).
The preliminary results of running the plant were reported elsewhere (Motyka and Szczypa, 1976). During the four years of exploitation, the refinement of the process along with changes in the equipment were carried out as reported recently. The most important changes were dealing with pH control and simplification of the removal of suspension of calcium sulphate dihydrate from the brine. Following extensive investigations the heat exchangers were converted to appropriate stainless steel tubing and improved layout with vertical position of the exchanger tubes. Also the influence of process parameters and of the quality of incoming brine on the purity of the product salt was studied. The process and equipment may now be characterized as mature and reliable. A series of plants of appropriate size may now be designed in order to utilize the most concentrated coal mine brines (Motyka, 1979).

In order to perfect the process of utilization, the application is studied of valuable components of the mother liquor extracted from the plant. Potassium chloride, magnesium salts, bromine and iodine may be recovered. To date most effort is devoted to the utilization of magnesium content in order to obtain magnesium oxide for the refractory industry (Motyka and Gunia, 1978).

Utilization of Coal Mine Brines in Electrochemical Industry

The research centered on use of the most concentrated coal mine brines, for the production of chlorine and caustic was initated in the Department of Applied Electrochemistry of the Silesian Technical University in Gliwice a few years ago. Recently, the investigations are under way concerning a coal mine with the output of some 1000 m^3 daily of brine with salinity of 140 gram salt per litre, in the close vicinity of an electrochemical plant.

The use of diaphragm cells is envisaged and a suitable pilot plant is under construction. At the same time the purification of the brines is being investigated. In order to get rid of excessive magnesium content the appropriate two-step purification scheme was worked out enabling easy sedimentation and filtration of magnesium hydroxide precipitate. Three variants of brine preparation, namely saturation with salt, concentration by evaporation and intermediate method were worked out (Gnot, Turek and Walburg, 1979).

Desalination of Brackish and Moderately Saline Mine Waters

The brackish and moderately saline waters do occur in large quantities in the whole Upper-Silesian Coalfield, most of them in the Vistula tributary. The volume and the salt load of these waters are expected to increase substantially in the next decade following the development of coal mining now under way. Therefore a research program on management and utilization of these waters was started in 1974, under supervision of the Central Mining Institute.

The general concept is to use the brackish waters for coal purifying and to apply them in other branches of industry. The moderately saline waters-above 3 g/l salt - are to be preconcentrated by membrane methods in order to obtain potable water and brine concentrate. The concentrate is to be utilized by evaporation as described above in order to obtain salt and distillate (Kępiński, Lipiński and Chlubek, 1976).

The commercial development of reverse osmosis at the time of starting the program was limited to brackish waters, mainly in the 2-3 g/l range of salinity with research in Israel up to 6 g/l. Therefore the investigation was started using commercial tube and hollow fibre B-9 modules in order to establish the performance at initial salinities up to 30 g/l and more. The commercial availability of B-10 modules enabled desalination of the waters of such salinity in one-step installation. Nevertheless, the use of the B-9 or similar modules may be advisable at lower salinities in view of their flux characteristic. A suitable pretreatment scheme was elaborated along with the desalination (Kępiński and Lipiński, 1978). A pilot plant of 200 m^3/d capacity was designed, and a feasibility study of the full scale plant was also carried out. Assuming the technical feasibility of the utilization scheme, full scale cost estimates indicate that the investment cost is about the same at 6 g/l and 30 g/l initial salinity, provided the right array of B-9 and B-10 modules is used in the first case, and one-step desalination using B-10 modules is applied in the second case (Kępiński and Lipiński, 1979).

The decisions on building the pilot plant and implementing the desalination and utilization scheme of moderately saline waters are pending.

At the same time some action was taken in order to prevent or reduce the salt burden drained to the most handicapped small effluents of the Vistula by building distribution reservoirs and planning the pipelines (Gisman and Szczypa, 1979).

REFERENCES

Delyannis, A., and E. Delyannis (Eds.), (1974). Gmelin Handbuch. Sauerstoff. Anhangband. Water Desalting - Wasserentsalzung. Springer, Berlin.

Gisman, S., and H. Szczypa (1979). Protection of environment against the harmful influence of saline mine waters. Sci. Papers of the Szczecin Technical Univ., N° 123, 7-14.

Gnot, W., M. Turek and Z. Walburg (1979). Utilization of coal mine brines for chlorine production. Sci. Papers of the Szczecin Technical Univ., N° 123, 51-60.

Kępiński, J. and K. Lipiński (1978) Pretreatment of brackish mine waters for desalting by reverse osmosis. Proc. 6-th Int. Symp. Fresh Water Sea (Las Palmas). Athens, Vol. 3, 391-394.

Kępiński, J. and K. Lipiński (1979). Technical and economical aspects of desalination of brackish mine waters by reverse osmosis. Sci. Papers of the Szczecin Technical Univ., N° 123, 75-84.

Kępiński, J. K. Lipiński and N. Chlubek (1976). Desalination of brackish mine waters by reverse osmosis. Proc. 5-th In. Symp. Fresh Water Sea. (Alghero). Athens, Vol. 4, 347-352.

Motyka L, (1979). Improvement of process and equipment for utilization of coal mine brines based upon the operation of a demonstration plant. Sci. Papers of the Szczecin Technical Univ., N° 123, 15-27.

Motyka, L and W. Gunia (1978). Wasteless utilization of saline mine waters. Proc. 6-th Int. Symp. Fresh Water Sea (Las Palmas). Athens, Vol. 1, 123-134.

Motyka, L and H. Szczypa (1976). Desalination of mine waters in the Polish coal mining industry. Proc. 5-th Int. Symp. Fresh Water Sea (Alghero) Athens, Vol. 1, 37-44

Schultz, J., and G.A. Newly (1966). Desalination by reverse osmosis. In: Membrane Processes for Industry, Southern Res. Institute, Birmingham Alabama.

Sergeenko, I.L. (1970). Proc. 3-rd Int. Symp. Fresh Water Sea (Dubrovnik). Vol. 3, 491-500.

Spiegler, K.S. (1977). Salt-Water Purification. Plenum Press, New York.

Wilson, J.R., (Ed.). (1960). Demineralization by electrodialysis. Butterworths, London.

ION SELECTIVE ELECTRODES IN WATER AND WASTEWATER ANALYSIS

K. Sykut

Department of Analytical Chemistry and Instrumental Analysis, Institute of Chemistry, Maria Curie-Skłodowska University, Lublin, Poland

ABSTRACT

Ion selective electrodes (ISE) have been the subject of rapidly increasing interest to analysts because of several advantages among these are: the simple way of measurement and preparing of samples, short time of analysis, wide range of applications and low cost equipment.
Nowadays you can estimate about 25 species i.e. cations and anions. When the direct measurement method is applied, the response range for ISE is from $1 \cdot 10^{-5}$ to $0,1M$ (ca $0,5-5000$ ppm) with the most frequently the 1-1000 ppm one.
The estimation, according to the kind of sample, may be performed by the method of calibration curves, the standard addition substraction or titration method.
The best results can be achieved while using the special analysers with microprocessor.
The ISE have been applied in the environmental protection, to the pollution control of water, analysis of wastes and food products with good results.

KEYWORDS

Ion selective electrodes; wastewater analysis; water analysis.

Analytical problems are very important in the natural environment protection. They include pollution control of biosphere and geosphere as well as waste neutralization control, purification and water recovery processes and others.

Among instrumental analytical methods, electrochemical are of great importance because of their wide range of applications, high sensitivity, low costs of equipment, simplicity in use and possibility of out of laboratory measurement places. Table one gives the electrochemical methods already used in studies on natural environment protection. It is hard to say which of them is the best but it is

possible to state which is the best in a given problem solution. For example only stripping voltammetry can be used to determine metal contents in sea water because of its high sensitivity.
Ion Selective Electrodes have been the subject of rapidly increasing interest to analysts because of several advantages, such as a simple way of measurement, short time of analysis, wide range of applications and low cost of equipment. The theoretical background of ISE performance, measurement techniques and applications are included in the monographs by Darst (1969), Koryta (1972), Moode (1971), Camman (1973), Lakshminarayanaiah (1976), Balley (1976), Wessely (1977).

Table 1. Analytical Electrochemical Methods.

	detection limit		Main applications
	M/l	ppm	
DC-polarography	$1 \cdot 10^{-6}$	0,1	metals (cations), some organic compounds
AC-polarography	$1 \cdot 10^{-5}$	1	surfactants, detergents
SW-polarography	$1 \cdot 10^{-7}$	0,01	metals (cations)
Pulse polarography	$1 \cdot 10^{-8}$	1 ppb	" "
Stripping Voltammetry	$1 \cdot 10^{-11}$	1 ppt	" "
Ion Selective Electrodes	$1 \cdot 10^{-4}$ $1 \cdot 10^{-7}$	109 to 0,01	cations, anions, some organic compounds

Development of the ISE research, achievements in new types of electrode and measuring apparatus construction are discussed in the review articles by Buck (1972-1978), Pick, Hulanicki (1972).
ISE are succesfully employed in the investigations concerning the natural environment protection as their usage simplifies the complicated, time and labour consuming analytical procedures.
Detection of NH_3, NO_3^-, NO_2^- in water and natural products (cucumbers, tomatoes, carrots, e.t.c.) may serve as a good example.
Ion selective electrodes being in serial production and commercially available are given in Table 2.

Table 2. ISE Being in Serial Production

Type of electrode	Solid membrane				
electrode	Cd^{2+}	Cu^{2+}	Ag^+	Pb^{2+}	Na^+
detection limit ppm	0,011	$6,3 \cdot 10^{-4}$	0,01	0,02	0,023
electrode	Br^-	Cl^-	CN^-	F^- J^- S^{2-}	SCN^-
detection limit ppm	0,4	1,7	0,02	0,02 0,02 0,003	0,3

Type of electrode	Liquid membrane				
electrode	Ca^{2+}	K^+	BF_4^-	NO_3^-	ClO_4^-
detection limit ppm	0,4	0,4	0,26	0,37	0,20

Type of electrode	Gas electrodes				
electrode	NH_3	CO_2	H_2S	NO_x	SO_2
detection limit ppm	0,017	4,4	0,034	0,022	0,192

They are classified into three groups according to their construction: solid state sensor electrodes of single crystals or pressed polycrystalline powder, liquid-membrane electrodes and gas sensing probes. To appreciate their application, the detection limit in ppm was given. The table does not include all types constructed so far. Some new electrodes have been constructed (literature reports) to determine surfactants, enzymes and some chemical compounds. The development of the ISE research provides the evidence for electrode construction of the required properties i.e. selective in relation to a particular chemical species. To appreciate the full application of ISE one must get familiar with their properties and parameters (Fig. 1).

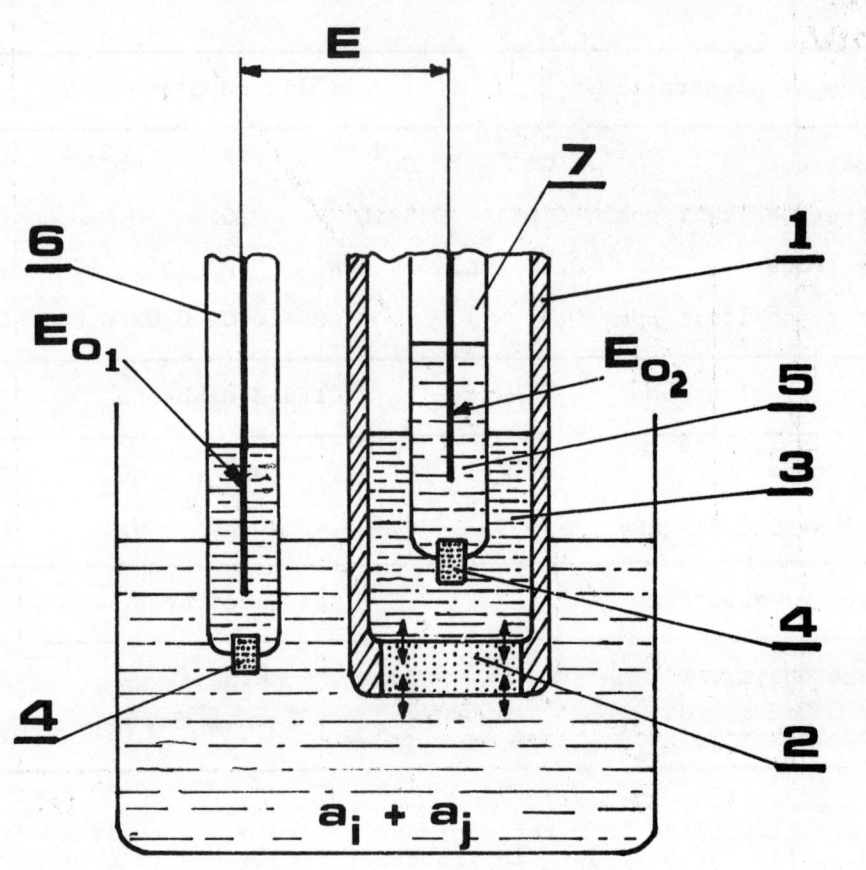

Fig. 1. Measuring cell with an ion selective electrode.
1 - ion selective electrode; 2 - membrane;
3 - internal solution; 4 - electrolytic key;
5 - internal solution of the electrode;
6,7 - reference electrodes;
a_i - main ion; a_j - interfering (disturbing) ion.

ISE, a reference electrode and an examined solution form a measuring cell of which characteristics is given on the plot with the coordinates; potential - and logarithm of the main ion activity (Fig. 2).
The characteristics is given from Nikolski-Eisenman equation:

$$E \ E \ Est \pm \frac{RT}{nF} \ln \left(a_i + k_{ij} \, a_j^{z/n} + 1 \right)$$

where a_i is the main ion activity
a_j is the interfering ion activity.

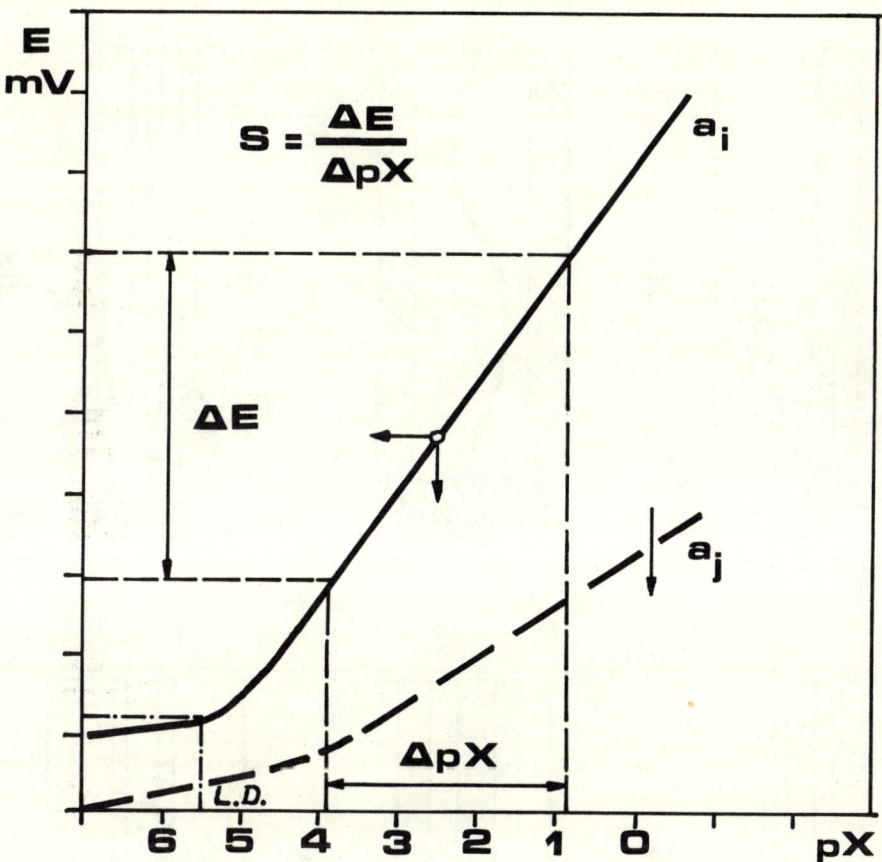

Fig. 2. Characteristics of an ion selective electrode.
k_{ij} is the activity coefficient
l_{ij} is the limit value.

Ion selective electrode constructions are given in Fig. 3. While measuring with ISE one must remember that the electrode potential is the function of the main ion activity.
Using solutions of the constant ion strength concentration calibration curves, very useful in practice, are obtained. Studying samples of the unknown ion strength, "known addition" methods giving reliable results are used. Development of measuring apparatus from a simple compensating potentiometer through a digital milivoltmeter to the universal apparatus equipped with a microprocessor enables the automatic calculations in any measuring technique. That makes the analysis with ISE quicker and more precise. With the direct method of measurement the response range of most ISE is from $1 \cdot 10^{-5}$ to 10^{-1} M/l i.e. about 0.5 to 5.000 ppm.

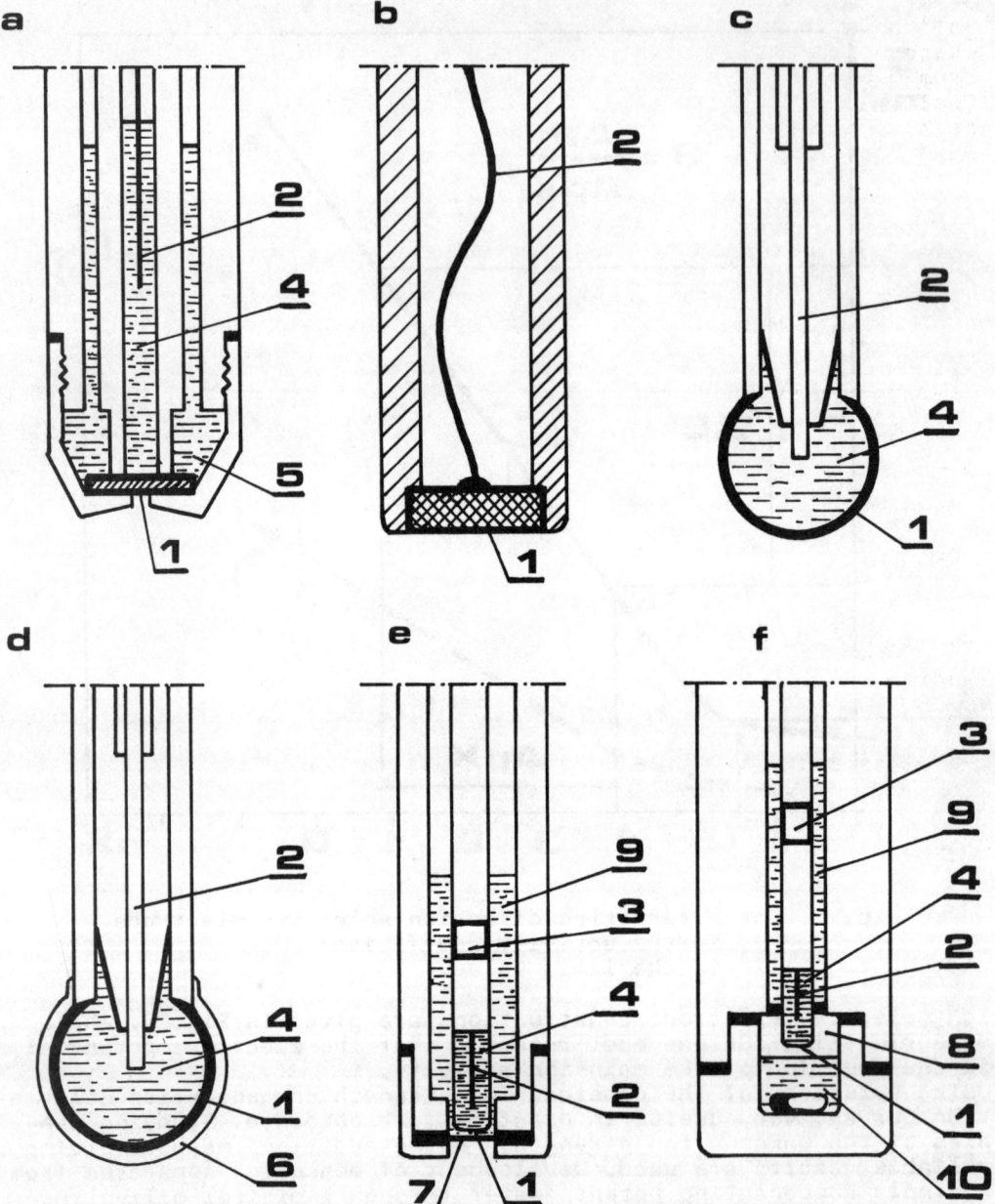

Fig. 3. Types of Ion Selective Electrodes: a - liquid membrane electrode; b - solid membrane electrode; c - glass electrode; d - enzyme electrode; e - gas electrode; f - gas electrode with an air gap.
1 - ion selective membrane; 2 - internal electrode; 3 - reference electrode; 4 - internal solution; 5 - liquid ion exchanger; 6 - enzyme layer; 7 - membrane permeable to gas; 8 - air gap; 9 - reference electrode solution; 10 - sample.

Determinations are generally carried out in the concentration range from 1 to 1000 ppm. The admissable toxic concentration of substances in water, air, food products (meant for consumption) is from 0.01 to 5 ppm i.e. close to the limit sensitivity of ISE. Quality and characteristics of an electrode as well as measurement methods are very important. Ion selective electrodes with solid membranes and gas sensing probes are the best in very low concentration measurement of strongly toxic substances. Other instrumental methods possess more advantageous detection limits. Table 3 gives the comparison of ISE and absorptiometric methods in water pollution determination.

Table 3. Comparison of Water Pollution Limit Detection (ppm) for Absorptiometric and Potentiometric - ISE Methods.

cation	NH_4^+	Cd^{2+}	Cu^{2+}	Pb^{2+}
admissable concentration	0,5	0,01	1	0,05
absorptiometric methods	0,001	0,05	0,1	0,04
ISE method	0,017	0,011	$6,3 \cdot 10^{-4}$	0,025
anion	CN^-	F^-	NO_3^-	Cl^-
admissable concentration	0,2	2	10	250
absorptiometric methods	0,02	0,01	0,02	0,04
ISE method	0,026	0,019	0,37	1,7

It comes out that ISE methods are sensitive enough in these determinations. Yet, for an experienced analyst, limit detection numbers are not the only criterion for the choice of a method. Sample composition, in particular, contents and kind of interfering substances determine the proper analytical method.
In these cases ISE application often gives very good effects. Water is a suitable object for using ISE methods because of low ion strength of samples and almost neutral reaction. There are suitable electrodes for all ions (25) which can contaminate water. It is easy to use monitors (continuous action analysers) for water being in communal use. The addition method is recommended in water analysis. Wastewater is a more complicated object but as its concentration is greater before purification, requirements for the ISE are smaller. In practice the known qualitative composition of industrial wastes makes the choice of an analytical procedure easier.
The electrode properties (selectivity) can be "improved" by a proper choice of adjustors and condisoles i.e. solutions determining pH and ion strength values as well as masking interfering substances in a sample.

Analytical procedure preparation is not difficult but requires knowledge of ISE properties. In practice it consists in mixing a sample and solution in a proper ratio and EMF measuring. It does not require highly qualified personnel.

Concentration measurement accuracy with ISE depends on a method. Determination with a (single) standard addition to a sample or inversely and indirect determination have the relative error from 2 to 10 per cent for n = 1 and from 4 to 10 per cent for n = 2.

The relative error of the direct measurement method is from 1 to 5 per cent for n = 1 and from 2 to 10 per cent for n = 2.
Gran's and titration methods are the most accurate because the relative error is about 0.5 per cent. These methods are more time and labour consuming and do not give good results for very low concentrations close to the detection limit.
Though there are more and more papers devoted to ISE applications in water and wastewater analysis, they are not widely used as newly introduced into practice they have not become very popular. Besides they are relatively expensive.

REFERENCES

Durst, R.A. (1969). Ion Selective Electrodes, NBS, Spec.Publ.3,4.
Moody, G.J., J.D.R.Thomas (1971). Selective Ion Sensitive Electrodes,Merron.
Koryta, J. (1972). Jontowe selektiwni membranowe elektrody. Academia - Praha.
Cammann, K. (1973). Das arbeiten mit Ionoselektiven Elektroden, Springer - Verlag, Berlin.
Lakshminarayanaiah, N. (1976). Membrane electrodes, Academic Press, New York.
Bailley, P.L. (1976). Analysis with ion - selective electrodes, Heyden - London.
Wessely, J.D., D.Weiss, K.Stulik. (1977). Analysis with ion - selective electrodes.
Buck, (1972, 1974, 1976, 1978). Anal.Chem.Revievs.
Pick, J. Hung.Scient.Instruments. 34, 41, 42, 43.
Hulanicki, A. (1972). Chem.Anal. 17, 217-242.

PURIFICATION OF WASTEWATERS FROM A PIG FARM BY MEANS OF ELECTROLYSIS

Z. Drabent, J. Dziejowski and L. Smoczynski

Institute of Agricultural Chemistry, AR-T, 10-728 Olsztyn, Poland

Abstract - A laboratory test for purification of wastewaters by means of electrolysis in a static system and continuous flow of wastewaters between the electrodes has been described. The electrolysis was conducted with the application of iron electrodes and current density 24 mA/cm^2, during 5 and 10 minutes. For comparison chemical coagulation of wastewaters was conducted by using $FeSO_4$. It was found that as a result of electrolysis the pH of wastewaters increases simultaneously with the reduction of nitrogen phosphorus, calcium and COD. From the calculated amount of energy consumption together with the purification effects /also visual/ obtained, this method seems to be useful in purification of wastewaters from pig farms.

Keywords: animal wastes, electrolysis, iron electrodes, coagulation

INTRODUCTION

Farms are a source of a considerable amount of wastewaters and cause a threat to the natural environment. Most of the farm wastewaters are utilized in agriculture, but the rest is put through a special treatment in order to remove the undesirable properties resulting from the character of the waste /Baader and others 1972, Loehr 1974, Taiganides 1977/. Animal wastes are characterized by a considerable amount of solid substances and suspensions, organic and inorganic compounds showing considerable toxity in some cases /Wadekind and Koriath 1969, Middlebrooks 1974/. They are a carrier of pathogenic organisms, and their improper utilization is dangerous for human and animal health /Strauch 1972/. In larger quantities they unfavourably affect soil, contaminate the surface and ground waters and atmosphere round the farm /Bardtke and Jeserich 1972, Pratt 1979/. The hitherto applied technologies of wastewaters treatment do not meet sufficiently the demands put forward by agrotechnics, zootechnics, economics and water protection /Baader and others 1972, Taiganides 1977, Mazur and Mackowiak 1978, Wilkinson /1979/.

ELECTROLYSIS

The technologies of animal wastes treatment can be divided into:

physico-chemical, biological and combinations of the above technologies. The method based on the flow of electric current through the waste is of particular interest /Belau and co-workers 1976, Cross 1966, Nurnberger and co-workers 1966, Kalisch and co-workers 1970, Schwab and co-workers 1975/. In an animal waste colloidal micelle, ions of organic and inorganic compounds are displaced to the corresponding electrodes under the influence of direct electric current, where they can undergo suitable physico-chemical changes /coagulation, redox processes, precipitation of flocks, etc./. The variety of phenomena occuring in the process of current flow through animal wastes /electrolysis, redox processes, electro-dialysis, electro-osmosis, electrophoresis, coagulation, precipitation of flocks, various secondary reactions etc./ is the cause of difficulties in accurate determination and prediction of the most favourable conditions for the precipitation of flocks and the improvement of wastewaters treatment. Kalisch and others /1970/ and Belau and others /1976/ reported that wastewaters from a pig farm purified by electrolysis are characterized by a reduction in contamination charge determined by COD and BOD. Electrolysis causes almost a complete removal of phosphorus, reduction of iron and calcium content and reduction of the amount of bacteria. However, Schwab, Strauch and Muller's /1975/ investigations showed that this method cannot be applied in practice to the sterilization and disinfection of wastewaters. While examining the purification of animal wastes by electrolysis, Kalisch and others /1970/ used various electrodes /carbon, platinum, aluminium and iron/ of different surfaces. They found that the use of iron electrodes was the most feasible. It is obvious that in the course of electrolysis of animal waste the process of anodic dissolution takes place and simultaneously the temperature of the system increases. Therefore a question arises if it is possible to achieve the same practical effect without using current by means of the selection of suitable ferrous ion concentration, pH and temperature, i.e. by a chemical method only. From scientific data it appears clear that before the introduction of electrolysis in the treatment of animal wastes numerous investigations should be carried out on a laboratory scale

This paper presents the results obtained from the examinations of comparative changes taking place in the chemical composition of wastewaters purified by means of electrolysis with the use of iron electrodes and purified by coagulation by means of $FeSO_4$ /ferrous sulphate/. The electrolysis of wastewaters was conducted by two methods: in a static system /ES/ and by continuous flow of current through the space between the electrodes /ED/.

EXPERIMENTAL

Pig wastes were collected from the purification plant located at the farm of the type "Agrocomplex" with an annual production of 15000 pigs and an average stock of 10000 head. The examined wastes were passed through the vibrating screen before they were used. The biologically purified wastes were collected from a settling tank connected with an aeration tank. The purification of waste was carried out on a laboratory scale in the device shown in fig. 1. The wastes continually mixed by a magnetic stirrer were introduced from the bottom to an electrolyser by means of a dosing micropump. The iron electrodes were supplied with a direct and variable current feeder of $I_{max}=5A$. In the course of electrolysis the direct current of 2A was maintained by means of a feeder potentiometer.

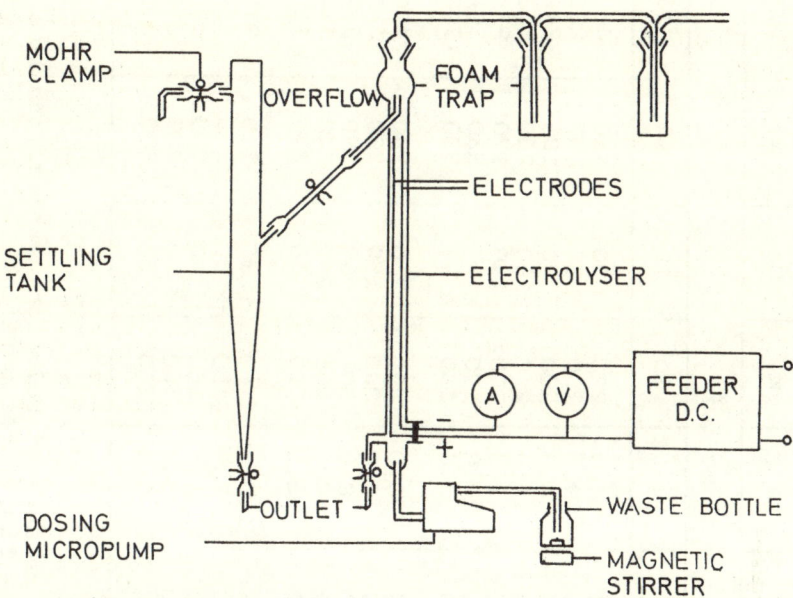

Current density was 24 mA/cm^2 /$\bar{U}$=10V/

Electrical energy consumption:
- in the course of 5 minutes electrolysis 8,3 kWh/m^3 of wastes
- in the course of 10 minutes electrolysis 16,6 kWh/m^3 of wastes

Consumption of iron was respectively:
- 0,87 kg anodic iron/m^3 of wastes
- 1,74 kg anodic iron/m^3 of wastes

Average increases of temperature were:
- 5 minutes ES - +13K , 10 minutes ES - +13,5K
- 5 minutes ED - +8K , 10 minutes ED - +12,0K

<u>Fig. 1</u>   A device for electrolytic purification of wastewaters

The changes of the potential were also noted /U=10±2V/. Iron electrodes of size 40·1·0,05 /cm/ were used. The distance between electrodes was about 1 cm. The capacity of electrolyser to the level of "overflow" was 0,2 dm^3. In a static system the electrolyser was filled up with the wastewaters to the overflow level at the closed inflow and outflow. In a dynamic system the outflow to the settling tank was open and the wastes were proportioned in order that the exchange time of the whole electrolyser volume was 5 or 10 minutes, which corresponds with the electrolyser load of volume 1 dm^3, respectively, 12 dm^3/h or 6 dm^3/h. The temperature was measured before and after purification. After electrolysis the whole volume of the electrolyser /ES/ or the settling tank /ED/ was used for analyses, however, collecting the solution in the settling tank took place after 30 minutes of flow, i.e. after the settling of stable parameters which characterized the work of the electrolyser. Each

Table 1   Average results of investigations concerning purified animal wastes

Number of measuring series	Method of purification	t /s/	pH	T /K/	N_K	N_{NH_4}	N_{org}	P	K	Ca	Na	COD
							mg/dm^3					
1	2	3	4	5	6	7	8	9	10	11	12	13
Waste No I crude 01.79	-	-	7,50	291,5	1260	665	595	727,0	220	620	310	4880
	ES	300	9,40	306,0	660	525	135	19,0	200	80	200	2860
	ES	600	9,60	309,0	565	520	45	4,5	200	70	180	2120
	ED	300	9,00	297,0	715	580	135	14,5	210	160	200	3630
	ED	600	9,40	301,5	660	570	90	2,0	200	140	170	2740
	K5	-	8,50	-	695	570	125	20,0	190	-	230	3940
	K10	-	8,50	-	590	530	60	33,0	190	-	230	2960
Waste No II crude 11.78	-	-	7,30	291,5	1490	810	680	790,0	290	780	240	4500
	ES	300	9,50	304,0	855	615	240	15,0	280	150	260	2420
	ES	600	9,40	311,0	820	750	70	8,0	300	60	280	2400
	ED	300	9,20	299,0	1025	695	330	14,5	290	200	240	3340
	ED	600	9,30	303,5	870	780	90	3,0	280	130	270	2850
Waste NoIII 11.78 B.P.	-	-	7,70	290,0	735	370	365	120,0	150	230	210	2770
	ES	300	9,30	302,0	495	380	115	3,0	150	50	200	2080
	ES	600	9,50	303,0	430	360	70	1,0	160	40	200	1900
	ED	300	9,10	300,5	510	380	130	5,0	160	80	220	2150
	ED	600	9,40	303,5	465	375	90	1,0	150	60	200	2090

In the case of coagulation calcium was not determined because of solution alkalization by lime milk! K /5,10/ - chemical coagulation , B. P. - biological purified operation was repeated three times and average results from the three operations are shown in TABLE 1. For each repetition of the electrolysis the direction of the current on the electrodes was changed. After two hours the solution which appeared over the precipitate was collected for chemical determination. The analyses were carried out according to the standard methods /Hermanowicz and others 1976/. The coagulation of wastewaters by means of $FeSO_4$ was conducted in the following way: the corresponding amounts of $FeSO_4$ calculated according to Faraday's law were introduced to 200 cm^3 of waste with continuous mixing. The $FeSO_4$ amounts were equivalent to the amounts of Fe evolved in 5 and 10 minutes of electrolysis. Afterwards the solution was alkalized by lime milk to pH = 8,5 and further operations were carried out analogically to the experiments with electrolytic coagulation. The obtained results are presented in TABLE 1.

DISCUSSION

In the experiments the crude, grey and green, turbid wastewaters showed incomplete clarity after 5 minutes electrolytic purification whereas after 10 minutes complete clarity was observed. The solution over the flocks, dark blue and green precipitate /after some time becoming russet showed complete clarity of light-honey colouring. The flocks of the precipitate which appeared in the course of electrolysis settled quickly. Enormous amounts of foam were formed on the surface of the wastes, and wastes pH increased to 9 - 9,6. The loss of organic nitrogen within the range of 79 - 90% was observed in the case of chemical coagulation and about 90% during 10 minutes of electrolysis with a slight decrease in ammonia nitrogen content. Phosphorus was removed from the wastes almost completely and the calcium content in the waste purified by means of electrolysis was maintained at a low level of about 8 - 23% in comparison with the initial sample. The decrease of contamination /determined in COD/ by about 19 - 30% was observed in the course of coagulation whereas for 10 minutes static electrolysis was 46 - 53%, and for 10 minutes ED 37 - 40%. The loss of the mentioned compounds after purification in the solution over the precipitate indicates that they are accumulated in the precipitate enriching thus the potential fertilizer with such compounds as N, P, Ca. The obtained results compared with those of other authors /Belau and co-workers 1976, Cross 1966, Kalisch and co-workers 1970, Nurnberger and co-workers 1966, Schab and co-workers 1975/ indicate that the electrolytic method leads to a more efficient removal of phosphorus, organic nitrogen and impurities determined in COD of wastes, as compared with chemical coagulation. The comparison of the results obtained after the electrolytic purification of wastes in the static and dynamic system shows negligible differences in the results of purification in favour of the static method. The specification of the data on crude wastewaters and wastewaters purified biologically does not permit to draw a final conclusion on the subject concerning the application of biological purification before electrolysis. However, it should be emphasized that the absolute content of N_{org}, P and Ca, and also impurities determined in COD, is lower in the wastewaters previously purified biologically than in crude wastewaters. The presented results suggest that further experiments in this field should be continued.

REFERENCES

Baader W., R. Thaer, H. Traulsen /1972/. Verfahren zur Behandlung von Abfällen der tierischen Produktion. Berichte über Landwirtschaft. 50, 612-627.

Bardtke D., G. Jeserich /1972/. Einfluss von Abfällen und Ausscheidungen der tierischen Produktion auf Wasser und Gewässer. Berichte uber Landwirtschaft. 50, 666-674.

Belau L., H. G. Hummel, W. Langecker /1976/. Gegenwärtiger Stand der elektrischen Gülleaufbereitung. Wasserwirtschaft - Wassertechnik. 3, 90-93.

Cross O. E. /1966/. Removal of moisture from poultry waste by electro-osmosis /Part 1/. In: Management of farm animal wastes. ASAE Publ. No. SP - 0366/1966. Amer. Soc. Agric. Engrs., St. Joseph Mich., 91-93.

Hermanowicz W., W. Dożanska, J. Dojlido, B. Koziorowski /1976/. Physical and Chemical Examination of Water and Wastewaters. Warszawa - Arkady.

Kalisch H., K. Krannich, L. Türpitz /1970/. Versuche zur Behandlung der flüssigen Phase von Schweinegülle mit Gleichstrom. Deutsche Agrartechnik. 20, 181-183.

Loehr R. C. /1974/. Agricultural Waste Management. Academic Press, New York and London.

Mazur T., Cz. Mackowiak /1978/. Liquid Manure Fertilization. PWRiL Warszawa.

Middlebrooks E. J. /1974/. Animal Wastes Management and Characterization. Water Research. 8, 697-712.

Nurnberger F. V., C. J. Mackson, J. Davidson /1966/. Removal of moisture from poultry waste by electro-osmosis /Part 2/. In: Management of farm animal wastes. ASAE Publ. No. SP - 0366/1966, Amer. Soc. Agric. Engrs., St. Joseph Mich., 93-95.

Pratt P. F. /1979/. Management Restriction in Soil Application of Manure. Journal of Animal Science. 48 /1/, 134-143.

Schwab H., D. Strauch, W. Müller /1975/. Hygenische Beurteilung technischer Verfahren zur Flussigmistbehandlung. Berliner und Munchener Tierärztliche Wochenschrift. 88, 184-187.

Strauch D. /1972/. Hygienische Anforderungen an die Verfahren zur Behandlung tierischer Abfalle und Ausscheidungen. Berichte über Landwirtschaft. 50, 602-611.

Taiganides E. P. /Ed./ /1977/. Animal Wastes. Applied Science Publishers Ltd. London.

Wadekind P., H. Koriath /1969/. Substanz und Nahrstoffgehalt der Gulle. Feldwirtschaft. 7, 319-320.

Wilkinson S. R. /1979/. Plant Nutrient and Economic Value of Animal Manures. Journal of Animal Science. 48 /1/, 121-133.

NEW COAGULANTS FOR TREATMENT OF PIGGERY WASTE-WATER

M. Rutkowski, H. Pielichowski, T. Tanowy* and
M. Korczak*

*Institute of Chemistry and Technology of Petroleum and Coal,
Wrocław Technical University, Poland*
*Institute of Environmental Development, Wrocław, Poland

ABSTRACT

The paper presents results of initial studies on possibilities of replacement of aluminium sulphate usually applied as a piggery wastewater coagulant by the new reactants. Research works were initiated by the fact that imported aluminium oxide is commonly used as a raw material to produce aluminium sulphate. The new reactants are manufactured from clay existing in the country and existent in large amount waste sulfuric acids after methyl methacrylate production. These new coagulants have a mixture of aluminium and ferric compounds acting as coagulant as well as active silica and carbonized organic compounds acting as adsorbent. Occurence of different amount of acid ammonium sulphate and of little amount of organic substance /below 2%/ in waste sulfuric acids requires not only determination of efectiveness of new reactants in comparison with aluminium sulphate but also determination of influence of new reactants on biological treatment of piggery wastewater. On the ground of initial laboratory studies was stated that employing twice bigger doses of new coagulants than aluminium sulphate it is possible to acquire the same effect of treatment.

KEYWORDS

Piggery wastewater; coagulation; basalt detritus; aluminium sulphate; activated sludge process.

INTRODUCTION

The protection of natural environment creates a constant demand for inexpensive and effective coagulants. In Poland mainly aluminium sulphate is used as a coagulant. The production of the coagulant is based on aluminium hydroxide, which is mostly imported. A domestic use of this expensive coagulant only for the treatment

of piggery wastewater comes up to several thousands tons per year and will increase considerably in near future.
Beside aluminium coagulants ferric ones can be applied for treating waters and sewage. To increase coagulation effects a mixture of aluminium and ferric coagulants is used very often.
In the technique of sewage treatment the use of additional reagents, which aids the flocculation and sedimentation of a sludge being precipitated is known. Among inorganic reagents these are clays rich in illite, kaolinite, montmorillonite or diatomite and silicate earth.
Very good results are obtained when chemically activated silica is applied. By introducing the activated silica together with the solution of aluminium and ferrum salts, the optimum range of coagulation pH is expanded, lower coagulation temperatures are possible and large easily settling floccules are formed faster. The volume of sludge is much lower than while using only metal salts. Contamination, which is not removed during the coagulation must be additionaly treated with activated carbon.
As a result of research carried out at the Institute of Chemistry and Technology of Petroleum and Coal on the usability of some domestic mineral raw materials for the industry, it has been found that samples of basalt eluvium deposit of Dunino by Legnica after treatment with sulphuric acid, provides active silica and rich solutions of aluminium and ferrum salts as well. Further research has shown that aqueous suspension of these salts with silica is an exceptionally good coagulant matching the expensive aluminium sulphate.
Detailed examination resulted in the development of a simple process for obtaining a coagulant containing aluminium sulphate, ferric sulphate and active silica.
Aluminium and iron contained in the deposit is transformed in about 95% a form soluble in the coagulant.
Later works dealt with the possibilites of applying a waste sulphuric acid obtained during the production of methyl methacrylate at the Chemical Works in Oświęcim, instead of the technical grade acid. This waste product contains about 10% of free sulphuric acid, about 50% of acid ammonium sulphate and about 2% of organic substances.
It was found that the basalt detritus of Dunino is especially good for the treatment with this acid, giving coagulants of alum type. Good solubility of the basalt detritus is caused by physical and chemical properties of its mineral components, especially halloysite, haematite and fine grain forms of kaolinite.
The method of producing the coagulant consists in dissolving the powdered basalt detritus in the waste sulphuric acid at the weight ratio 1:3 at 120°C for 3 hours. The suspension is neutralized an additional amount of the same detritus, depending on a desirable acidity.
For example, a coagulant neutralized with 33% of powdered basalt detritus had the following chemical composition:

Al_2O_3	-	6.4 to 6.6%
Fe_2O_3	-	4.0 to 4.1%
H_2SO_4	-	3.2 to 3.5%
insoluble components	-	18.0 to 18.5%

The coagulant contains about 5.0% of ammonia nitrogen. This nitrogen introduced with the coagulant to piggery wastes is a small percent of total nitrogen quantity and does not affect ditrimentally a further biological waste treatment. The above coagulant was marked as 3F-I for conveniency.
Further heating of 3F-I coagulant up to 350°C results in a significant removal of ammonia nitrogen and at a time carbonization of organic compounds contained in the coagulant. That means a formation of substances having additional absorptive properties.
This type of the coagulant was marked with FS symbol.

EXPERIMENTAL PROCEDURE

Besides aluminium sulphate /technical grade, 14% Al_2O_3/, two different coagulants were used: simple three-component 3F-I coagulant and its improved FS form. The latter was obtained by removing some ammonia nitrogen from the simple coagulant and by a carbonization of organic compounds. Piggery wastewater was taken from farm waste treatment plant in Rokitniki and Żródła.
The investigation of coagulation process consisted of two parts. At first optimum doses of coagulants were defined by means of jar test apparatus. Doses applied in the tests were as follows:

$Al_2/SO_4/_3$ - from 500 up to 1500 g/m^3
3F-I - from 1500 up to 3500 g/m^3
FS - from 500 up to 3500 g/m^3.

Then the coagulation of the piggery wastewater was carried out with the application of optimum reagent doses found. The original piggery wastewater and that coagulated was filtered and analyzed /for reaction, colour, turbidity, ammonia nitrogen, BOD_5 and COD/, post-coagulation sludge settling was defined in Imhoff's funnel.
An influence of new coagulants on the activated sludge was tested in batch system activated sludge process during 40 days.
Three tanks with the activated sludge were fed with wastes coagulated with aluminium sulphate, 3F-I and FS respectively. The effect of treatment was defined by the physical and chemical analysis as described above. Process conditions /dry suspended solids of activated sludge, volume index/ were checked and the sludge of every tank was analyzed under a microscope.

COAGULATION PROCESS

Basing on the jar test, the optimum coagulants doses were found depending on the degree of wastewater permanganate oxygen demand decrease as well as colour, turbidity and reaction /Fig. 1./.
The following doses are optimum ones:

$Al_2/SO_4/_3$ - 1250 g/m^3
3F-I - 3000 g/m^3
FS - 2500 g/m^3.

The coagulation with 1250 g/m^3 $Al_2/SO_4/_3$ dose resulted in a decrease of permanganate oxygen demand of wastewater by 35% and its turbidity by 95%. Its colour, however, increased by about 70%.

Respectively higher doses of the new coagulants enabled to achieve a similar degree of permanganate oxygen demand and turbidity decrease /Fig. 1./. The coagulants caused also decolorization by about 30%.

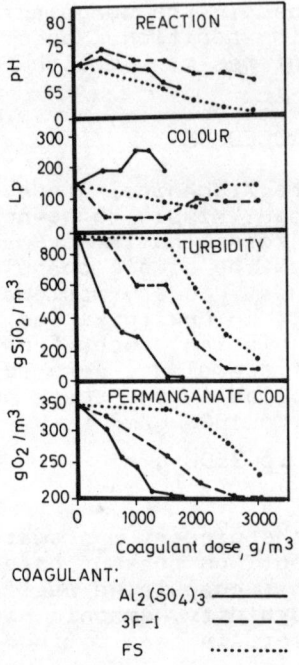

Fig. 1. Results of coagulation jar test.

The jar tests enabled a preliminary choice of doses and treatment effect for the new coagulants. The test results were used at the next research stage on the influence of the new reagents on the activated sludge process.

Average coagulation results of the piggery wastewater preliminary treated in order to feed the activated sludge, were calculated as shown in Table 1.

A percentage of organic compounds removal expressed by means of dichromate COD and BOD_5 was similar after the application of three coagulants under test, however, it was the highest when FS was added /51% COD removal and 59% BOD/.

A decrease in turbidity was also observed; it was the highest when aluminium sulphate was applied /47%/. In the tests after the coagulation by means of the new coagulants an increase of ammonia nitrogen content was observed by 49 g $N-NH_4/m^3$ for 3F-I and by 11 g $N-NH_4/m^3$ for FS on the average. The dosage of reagents /aluminium sulphate among others/ resulted also in a colour increase of the wastewater, in contrast to jar tests.

In spite of high doses of reagents the piggery wastewater reaction did not change essentially and was within 7.4 to 7.8 pH.

TABLE 1 Results of Piggery Wastewater Treatment by Coagulation

Measurements[x]	Unit	Raw Wastewater	Wastewater after Coagulation					
			Composition			% of removal		
			I	II	III	I	II	III
COD	gO_2/m^3	5872	3502	3059	2900	40	48	51
BOD_5	gO_2/m^3	3217	1550	1385	1332	52	57	59
Ammonia nitrogen	$gN-NH_4/m^3$	592	447	631	593	23	-	-
Turbidity	$gSiO_2/m^3$	1500	797	900	915	47	40	39
Colour	Lp	114	200	160	207	-	-	-
Reaction	pH	7.37	7.80	7.43	7.42	-	-	-

[x]Analysis done in filtered samples.
I - samples after the coagulation with $Al_2/SO_4/_3$
II - samples after the coagulation with SF-I
III - samples after the coagulation with FS

A graph of post-coagulation sludge settling and its volume measured in Imhoff's funnel is shown in Figure 2.

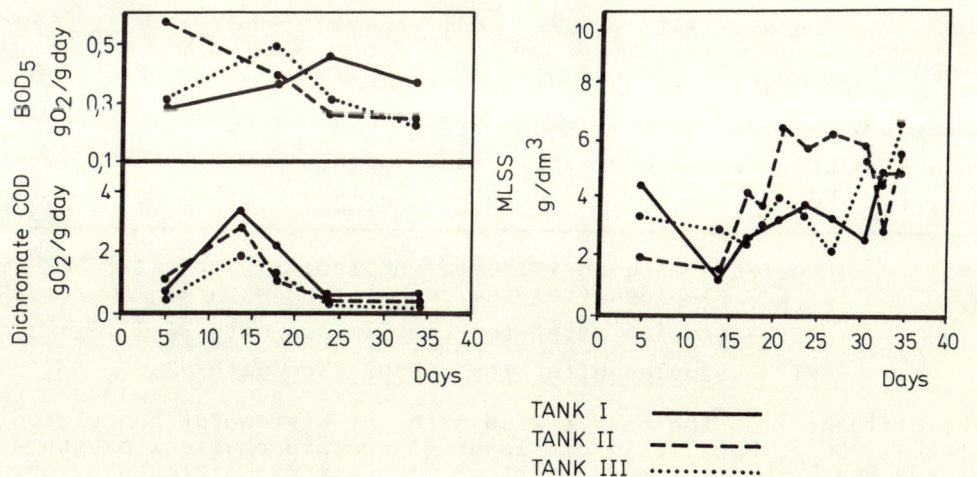

TANK I ———
TANK II – – – –
TANK III ···········

Fig. 2. Sedimentation curves of post-coagulation sludge.

The setting velocity of sludge was the highest after using 3F-I coagulant. After 120 min of settling the sludge volume was the least - 120 cm^3/dm^3. The FS coagulant gave also a smaller sludge volume than aluminium sulphate. However, during three of eight coagulation tests a flotation of certain sludge amount occurred on the wastewater surface.

ACTIVATED SLUDGE PROCESS

When testing the influence of the new coagulants on the piggery wastewater treatment process by means of activated sludge, three aspects were considered:
a/ effect of wastewater treatment
b/ technological properties of the sludge
c/ sludge biocenosis.
The effect of wastewater treatment because of biological decomposition of contaminations by activated sludge organisms was similar in all the three tanks /Table 2/.

TABLE 2 Results of Piggery Wastewater Treatment by Activated Sludge Process[x]

Measure-ments	Unit	Raw Waste-water	Wastewater after Biodegradation					
			Composition			% of removal		
			I	II	III	I	II	III
Dichromate COD	gO_2/m^3	5872	636	679	705	82	78	76
BOD_5	gO_2/m^3	3217	677	445	434	60	68	67
Ammonia nitrogen	$gN-NH_4/m^3$	592	291	461	377	35	27	36
Turbidity	$gSiO_2/m^3$	1500	360	712	562	55	21	38
Colour	Lp	114	222	145	128	-	9	-
Reaction	pH	7.37	8.18	8.27	8.22	-	-	-

[x] Analysis done in filtered samples.
    I - samples after the coagulation with $Al_2/SO_4/_3$
    II - samples after the coagulation with SF-I
    III - samples after the coagulation with FS

The effluent from the tank I /fed with the wastewater coagulated with $Al_2/SO_4/_3$ had little bit lower dichromate chemical oxygen demand and turbidity, and comprised less ammonia nitrogen by about 86-170 g/m^3.
However, the effluents from the tanks II and III /fed with the wastewater coagulated with SF-I and FS reagents/ included less easily biologically decomposible compounds /lower biological oxygen demand /BOD_5/ by 182-193 g O_2/m^3/ and had a lower colour. The removal of contamination expressed by means COD and BOD as a result of biodegradation, can be described with an equation taking into account the kinetics of the phenomena.

From Eckenfelders' equation /Adams, 1974/ the degradation rate of organic compounds /K constant/ can be calculated. The equation is as follows:

$$S_e/S_c = e^{-KX_v t/S_c}$$

where:
- S_c - concentration of contaminations in raw wastewater, g/m^3
- S_e - concentration of contaminations in effluent, g/m^3
- X_v - weight of activated sludge, g
- t - sludge aeration period, days

An average value of degradation rate constant - K found for dichromate chemical oxygen demand indicator, was respectively:

Tank I - $K = 2.60 \, d^{-1}$
Tank II - $K = 2.06 \, d^{-1}$
Tank III- $K = 1.57 \, d^{-1}$

It means that the highest rate of organic compounds biodegradation occurred in the tank fed with wastewater coagulated with alumina sulphate. In the remaining tanks the rate was lower. Considering, however, the long aeration time /22 hours/ in this case the differences of K constant did not affect the treatment results.
The ability of sludge settling measured as a volume index by Mohlmann, is an important parameter considering the process effectiveness.
Average values of volume index in the tested tanks I, II and III were equal 290 cm^3/g, 164 cm^3/g and 255 cm^3/g respectively. The unsatisfactory ability of sludge settling is not related to the applied coagulants but with wastewater nature and the experiment procedure. As observed in other investigations the sludge volume index under laboratory conditions assumes high values in the first phase of process. On a technical scale the index periodically achieves 1000 cm^3/g in piggery wastewater treatment plant, which is considered a failure of the plant.
The remaining parameters such as solids, a BOD_5 and dichromate COD load of the sludge were similar for all the tanks under comparison /Fig. 3./.
Biocenosis of activated sludge was checked during hydrologic tests by four series of microscopic analysis.
Species composition of the biocenosis for the tested tanks was very similar, a difference was observed only in a quantitative development of individual species. It appeared in the development of species of Opercularia genus, which dominated in the activated sludge used for the culture inoculation achieving 87.7% of the total number of observed organisms.
In the middle phase of the experiment these species were not found. After 40 days of the culture the percentage content in the total number of organisms was 23.6% in the tank I, 16.3% in II and 29.3% in III.
The described changes in the biological nature of sludge in three tanks tested parallely do not suggest a negative influence of individual coagulants dosage on the development of the activated sludge biocenosis.

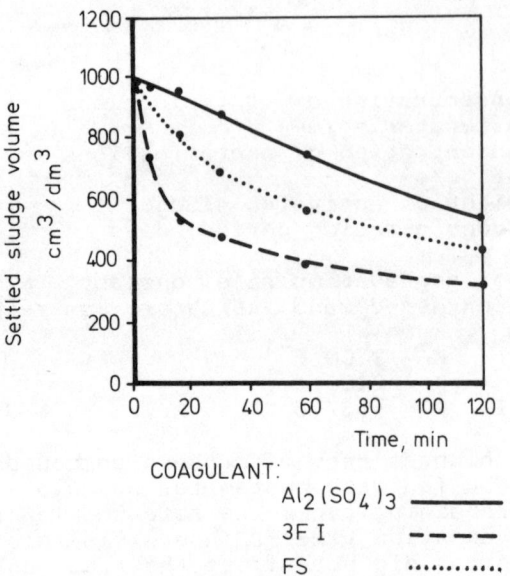

Fig. 3. Activated sludge characteristic.

SUMMARY AND CONCLUSIONS

1/ Doses of the new coagulants for the piggery wastewater coagulation, defined under laboratory conditions are more than twice higher than doses of aluminium sulphate.
2/ The new coagulants have at least the same ability of organic compounds removal from piggery wastewater as the aluminium sulphate.
3/ The advantage of the new coagulants is a small volume of the post-coagulation sludges. Their volume after the coagulation with the 3F-I reagent is about twice lower than after the coagulation with aluminium sulphate. After the coagulation with FS the sludge volume is on average by 17% lower as compared with aluminium sulphate /periodically a partial sludge flotation was also observed/.
4/ An anticipated increase of ammonia nitrogen amount in the wastewater coagulated with the 3F-I reagent was on the average 49 g $N-NH_4/m^3$. The dosage of FS increased the contents of that nitrogen form only by 11 g $N-NH_4/m^3$.
5/ The effect of biological treatment of the wastewater coagulated with aluminium sulphate and the new coagulants, under the activated sludge method, was similar. The manure treated in three separate tanks had a similar physical-chemical composition.
6/ The biological nature of the sediments in the three tanks tested parallely does not suggest a negative influence of the new coagulants dosage on the biocenosis of activated sludge.

7/ The positive results of the tests obtained on a laboratory
   scale entitle the continuation of experiments on a technical
   scale, both in the range of new coagulants production and
   their application.

REFERENCES

Adams, E. C., and W. W. Eckenfelder /1974/. <u>Process Design Technique for Industrial Waste Treatment</u>. Enviro Press, U.S.A.

THE OZONIZATION OF DIHYDROXY-BENZENES IN THE MODEL SOLUTION

D. Leszczyńska and A. L. Kowal

Institute of Environmental Protection Engineering of Wrocław Technical University, 50-377 Wrocław, Plac Grunwaldzki 9, Poland

ABSTRACT

The result of the ozonization of catechol, resorcinol and hydroquinone were discussed. The research were carried out on the model solution with initial concentrations of the compounds 400 mg/dm^3 and constant pH level equal to 10 or 3. It has been established that the direction of the ozonization reactions depend on the pH of the solution. The catechol, resorcinol and intermediate products of their oxidation were decomposed to a higher degree in the acid than the basic solutions. Hydroquinone was better oxidized in the basic solution. The comparison was made of the effectiveness of the ozonization of dihydroxybenzenes. It has been proved that in basic solution the ozone consumption was approximately equal for all compounds. In the acid solutions the ozone consumption increases as follows:
  catechol> resorcinol> hydroquinone .

KEYWORDS

Ozonization; pH level; catechol; resorcinol; hydroquinone; ozone.

INTRODUCTION

The quality of the water resources goes from bad to worse because of industrialization, chemization of agriculture and insufficient control over the physical-chemical composition of the wastewater discharged to natural water resevoirs. The phenolic compounds are one of the most troublesome group of pollutants in water.
The term "phenolic compounds" includes a large number of homologous compounds of phenol. Phenol, catechol, resorcinol and hydroquinone, can be considered as the simplest representatives of these compounds. Table 1 presents the list of industrial plants with these compounds in the wastewater /Chojnacki, 1966/. The wastewaters originated from these plants are especially hazardous because of their toxic properties, high oxygen demand and low biodegrability. It is assumed /Chojnacki, 1966/, that toxicity of the dihydroxybenzenes is greater than toxicity of the phenol.

TABLE 1 Phenolic Compounds in Industrial Wastewater

Chemical compound	Formula	Industrial plants
Catechol	(benzene ring with two OH groups, ortho)	Pharmaceutical plants, coking plants, gas-works, low-temperature carbonization plants, coal carbonization plants.
Resorcinol	(benzene ring with two OH groups, meta)	Synthetic dyestaff plants, coking plants, gas-works, wood distillation plants, low-temperature carbonization plants.
Hydroquinone	HO—(benzene ring)—OH	Photographic film plants, dyestaff plants, low-temperature carbonization plants.

The methods used for phenolic wastewater treatment can be divided two groups:
- recovery methods, which are economically justified when concentrations of phenols are high;
- destructive methods, used when concentration of phenols are about 1 g/dm^3 and less.

Chemical oxidation with ozone, used at low concentration of phenol, leads to irreversible destruction of molecular structure of phenol. Ozone is the strongest oxidizing agent. Its oxidizing potential amounts to 2.07 V in acid solutions and 1.24 V in basic solutions /Peleg, 1976/. The comparative oxidizing potential of chlorine and chlorine dioxide are 1.36 and 1.275 volts versus hydrogen, respectively /Diaper, 1975/. Mechanisms and kinetics of dissociation of ozone were investigated by a great number of workes but these processes are not precisely described, yet. Hence, it is impossible to determine the direction of reaction path way during ozonization of organic compounds in water in spite of the fact that the initial parameters of the process are not well known. It was found /Peleg, 1976/, that under given conditions of ozonization process, water contains beside molecular ozone, OH• and HO_2• radicals, O^-, O_3^-, O_2^- ions, and molecular oxygen. The occurence of these molecules and their concentrations depend on pH of the water. The reactivity of the ozone dissociation products depends on the pH of the solution.

The ozone can react with organic substances in water by the following ways /Hoigne, 1976/ :
- direct reaction between ozone and dissolved organic compounds;
- secondary reaction of ozone decomposition products with organic compounds, proceeding at a pH over a critical level.

The products formed during reaction of the ozone with water can accelerate or inhibit the rate of ozone decomposition /Hoigne, 1976/. This paper presents the results of ozonization of dihydroxybenzenes.

METHODS

Solutions of catechol, resorcinol and hydroquinone at initial

concentrations 400 mg/dm^3 were investigated. The pH of water was controled and maintained at constant level either at pH equal to 10 or 3.
The scheme of experimental installation is shown in Fig. 1 .

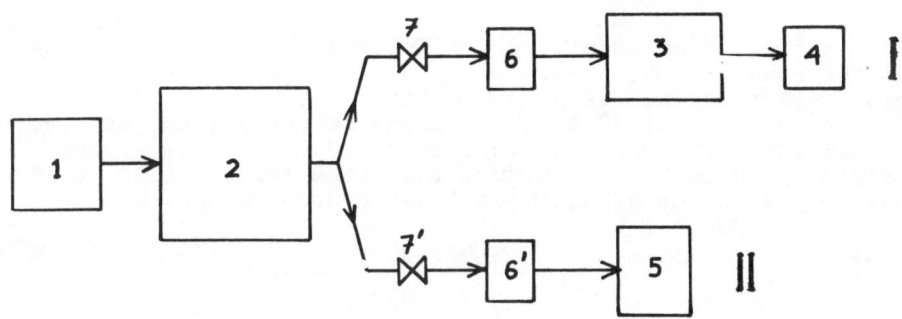

Fig. 1. The experimental installation used for ozonization.

It consisted of two parallel sets working simultaneously. Set I was used for ozonization process and set II for measuring of initial amount of ozone introduced to the reactor /3/ of set I.
The technique used was the following /Leszczyńska, 1977/ : the air was supplied by compresor /1/ at flow rate 130 dm^3/h through rotameter installed on ozonator's inlet.
The air stream with ozone leaving ozonator was directed to both sets:
- set I : the gas measured by rotameter /6/ controling the flow rate /100 dm^3/h /, supplied the reactor /3/ and subsequently wash bottle /4/ containing of 2% solution of potassium iodide where absorption unreacted ozone took place ;
- set II: the gas flow through rotameter /6'/, where the flow rate was controlled and maintained at 20 dm^3/h, to the wash bottle /5/, where ozone was absorbed in potassium iodide solution.
The amount of absorbed ozone in the wash bottle /5/ was taken to calculation of the initial concentration of ozone in the gas introduced to the reactor /3/. The amount of absorbed ozone in the wash bottle /4/ was taken when the amount of unreacted ozone was calculated. The difference between these two quantities was taken as the amount of reacted ozone in a given sample of wastewater The concentration of ozone was determined according to the Standard Methods /1960/. The final concentration of the oxidized compounds were determined by spectrophotometric method based on the coupling reaction between phenols and p-nitroaniline in the basic solution /Korenman, 1973/. Variations of the colour intensity in the ozonated solution were continously observed and measured using spectrophotometer at wavelength of 500 nm /Leszczyńska, 1977/. The pH was recorded with N 5122 pH-meter. The COD based on permanganate value was also determined according to Hermanowicz /1976/. The pH was adjusted by either sulphuric acid solution dilluted with distilled water at ratio 1:5 or with 1 M solution of sodium hydroxide. Constant pH was achieved by the following way: after a given period of time from the begining of the process the ozonator was stopped and wastewater was determined

the phenolic compound concentration, COD, pH, colour, and initial
and unreacted amount of ozone. Subsequently, pH was adjusted to the
initial value and a new colour intensity was measured and ozonator
was put to work again.

RESULTS

The ozonization of catechol, resorcinol and hydroquinone was carried
out at the constant pH level.
Experiments were conducted under conditions:
- maintaining pH about 10 at which ozone decomposes to reactive
  radicals and ions /Peleg, 1976/, and
- maintaining pH about 3 to assure the ozone molecules to be stable
  in water /half-time of decomposition is more then 41 minutes
  /Stumm, 1958 /.

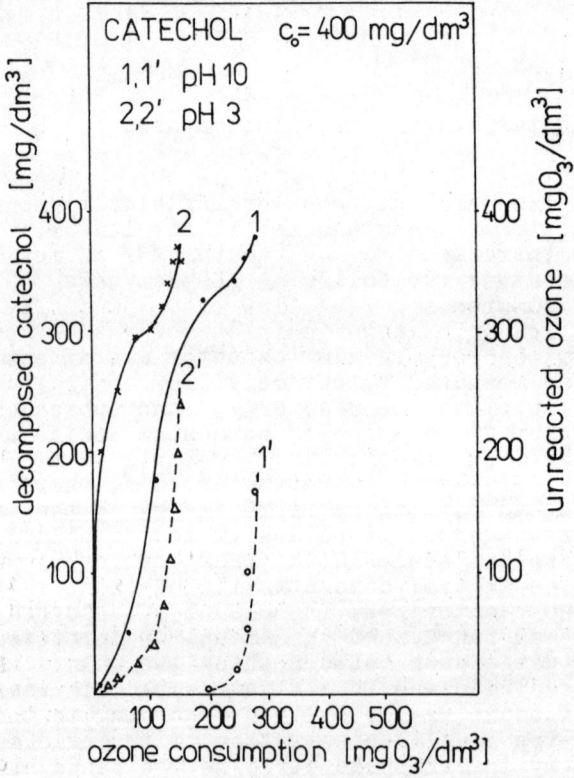

Fig. 2. The results of the ozonization of catechol at constant pH.

Figure 2 presents the relationship between the amount of decomposed
catechol and ozone consumption during ozonization which was carried
out at constant pH 10 /curve 1/ and pH 3 /curve 2/. From Fig. 2. it
follows that the lowest ozone consumption occurs at pH 3. For example
in order to conwert 75% of catechol at pH 3 100 mg/dm^3 ozone was
needed, whereas at pH 10 170 mg/dm^3 O$_3$ was required.

The breakdown point, corresponding to the appeerence of ozone in the products of reaction was dedected in acid solution at the begining of experiment in basic solution it appeared after 200 mg/dm³ ozone has reacted.

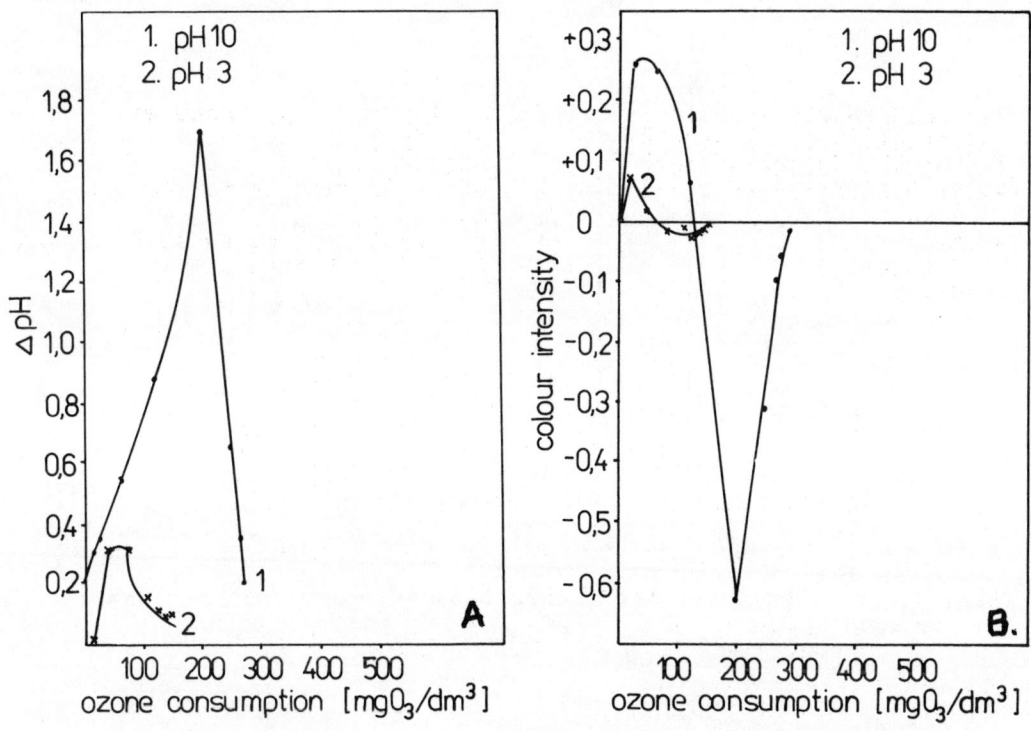

Fig. 3A, 3B. Variations in pH and colour intensity during ozonization of catechol at constant pH.

Variations in pH and colour intensity related to the quantity of ozone consumed are given in Fig.3A and 3B, respectively. In the course of ozonization in basic solution the maximum increase in the concentration of hydrogen ions coresponded to maximum decrease of colour /Fig.3A curve 1 and Fig.3B curve 2/. This relationship was not observed during ozonization in acid solution /Fig. 3A,B curves 2/. The reaction curves of resorcinol Fig. 4. were similar up to decrease of its initial concentration, with different doses of ozone. The reaction between resorcinol and ozone in basic solution was complete up to 96% decrease of resorcinol. In acid solution the unreacted ozone was observed instantaneously at the begining of the experiment /Fig. 4 curve 1' and 2'/. At pH 3 the ozone consumption for the oxidation of 80% of resorcinol was twice higher than at pH 10.
The variations of pH and colour intensity in resorcinol solution versus ozone consumed are given in Fig. 5A and 5B, respectively. In the course of the ozonization carried out at pH 10, maximum increase of hydrogen ions concentration and maximum decrease of colour intensity were observed at the point 70 to 130 mg/dm³ O_3 consumed Fig.5A,B curves 1/. This was not stated in acid solution, which was

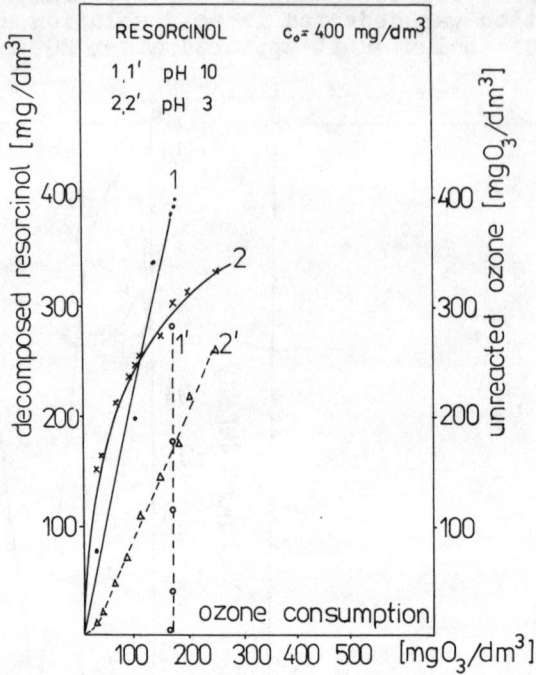

Fig. 4. The results of the ozonization of resorcinol at constant pH.

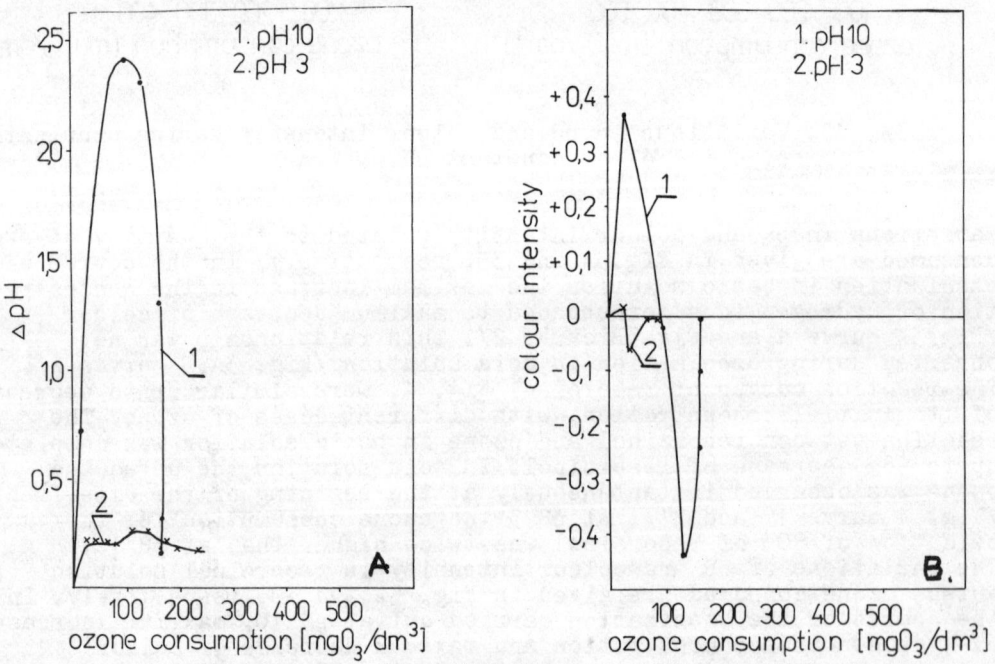

Fig. 5A, 5B. Variations in pH and colour intensity during ozonization of resorcinol at constant pH.

almost colour less like in ozonization of catechol carried out under the same conditions /Fig. 5A,B curves 2/.

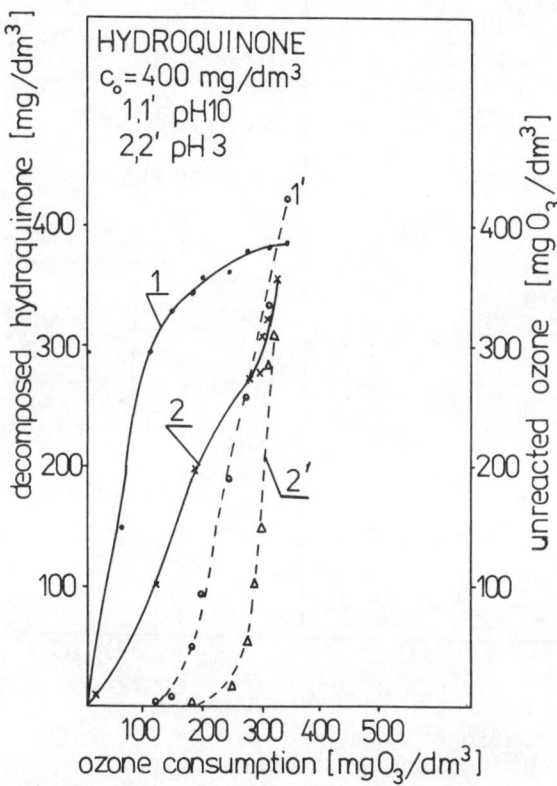

Fig. 6. The results of the ozonization of hydroquinone at constant pH.

The relationships between the amount of hydroquinone decomposed and ozone consumed at pH 10 and pH 3 are shown in Fig. 6. curves 1 and 2 respectably, were pH dependent. For hydroquinone to be decomposed in 50% of initial concentration, 70 mg/dm^3 O_3 was required at pH 10, and 180 mg/dm^3 O_3 at pH 3. The breakdown point of ozone was observed when the concentration was over 100 mg/dm^3 O_3 in both acid and basic solution /Fig. 6. curve 1' and 2'/. Nevertheless, the decrease initial amounts of hydroquinone in the breakdown point of ozone were 75% and 19% in the basic and acid solution, respectively.
The variations of pH and colour intensity versus the amount of ozone consumed are given in Fig. 7A and 7B. During ozonization of hydroquinone at pH 10 the maximum decrease of pH corresponded to maximum increase of solution colour intensity /Fig. 7A,B curves 1/. These relationships were considerably different when hydroquinone was ozonizated in the acid solution. The colour intensity of the solution during ozonization of hydroquinone in basic medium was considerably higher as compared to catechol and resorcinol ozonization.

DISCUSSION

The investigations results indicate that the ozonization of

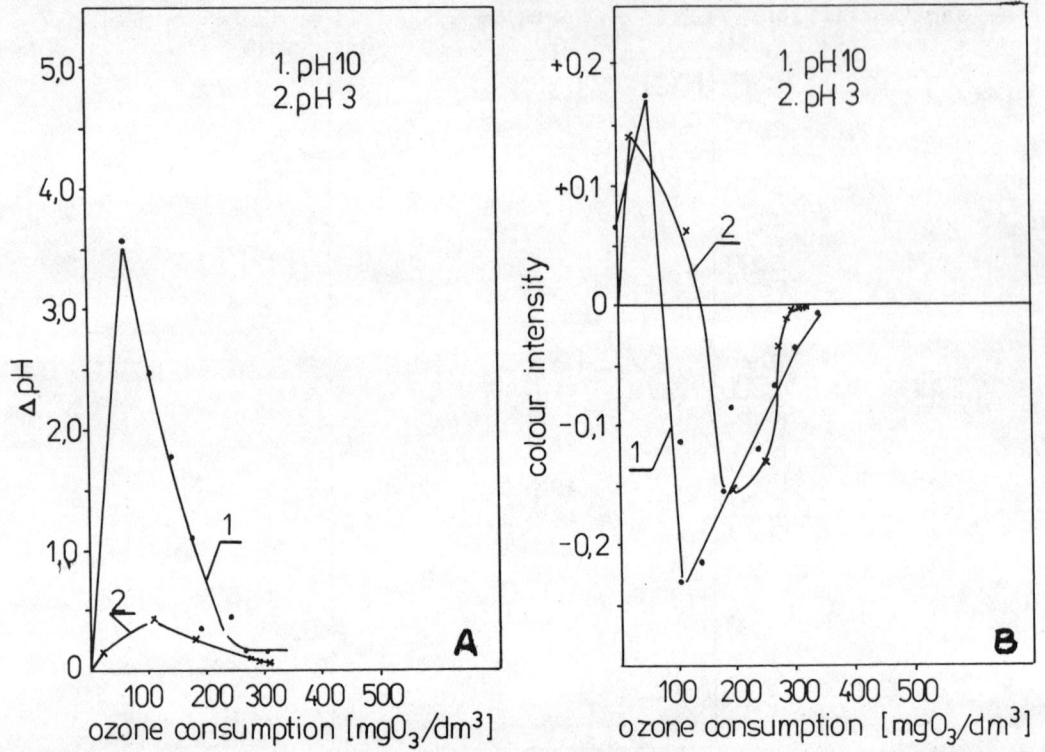

Fig. 7A, 7B. Variations in pH and colour intensity during ozonization of hydroquinone at constant pH.

dihydroxybenzenes were pH dependent. The process proceeds through various intermediat products.
The following factors determine the way of the reaction:
- the quantity of the compound to be decomposed;
- the quantity of consumed and unreacted ozone;
- increase of hydrogen ions concentration, indicating an acidic character of intermediaty products;
- variations of colour intensity. Colour is an indicator of quinones presence.

The data from ozonization of catechol suggest that ozone introduced to water firstly decomposed into products which reacted with catechol molecules. This led toward forming of o-benzoquinone, which may be proved quantitative reaction with ozone, and rapidly increased colour of the solution and rapid decrease of the initial concentration of catechol /about 50%/. Further oxidation of quinons break of the benzene ring at the same time, an increase of hydrogen ions concentration was observed with simultaneous decrease of colour intensity. Decrease of residual catechol required much higher concentration of ozone. The experiments carried out in acid solution shown that under these conditions the mechanism of oxidation of catechol was different than in the basic solution and lead to anyother intermediat products. In this case molecular ozone was a major oxidizing agent. Catechol was mainly decomposed by direct breaking of the aromatic rings. The light colour of the solution indicated that only a small part of

catechol was decomposed through quinons as intermediat products. The ozonization of catechol in acid solution is more than as compared to basic solution. The amount of ozone required during decomposition of catechol was much smaller too, and in the acid solution oxidation reaction was not stopped at the quinons formations. The considerable increase of hydrogen ions concentration indicates the formation of simple products from catechol decomposition.
The data obtained during ozonization of resorcinol and hydroquinone in basic solution indicate that the reaction run through quinons, too. The amounts of ozone being utilized were nearly the same without respect to the oxidized compound. The assumption of free radical oxidation is being proved additionally by that fact.
The resorcinol oxidized in the acid solution practically without formation of quinon. Decrease of resorcinol concentrations resulted from direct oxidation of the ring and its subsequent decomposition.
The oxidation of hydroquinone in the acid solution took a different course. Because of its degrability, oxidation of hydroquinone consisted in both reaction toward formation of quinons and indirect breaking of rings with subsequent oxidation of the decomposed products.
The ozone consumption varied dependently on particular dihydroxybenzene to be oxidized due to different degrability of intermediat products formed during rings decomposition. It proves the assumption of electrophilic mechanism of oxidation going on through unstable ozonides.

CONCLUSIONS

The way of the ozonizations of dihydroxybenzenes is dependent on the pH of the solution.
Ozonization of the catechol in the acid solution is recommended for the decomposition of aromatic rings into simple chemical compounds.
Ozonization of resorcinol both in acid and basic solution resulted in the similar decrease of its initial concentration.
For oxidation of hydroquinone the basic solution is recommended.
Effectivenes of dihydroxybenzenes decomposition in the basic solution is similar / the same concentration of compound tested and ozone consumed/.
The reduction of initial concentration of dihydroxybenzenes in the course of ozonization at low pH is in good egreement with their theoretical susceptibility to degradation.
The oxidation of dihydroxybenzenes proceeds through intermediat products and may be stopped total decomposition takes place.

REFERENCES

Chojnacki A. /1966/ The technology of the industrial wastewater , Part II, PWN, Warszawa.
Diaper E.W.J. /1975/ Disinfection of water and wastewater using ozone Ann Arbor Science Publishers Inc. , 211.
Hermanowicz W. /1976/ The physicochemical investigation of water and wastewater , Arkady , Warszawa.
Hoigne J.,Bader H. /1976/ The role of hydroxyl radical reactions in ozonation processes in aqueous solutions, Water Research, 10, 377
Korenman I.M. /1973/ The photometric analysis, WNT, Warszawa.

Leszczyńska D. /1977/ The degradation of the phenols by ozone in the model solution and in the industrial wastewater, Theses, Politechnika Wrocławska, Wrocław.

Peleg M. /1976/ The chemistry of ozone in treatment of water, Water Research, 5, 361.

Standard Methods for the Examination of Water and Wastewater, /1960/ American Public Health Association INC.

Stumm W. /1958/ Ozone as a disinfectant for water and sewage, J.Boston Soc.Civil Engr., 45, 68.

STUDIES OF THE INFLUENCE OF THE COAGULATION PROCESS ON CERTAIN PESTICIDES IN WATER

W. Sztark

Institute of Inorganic Chemistry and Technology, Technical University of Cracow, Cracow, Poland

ABSTRACT

The effect of coagulation on the level of nabam in aqueous solutions has been investigated. Nabam stands for sodium ethylenebisdithiocarbamate /terminology accepted by International Organisation for Standarization, ISO/. Dithiocarbamates are increasing in importance as fungicides. The coagulation process was carried out with the aid of aluminium sulphate, ferrous sulphate, chlorinated ferrous sulphate and ferric chloride in the presence of calcium hydroxide. The fate of the nabam was followed spectrophotometrically /using the UV absorption bands/.
In the case of coagulation brought out by aluminium sulphate in solutions of pH 6 to 10 no decrease of nabam concentration or appearance of its decomposition products have been observed. This can indicate the absence of any influence of the coagulant on the investigated compound. In acid solutions /pH = 3/ aluminium sulphate catalyses the decomposition of nabam. In the presence of acid after two hours neither nabam nor its decomposition products could be found.
In the coagulation process brought about by ferrous and ferric salts under alkaline condition, catalytic decomposition of nabam to ethylenethiuram monosulphide /ETM/ and the formation of a transient complex between iron ions and ethylenebisdithiocarbamate /EDTC/ ion have been found. The coagulation brought about by ferrous and ferric ions causes the decomposition of nabam but does not remove its decomposition products from aqueous solutions. In acid medium iron ions remove nabam from solution completely by forming a comparatively stable insoluble compound with ethylenebisdithiocarbamate anion.

KEYWORDS

Coagulation ; dithiocarbamates ; nabam decomposition ; pesticides removal.

INTRODUCTION

Nabam - sodium ethylenebisdithiocarbamate /EDTC/ belongs to a group of fungicides, which is of increasing importance because of a high biological activity and comparatively low toxicity towards higher

organisms. Nabam is soluble in water, labile and in aqueous solution undergoes slow decomposition. Its decomposition products show, in some cases, even stronger biological activity than the parent substance /Klöpping, van der Kerk, 1951/.
The purpose of the present work was to investigate the changes caused by nabam when coagulation takes place in a system. For the determination of nabam concentrations the technique of UV spectrophotometry was adopted. Nabam gives a characteristic spectrum in the UV range with two absorption maxima /λ = 260 and 287 nm/.
The decomposition of nabam was studied as a function of pH. One set of samples was made alkaline by adding sodium hydroxide, the other was acidified with hydrochloric acid. After 2 hours the nabam concentration dropped by about 60 % in the acid solutions. In the pH range 5 to 8 the decomposition is slower and at pH 9 to 10 the decomposition is undetectable which is probably brought about by a depression of the hydrolysis of sodium EDTC. Coagulation studies were carried out for some commonly used coagulants such as aluminium sulphate, ferrous sulphate, chlorinated ferrous sulphate, ferric chloride.

COAGULATION PROCESS – AQUEOUS SOLUTION OF NABAM IN THE PRESENCE OF ALUMINIUM SULPHATE

The influence of various doses of coagulant on the given amount of nabam was investigated, keeping the other conditions constant ; they were the pH /9,5/ of the original solution, temperature /$20 \pm 2^{\circ}$C/, stirring rate /80 turns/min/, time of stirring /15 min/ and sedimentation time /2 hours/. UV spectra of the solutions were obtained after the removal of any precipitated sediment. In every case a control experiment was carried out in standard solution /S/. The experiments were carried out for nabam concentrations from 10 to 200 mg/dm^3. At pH 9,5 neither a decrease of nabam concentration nor an appearance of its decomposition products could be detected, which demonstrates that the coagulant does not influence the nabam present in the system.

COAGULATION PROCESS – AQUEOUS SOLUTION OF NABAM IN THE PRESENCE OF FERRIC AND FERROUS SALTS

Exploratory studies of the coagulation process in aqueous solutions of nabam in the presence of ferric and ferrous salts indicated that the coagulation took a different course than in the presence of aluminium sulphate. After addition of each coagulant : ferrous sulphate, chlorinated ferrous sulphate and ferric chloride a dark brown precipitate was separated. No nabam could be detected in the filtrates, but there was evidence of the appearance of a new compound, characterised by an absorption maximum at 280 nm. The compound was identified, on the basis of literature data, as ethylenethiuram monosulphide – ETM – /Ludwig, Thorn, Miller, 1954 ; Klisenko, Veksztein, 1971/. Several series of experiments were carried out for nabam solutions of various concentrations /10, 20, 50, 100, 200 mg/dm^3/, keeping the other conditions constant /time, temperature, rate and duration of stirring and sedimentation time/. For each concentration four paralell tests were carried out for three coagulants, the same amount /100 mg/dm^3/ being added in each case. The standard solution of nabam was made alkaline and diluted to the required volume without an addition of the coagulant. UV spectra were obtained /after filtration/. In all cases the same type of spectrogram of the decomposition product of nabam was obtained, the intensity of the peaks being dependent on the coagulant used.

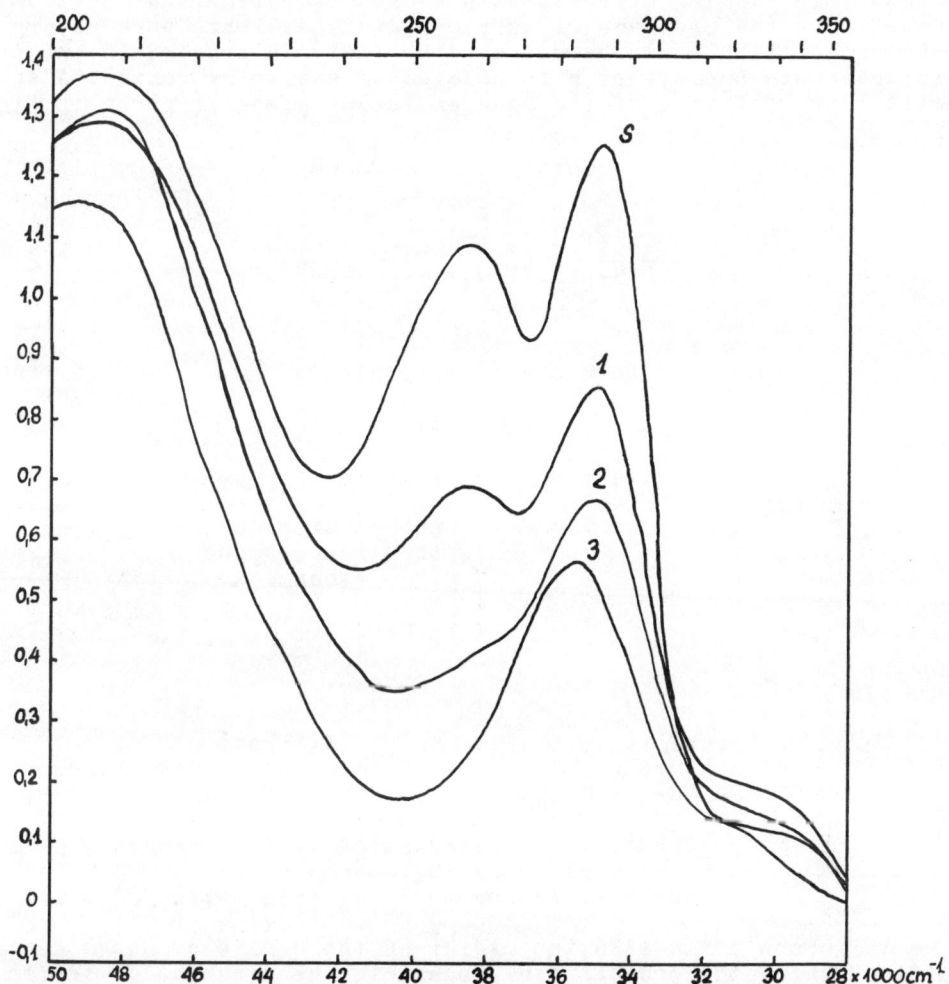

Fig. 1. Decomposition of nabam during coagulation in the presence of iron salts /pH = 9,5/ S standard solution ; 1 coagulant $FeSO_4$; 2 coagulant $FeSO_4$ + Cl_2 ; 3 coagulant $FeCl_3$.

UV spectrum of nabam shows two absorption bands ; the weaker one at 260 nm and the stronger one at 287 nm. Spectrum No. 1 obtained for nabam solution in the presence of ferrous sulphate shows two peaks: at 287 nm and at 260 nm which proves the presence of residual un - decomposed nabam. However, spectra 2 and 3 obtained in the case of coagulation in the presence of ferric ions show only one absorption band at 280 nm ; charakteristic of the decomposition product of nabam - ETM. The precipitate was analysed chemically and its IR -

spectrum was obtained. In the case of an alkaline medium /the sample with calcium hydroxide/only carbonate bands were found : these are much stronger than the adsorption bands of ferric hydroxide /a possible masking effect/. With sodium hydroxide the spectrum showed only the presence of ferric hydroxide. There were no organic substances which could be detected by their IR spectra in the precipitate obtained either with calcium or sodium hydroxide. A schematic representation of the changes taking place is given in Fig.2.

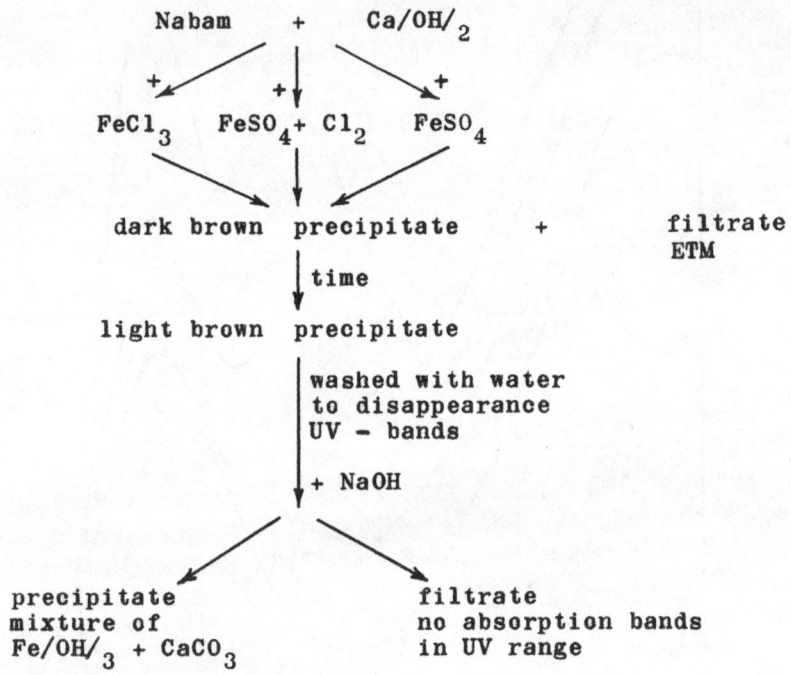

Fig. 2. Schematic representation of the changes taking place during decomposition of nabam in the presence of iron salts /pH = 9,5/.

The influence of an alkaline medium on the course of nabam decomposition during the coagulation process in the presence of ferric salts is illustrated in Fig. 3. The results obtained lead to the conclusion that calcium ions cooperate with ferric ions in bringing about the catalytic decomposition of nabam to a greater extent than with sodium ions.

A curious phenomenon was observed on adding a ferric or ferrous salt to a nabam solution without the previous addition of alkali. In a strongly acidic solution / pH= 3 / a black precipitate was separated, in which a new compound probably a combination of ferric ion with the EDTC anion was detected /Fig. 4/.

The IR spectrum 'b' and 'c' differ from the IR spectrum of nabam 'a'. The more complicated type of spectrum 'c' could be due to a more complex structure of the product of reaction of ferric ions with the EDTC anion. The changes are shown schematically in Fig. 5.

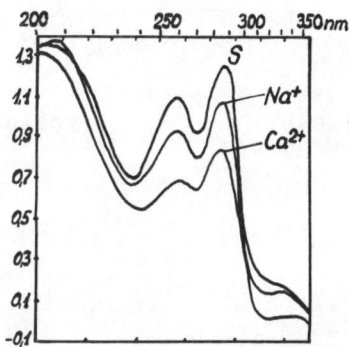

Fig. 3 Influence of calcium and sodium ions on the decomposition of nabam during coagulation in the presence of ferric salts.

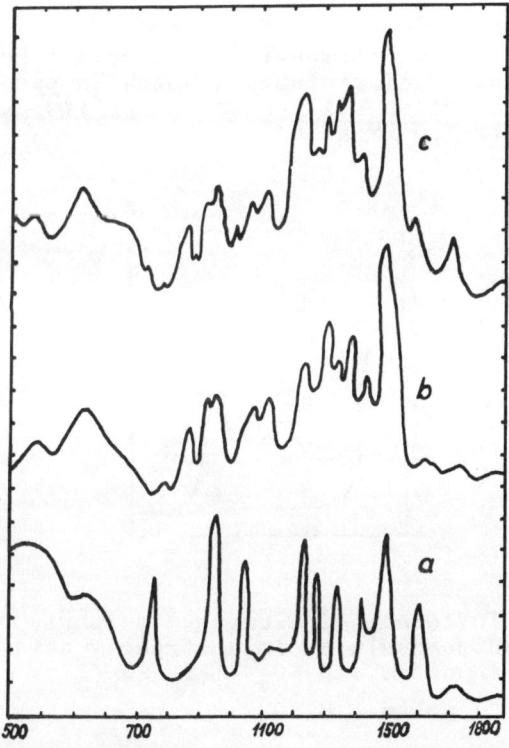

Fig. 4 IR spectra of : a nabam ; b combination of ferrous ions with the EDTC anions ; c combination of ferric ions with the EDTC anions.

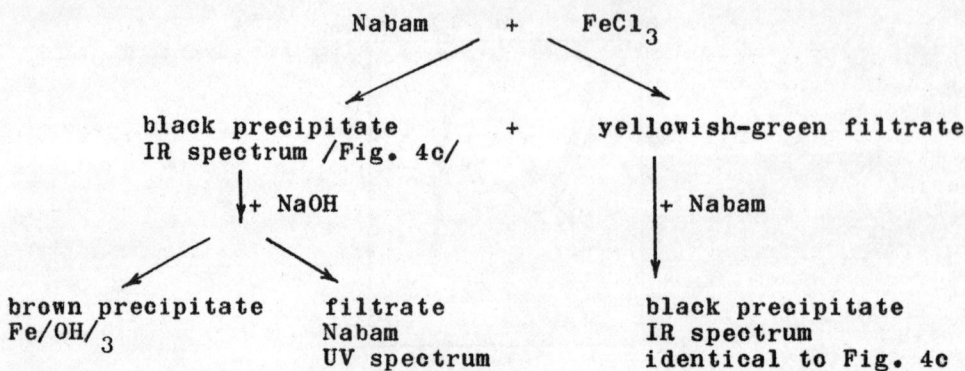

Fig. 5 Schematic representation of the changes taking place during decomposition of nabam in the presence of ferric salts in strongly acidic medium /pH = 3/.

Similar tests for nabam decomposition in acidic solution were carried out in the presence of aluminium sulphate. No precipitate was separated and in the solution neither nabam nor its decomposition products could be detected /Fig. 6/.

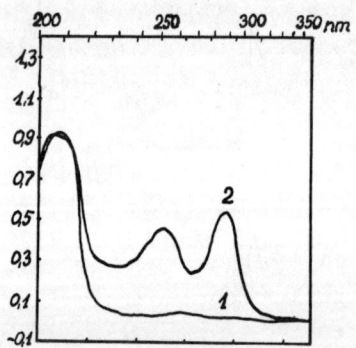

Fig. 6 Influence of aluminium sulphate on nabam decomposition in a strongly acidic solution,
1   nabam + H_2SO_4 + $Al_2/SO_4/_3$;
2   nabam + H_2SO_4

This can indicate that in this case the decomposition of nabam goes further, beyond the ETM intermediate. Thus the catalytic activity of aluminium sulphate in a strongly acidic medium is very strong while in weakly acidic, neutral and alkaline media this activity is insignificant.

CONCLUSIONS

1/ No changes of nabam concentration could be detected during coagulation brought about by aluminium sulphate under usual coagulation conditions.

2/ There is evidence of nabam decomposition in a strongly acidic medium /pH = 3/ in the presence of aluminium sulphate.

3/ During coagulation in the presence of ferrous and ferric salts /$FeCl_3$, $FeSO_4$, $FeSO_4$ + Cl_2/ catalytic decomposition of nabam takes place, to water soluble ETM and only slightly soluble /unstable in an alkaline medium/ combination of EDTC anion with iron ions.

4/ Acidification of aqueous solution of nabam causes the total removal of nabam from solution, with the formation of a relatively stable combination of iron ions with the EDTC anion.

5/ The ferric ions catalyse the process of nabam decomposition to a greater extent than ferrous ions because of the oxidising properties of the former.

REFERENCES .

Klisenko, M.A., and M. Sh. Veksztein /1971/. Decomposition rate and the nature of transformation products of dimethyl- and ethylenebisdithiocarbamates in various farm cultures. Vop. Pitan, 30, 79-81.

Klöpping, H. L., and G. J. M. van der Kerk /1951/. Investigation on organic fungicides. V. Chemical constitution and fungistatic activity of aliphatic bisdithiocarbamates and isothiocyanates. Recueil, 70, 949-961.

Ludwig, R. A., G. D. Thorn, and D. M. Miller /1954/. Studies on the mechanism of disodium ethylenebisdithiocarbamate /Nabam/. Can. J. Botany, 32, 48-51.

A COMPARISON OF PACKED-BED AND EXPANDED-BED ADSORPTION SYSTEM

S. W. Hermanowicz and M. Roman

*Institute of Water Supply and Hydraulic Constructions,
Warsaw University of Technology, Warsaw, Poland*

ABSTRACT

A mathematical model of fixed-bed and expanded-bed adsorption systems was presented. The model consists of two partial differential equations: the mass balance equation and the kinetics equation. Fluid dispersion for packed-bed and expanded-bed and solids mixing for expanded-bed were considered. To verify the model adsorption of phenol on granular activated carbon was examined. Model parameters: film transfer coefficient k_f and dispersion coefficient D_F were estimated. Using molar viscosity coefficient introduced by Kurgaev a satisfactory correlation of mass transfer factor j_d and modyfied Schmidt number Sc' with modyfied Reynolds number Re' was established.

KEYWORDS

Adsorption; granular activated carbon; expanded-bed; fixed-bed; mathematical model; fluid dispersion; molar viscosity.

NOTATION

a - concentration of sorbate in solid phase, mg/g ;
a_o - concentration of sorbate in solid phase in equilibrium with the concentration c_o , mg/g ;
A - ultimate concentration in the Langmuir isotherm equation, mg/g;
b - energy constant in the Langmuir equation, 1/mg ;
c - concentration of sorbate in fluid phase, mg/l ;
c_o - concentration of sorbate in fluid phase at the inlet, mg/l ;

c_{eq} - concentration of sorbate in fluid phase in equilibrium with the concentration a, mg/l ;
d - sorbent particle diameter, cm ;
D - molecular diffusivity, cm^2/s ;
D_F - fluid dispersion coefficient, cm^2/s ;
D_S - solid particles dispersion coefficient, cm^2/s ;
$j_d = \dfrac{k_f}{u} Sc^{2/3}$ - undimensional mass transfer factor ;
k_f - film mass transfer coefficient, cm/s ;
$Re = \dfrac{u\,d}{\nu \cdot (1-\varepsilon)}$ - undimensional Reynolds number ;
$Re' = \dfrac{u\,d}{\nu_M}$ - modyfied Reynolds number ;
S - external surface area of sorbent particles on bed volume unit, cm^2/cm^3 ;
$Sc = \nu/D$ - undimensional Schmidt number ;
$Sc' = \nu_M/D_F$ - modyfied Schmidt number ;
t - time, s, hr ;
u - fluid velocity /based on empty column cross area/, m/hr, cm/s ;
x - axial distance in column, cm ;
ε - bed porosity ;
ν - kinetic viscosity coefficient, cm^2/s ;
ν_M - molar viscosity coefficient, cm^2/s ;
ϱ_b - bulk density of bed, g/l ;
ϱ_s - density of sorbent, g/l ;
ψ - undimensional particle shape coefficient /ratio of the surface area of a sphere of the same volume as the particle to the particle surface area/.

MODEL OF ADSORPTION

The first equation taken into consideration is the equation of mass balance in a column. For a fixed bed and for isothermic flow of incompressible fluid the equation gives:

$$\varrho_b \cdot \frac{\partial a}{\partial t} + \varepsilon \cdot \frac{\partial c}{\partial t} + u \cdot \varepsilon \cdot \frac{\partial c}{\partial x} - D_F \cdot \varepsilon \cdot \frac{\partial^2 c}{\partial x^2} = 0 \qquad /1/$$

In this equation the influence of axial dispersion is only considered and radial gradients of concentrations are neglected. In an expanded bed particles of sorbent may move and the fluid passing through the column mixes them. The adsorbed matter is transferred through the column together with the moving particles of the sorbent. Moreover, the sorbate is also transported due to fluid axial dispersion as it takes place in the fixed-bed system but the rate of transport is much higher. The mass balance equation for the expanded bed has the following form:

$$\varrho_b \cdot \frac{\partial a}{\partial t} + \varepsilon \cdot \frac{\partial c}{\partial t} + u \cdot \varepsilon \cdot \frac{\partial c}{\partial x} - D_F \cdot \varepsilon \cdot \frac{\partial^2 c}{\partial x^2} - D_S \cdot \varrho_b \cdot \frac{\partial^2 a}{\partial x^2} = 0 \qquad /2/$$

In wastewater treatment the expansion of a bed is about 10 to 20 per cent /Process..., 1973/. In these cases, however, solid mixing seems to be of minor importance and mass balance can be expressed by the eq.1.

The next equations in the model has to describe the rate of adsorption. There are three elementary processes controlling the rate of adsorption /Weber, 1967/:
- film diffusion, i.e. the transport of the solute across the liquid film surrounding the particle of the sorbent;
- pore diffusion, i.e. the transport of the solute within the pores of the particle;
- surface reactions, i.e. adsorption of the molecules of the solute on the sorbent surface.

The rate of the surface reactions is much greater than that of the diffusion. So the pore and the film diffusion are controlling the kinetics of adsorption /Keinath, Weber, 1968; Zuchovickij 1945/. The influence of these two processes differs and depends on pore size and structure, character of fluid flow, kind of solute, etc. Keinath and Weber /1968/ stated that a number $D_G = a_o \cdot \varrho_b / c_o \cdot \varepsilon$ can describe the influence of the pore diffusion on the overall rate. As D_G approaches to the unity, pore diffusion controls the rate of adsorption.

If it is assumed that the overall rate is controlled by film diffusion the kinetics equation is:

$$\frac{\partial a}{\partial t} = k_f \cdot S \cdot \frac{\varepsilon}{\varrho_b} \cdot (c - c_{eq}) \qquad /3/$$

Proper boundary and initial conditions must be set to make the model complete. They are as follows:
- for the beginning of the process: $t = 0$, $c = 0$, $a = 0$ /4/

- for the column inlet: $x = 0$, $c = c_o$ /5/

- for the column outlet: $x = L$, $\left.\dfrac{\partial c}{\partial x}\right|_{x=L} = 0$ /6/

For more general expression the following undimensional variables are introduced: $\varphi = c/c_o$, $\varphi_{eq} = c_{eq}/c_o$, $\vartheta = a \cdot \varrho_b/c_o \cdot \varepsilon$, $\tau = t \cdot k_f \cdot S$, $\xi = x \cdot k_f \cdot S/u$, $F_F = D_F k_f S/u^2$.

This yields to the following equations:

$$\dfrac{\partial \varphi}{\partial \tau} + \dfrac{\partial \vartheta}{\partial \tau} + \dfrac{\partial \varphi}{\partial \xi} - F_F \dfrac{\partial^2 \varphi}{\partial \xi^2} = 0 \quad /7/$$

$$\dfrac{\partial \vartheta}{\partial \tau} = \varphi - \varphi_{eq} \quad /8/$$

$$\tau = 0: \varphi = 0, \quad \vartheta = 0 \quad /9/$$

$$\xi = 0: \varphi = 1 \quad /10/$$

$$\xi = \xi_L = \dfrac{L \cdot k_f \cdot S}{u}: \quad \left.\dfrac{\partial \varphi}{\partial \xi}\right|_{\xi = \xi_L} = 0 \quad /11/$$

A numerical solution of eqs. 7 - 11 gives predicted breakthrough curves for a bed.

EXPERIMENTAL PROCEDURE

Dynamics of adsorption was studied for fixed-bed and for expanded-bed systems in columns of 5.4 cm in diameter. Activated carbon "Carbopol Z-4" was used as a sorbent. The particle diameter of carbon was d = 0.0641 cm and the uniformity coefficient $d_{60}/d_{10} = 2.3$ The density of carbon was investigated by the two methods: with pycnometr with water, and by a flow method proposed by Ergun /1951, 1952/. By the latter method the density based on the volume surrounded by the external surface of a particle can be estimated. The density of carbon mesured following Ergun was ϱ_s = 390 g/l, and with pycnometer - 1585 g/l. In all studies water solution of phenol C_6H_5OH was used as a sorbate. The concentration of phenol was determined by aminoantipyrine method without extraction according to W.Hermanowicz and co-workers /1958, 1976/. An isotherm was determined for the sorbate and the Langmuir equilibrium equation was fitted to the experimental data:

$$a = \frac{A \cdot b \cdot c_{eq}}{1 + b \cdot c_{eq}} \qquad /12/$$

and the parameters were estimated: $A = 18.01$ mg/g, $b = 11.18$ l/mg The molar diffusivity was found from the equation stated by Wilke and Chang /1955/: $D = 0.88 \cdot 10^{-5}$ cm^2/s. For both systems pressure drop was investigated for various flow rates as described by Roman /1961/. It was shown that the minimum fluidization velocity was 13.5 m/hr and the shape coefficient $\psi = 0.712$. The experiment data are presented in Tables 1 and 2.

EXPERIMENTAL RESULTS

For all columns breakthrough curves $\varphi = c/c_o = f/t/$ were made.

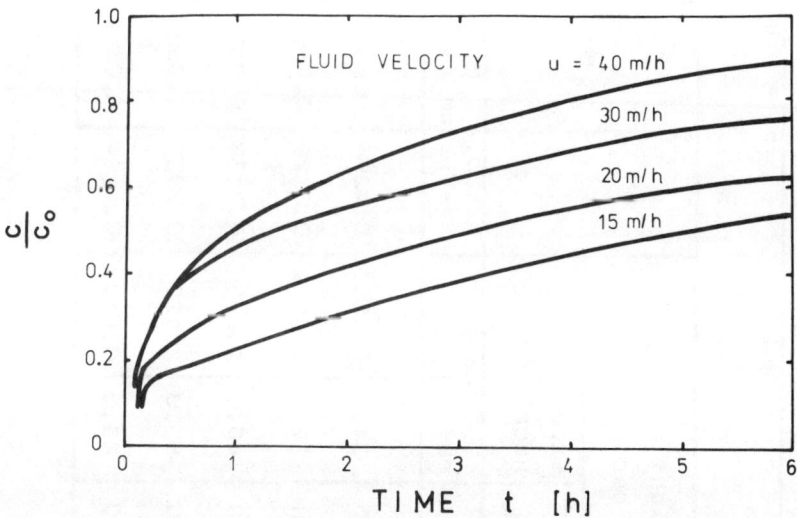

Fig. 1. Breakthrough curves for fixed beds

The breakthrough velocity /defined as $d\varphi/dt$ / is much greater for the expanded beds at the beginning but it decreases as the process develops /the breakthrough curves for expanded beds and for fixed beds become parallel/. This is probably due to greater fluid mixing in the expanded bed and the solute is transported through the column with fluid. Fluid mixing and perhaps solid mixing seem to be of the greatest importace among all phenomena

TABLE 1   Adsorption Parameters for Expanded-Bed System

Fluid velocity u	Mass of carbon	Concentration c_o	Height of bed initial	Height of bed expanded	Re'	ε	ν_M	D_F	k_f	Sc'	j_d
m/hr	g	mg/l	cm	cm	-	-	cm^2/s	cm^2/s	cm/s	-	-
15.0	107.5	2.12	20.0	34.5	0.1063	0.4254	0.2512	297.0	$7.718 \cdot 10^{-3}$	$8.456 \cdot 10^{-4}$	2.642
20.0	105.0	2.22	20.0	37.0	0.6002	0.6529	0.0593	139.0	$4.068 \cdot 10^{-3}$	$4.268 \cdot 10^{-4}$	1.045
30.0	102.5	2.28	20.0	40.0	1.245	0.7138	0.0429	261.3	$3.996 \cdot 10^{-3}$	$1.641 \cdot 10^{-4}$	0.6842
40.0	102.0	1.95	20.0	55.0	2.180	0.7929	0.0327	724.1	$4.023 \cdot 10^{-3}$	$0.451 \cdot 10^{-4}$	0.5166
20.0	58.0	1.60	12.0	23.0	0.8502	0.7184	0.0419	87.56	$8.610 \cdot 10^{-3}$	$4.784 \cdot 10^{-4}$	1.999

TABLE 2 Adsorption Parameters for Fixed-Bed System

Fluid velocity u	Mass of carbon	Concentration c_o	Height of bed	Re'	ε	ν_M	D_F	k_f	Sc'	j_d
m/hr	g	mg/l	cm	-	-	cm²/s	cm²/s	cm/s	-	-
15.0	102.0	2.23	22.5	0.1210	0.4434	0.2208	52.86	11.65·10⁻³	41.77·10⁻⁴	1.963
20.0	107.9	2.12	21.0	0.07054	0.3357	0.5048	115.2	9.976·10⁻³	43.84·10⁻⁴	2.562
30.0	102.0	2.00	20.5	0.1623	0.3891	0.3291	139.1	10.67·10⁻³	23.66·10⁻⁴	1.828
40.0	105.0	3.00	20.0	0.1644	0.3542	0.4333	223.1	6.244·10⁻³	19.42·10⁻⁴	0.8019

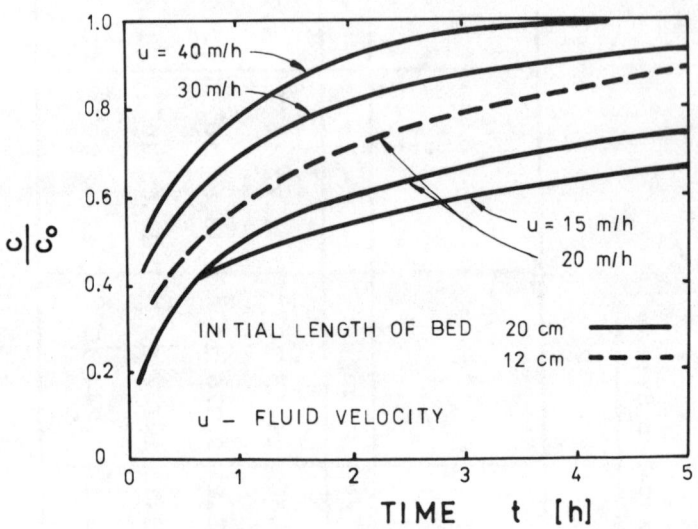

Fig. 2. Breakthrough curves for expanded beds

occurring in the column. They are, however, the most difficult to investigate and are dependent on a great number of various factors. The two parameters describe adsorption processes in the column. They are: film transfer coefficient k_f and fluid dispersion coefficient D_F.

In order to estimate these parameters the experimental breakthrough curves were approximated with numerical solutions of eqs. 7 - 11. The best fitting was found for the values of parameters presented in Tables 1 and 2. The estimated values were tested with the χ^2-test and for the confidence limit 0.99 there were no reason to reject them.

The values of surface area S were calculated from the modyfied formula stated by Letterman and co-workers /1971/:

$$S = \frac{6 \cdot (1 - \varepsilon)}{d \cdot \psi} \qquad /13/$$

The number D_G, as it was mentioned above, can describe the influence of pore diffusion on the rate of adsorption. For the beds in case the values of D_G were from 898 to 4289 for the expanded beds and from 3796 to 5425 for the fixed beds. Thus, the assumption that the film diffusion has controlled the overall rate is justified. The magnitude of the mass transfer coefficient k_f is the same for both types of bed, but it is lower for the expanded bed that for the fixed one at the same flow rate. The dispersion coefficient D_F is, on the other hand, greater for the expanded bed. Such result could be expected as the flow pattern for the expanded bed lies between piston flow and complete mixing. In the earlier works, Chu and co-workers /1953/ found a correlation between the mass transfer factor j_d and the Reynolds number Re. A similar correlation was established by Keinath and Weber /1968/. However, the previous works did not consider axial dispersion of fluid and thus the values of the k_f coefficient found in these investigations depended on fluid mixing. As these two phenomena were separated no reasonable correlation could be found between j_d and Re. As fluid flows through a bed /especially expanded one/ the viscosity of this heterogenous system is greater than of pure fluid. This is caused by eddies occuring in fluid and by the increased rate of momentum exchange between greater parts of fluid than between fluid molecules. To describe these phenomena Kurgaev /1977/ introduced a molar viscosity coefficient ν_M and gave the final formula for it:

$$\nu_M = \nu \left[1 + 2 \cdot (1 - \varepsilon) \cdot \left(\frac{2 - \varepsilon}{\varepsilon} \right)^2 \right] \qquad /14/$$

Using this coefficient a satisfactory correlation of the mass transfer factor j_d and the modyfied Schmidt number $Sc' = \nu_M/D_F$ with the modyfied Reynolds number $Re' = u \cdot d/\nu_M$ was found.
The following expressions were established /Fig. 3. and 4./:
- for expanded bed

$$Sc' = 1.540 \cdot 10^{-4} \cdot Re'^{-0.8890} \qquad /15/$$

$$j_d = 0.7833 \cdot Re'^{-0.5429} \qquad /16/$$

- for fixed bed

$$Sc' = 4.905 \cdot 10^{-4} \cdot Re'^{-0.8682} \qquad /17/$$

$$j_d = 0.2440 \cdot Re'^{-0.9110} \qquad /18/$$

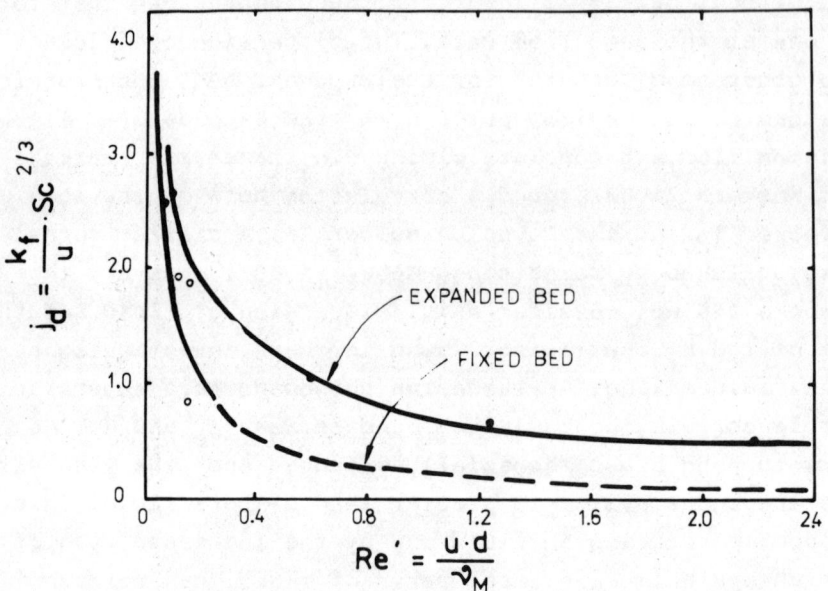

Fig. 3. j_d versus Re'

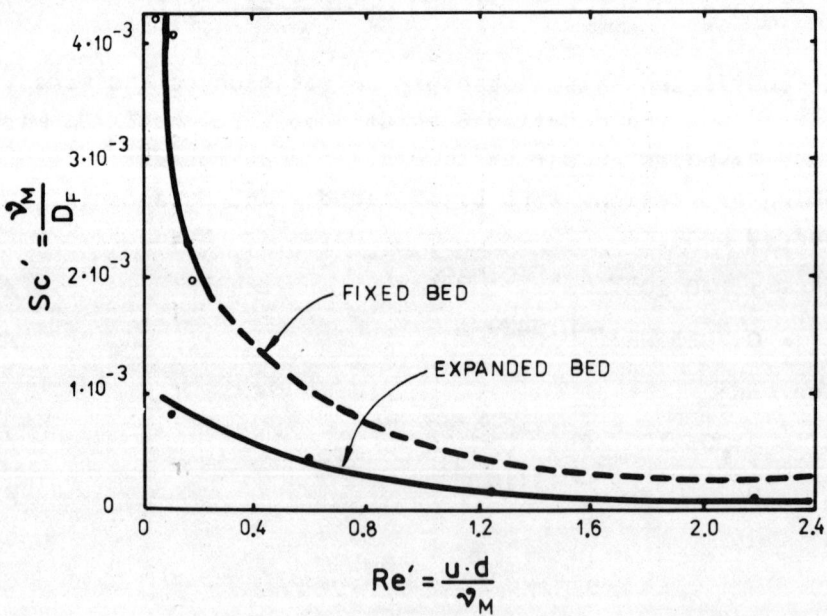

Fig. 4. Sc' versus Re'

CONCLUSION

Two phenomena are involved in the adsorption processes: axial dispersion caused by fluid mixing and adsorption itself. Fluid mixing is one of the factors of the greatest importance, especially for the expanded-bed systems.

The mathematical model incorporates the mass balance equation /eq.7/ and the kinetics equation /eq.8/ and provides breakthrough curves for both types of bed.

The two basic parameters: mass transfer coefficient k_f and dispersion coefficient D_F can describe adsorption processes on a bed.

Using the molar viscosity ϑ_M introduced by Kurgaev, a satisfactory correlation of the mass transfer factor j_d and the modyfied Schmidt number Sc´ with the modyfied Reynolds number Re´ was found and the expressions /15/ - /18/ were established.

REFERENCES

Chu, J., J.Kalil, and W.Wetteroth /1953/. Mass transfer in a fluidized bed. Chem.Eng.Prog., 49, 141.

Ergun, S. /1951/. Determination of particle density of crushed porous solids. Anal.Chem., 23, 151.

Ergun, S. /1952/. Determination of geometric surface area of crushed porous solids. Anal.Chem., 24, 388.

Hermanowicz, W. and H.Czarnodolowa /1958/. A comparative study on phenols determination in water. /Polish/. Gaz, Woda i Technika Sanitarna, 2, 60

Hermanowicz, W., W.Dożańska, J.Dojlido and B.Koziorowski /1976/. Physical and chemical examination of water and wastewater. /Polish/. Arkady, Warsaw. p.214.

Keinath, T. and W.Weber /1968/. A predictive model for the design of fluid-bed adsorbes. Jour. W.P.C.F., 40, 741.

Kurgaev, E. /1977/. Water clarifiers. /Russian/. Strojizdat, Moscow. pp. 5 - 14.

Letterman, R., J.Quon, and S.Gemmeld /1971/. Film transport coefficient in agitated suspension of activated carbon. Jour. Water Poll. Con.Fed., 43, 2536.

Process Design Manual for Carbon Adsorption /1973/. US Environmental Protection Agency.

Roman, M. /1961/. Hydraulics of adsorbers. /Polish/. In **Proceedings of the 4th Scientific and Technical Conference** "Technical Progress in Wastewater Treatment". NOT, Katowice.

Weber, W. /1967/. Sorption from solution by porous carbon. In S. Faust, J.Hunter /Ed./, **Principles and Applications of Water Chemistry**, John Wiley Sons, Inc., New York. pp. 89 - 123.

Wilke, C. and P.Chang /1955/. Correlation of diffusion coefficient in dilute solutions. **AIChE Jour.**, **1**, 264

Zuchovickij, A. /1945/. Gas adsorption from air stream. /Russian/. **Z.Fiz.Chim. /USSR/**, **19**, 253.

ELECTROCHEMICAL OXIDATION OF SODIUM SALT OF CETYL SULPHATE UNDER GALVANOSTATIC CONDITIONS

Zbigniew Gorzka, Anna Jóźwiak and
Bożenna Kozłowska

*Institute of General Chemistry, Technical University of Łódź,
Poland*

Investigations on the electrochemical oxidation of cetyl sulphate sodium salt have been carried out. The salt is a component of a detergent named Pretepon G, commonly used in textile industry. The investigations were carried out periodically on a laboratory scale under galvanostatic conditions. They concerned determinations of the effect of a number of basic parameters such as: charge 0.05 - 2.0 Ah, initial concentration of the investigated compound 50 - 1000 mg/dm^3, temperature 303 - 353 K, concentration of sodium chloride 5 - 30 g/dm^3 and current density 0.3 - 3.0 A/cm^2 on the course and efficiency of the process. Depending on the parameters used the degree of oxidation of the compound investigated was different and ranged from several to 85 %. Consumption of electric energy for the above mentioned parameters amounted from ca 40 to ca 500 Kwh/kg of the oxidized compound. Basing on the experimental results it was found that the quantity of the oxidized compound m expressed in mg can be estimated from the following equation :

$$m = A \; Q^a \; m_o^b \; T^c \; F^d$$

where: A - constant, Q - value of charge passing through the solution Ah, m_o - initial contents of the compound being oxidized in the investigated solution mg, T - temperature in an absolute scale K, F - anode surface cm^2, a,b,c,d - corresponding exponents.

KEYWORDS

Oxidation, neutralization

INTRODUCTION

Sodium salt of sulphuric acidester and cetyl alcohol of the formula $C_{16}H_{33}OSO_3Na$ is one of the basic components of a commercial preparation Pretepon G, being a detergent commonly used in textile industry. Properties of the salt are characteristic of anionic surface active substances of that type. Surface active substances (SAS) components of all kinds of detergents, emulgators and dispergators commonly used in industry to a higher and higher degree pollute water reservoirs (Pohl, 1964; Sturm, 1973; Bames, 1969; Lieber, 1969). The problem of rational methods for neutralization of such substances is a very actual one, and so far it has not been fully solved. It refers especially to industrial waste waters containing SAS and other toxic organic substances at high concentrations of hundreds and even thousands mg/dm^3. Investigations carried out so far have proved that one of the most effective methods for neutralization of such waste waters is the method of electrochemical oxidation (Mieluch, 1975; Kaczmarek, 1975; Gorzka, 1976; Easton, 1967; Miller, 1965). Application of this method on a technical scale requires a number of investigations on a laboratory scale, in order to determine the influence of basic parameters on the course and efficiency of electrochemical oxidation process. The parameters are: electric charge, current density, initial concentration of SAS, temperature, concentration of sodium chloride.

METHODS

The investigations were carried periodically out under galvanostatic conditions in a laboratory apparatus. The basic element of the apparatus was a cylindrical glass electrolyzer of a capacity $0{,}25$ dm^3 in which electrodes made of sheet platinum were placed: a cathode of a surface area 2 cm^2 and an anode of a surface area varying from $0{,}5$ to 4 cm^2. The distance between the electrodes was $2{,}5$ cm and electrolyte volume was 100 cm^3. During the process the solution was stirred using a magnetic stirrer at a constant speed. During the experiments a constant temperature of the solution was maintained. For the most of the measurements the time was constant - 1 hour.

The progress of the reaction was determined basing on the difference of substrate concentrations before and after the reaction. The contents of anionic active substance (AAS) was determined by the colorimetric method according to Jones with Longwelle's - Manience's and Abbot's modification (Longwell, 1955; Abbot, 1962) after removing free chlorine with sodium thiosulphate.

RESULTATS AND DISCUSION

Galvanostatic measurements followed potentiostatic investigations which allowed the determination of the range of current density values at which maximum degree of conversion can be obtained. The investigations gave also a number of data about the influence of electrode space separation on the values of the degree of oxidation. It was found that under the conditions of the investigations carried out the current of a density not lower than $0{,}025$ A/cm^2 should be used to obtain degree of conversion significant from the practical point of view. This current density value corresponds to the electrode potential equal 2,8 V in relation to a normal hydrogen electrode (N H E).

Results obtained within this range are presented in Table 1 and Fig. 1. For the electrode potential $\Delta\varphi \geqslant 2,8$ V NHE practically

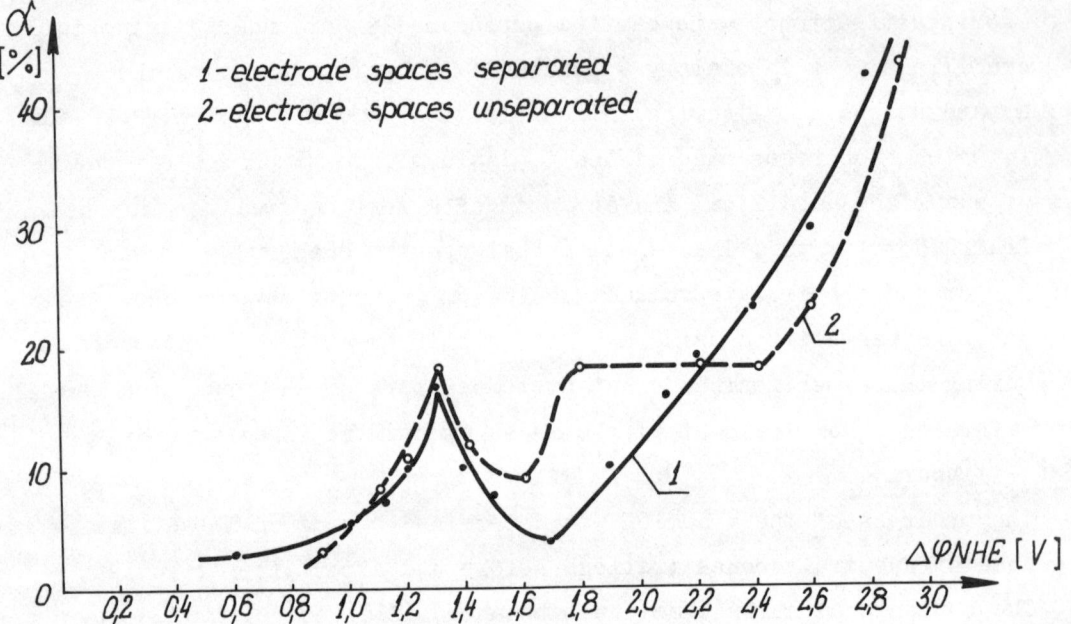

Fig.1 Dependence of the degree of oxidation of sodium salt of cetyl sulphate on electrode potential.

no influence of electrode spaces separation on the obtained degree of oxidation of the compound investigated was observed. That was the reason for which all the galvanostatic measurements were carried out in an electrolyzer with unceparated electrode spaces. As a part of galvanostatic investigation, experiments were made to determine the effect of basic parameters on the efficiency of electrochemical oxidation reaction. One of them is electric charge. The influence of this parameter on the degree of oxidation of AAS was investigated within the range 0,05 - 2,0 Ah for different initial concentrations of the compounds C_{AAS} investigated ranging from about 50 to 1000 mg/dm^3. The concentrations correspond

Table I. Degree of oxidation of the investigated compound depending on the electrode potential and separation of electrode potential and separation of electrode spaces/experimental conditions: initial concentration of AAS $5,8 \cdot 10^{-4}$ M, NaCl concentration 0,5 M, temperature 313 K, anode surface area 2 cm^2

NEW V	Separated electrode spaces		Unseparated electrode spaces	
	i A/cm^2	%	i A/cm^2	%
0,68	$1,350 \cdot 10^{-7}$	2,3	$1,400 \cdot 10^{-7}$	
0,93	$4,400 \cdot 10^{-7}$		$4,350 \cdot 10^{-7}$	1,8
1,03	$1,025 \cdot 10^{-6}$	5,5	$1,000 \cdot 10^{-6}$	
1,08	$1,300 \cdot 10^{-6}$		$1,300 \cdot 10^{-6}$	7,9
1,13	$1,775 \cdot 10^{-6}$	7,3	$1,800 \cdot 10^{-6}$	
1,18	$7,500 \cdot 10^{-7}$		$7,550 \cdot 10^{-7}$	
1,23	$8,950 \cdot 10^{-7}$	9,8	$9,050 \cdot 10^{-7}$	10,7
1,28	$3,395 \cdot 10^{-6}$		$3,405 \cdot 10^{-6}$	
1,33	$5,500 \cdot 10^{-5}$	16,4	$5,495 \cdot 10^{-5}$	16,5
1,38	$5,500 \cdot 10^{-5}$		$5,010 \cdot 10^{-5}$	
1,43	$1,938 \cdot 10^{-4}$	9,8	$2,004 \cdot 10^{-4}$	12,1
1,48	$3,680 \cdot 10^{-4}$		$3,620 \cdot 10^{-4}$	
1,53	$5,505 \cdot 10^{-4}$	7,5	$5,500 \cdot 10^{-4}$	
1,58	$5,965 \cdot 10^{-4}$		$5,905 \cdot 10^{-4}$	
1,63	$7,635 \cdot 10^{-4}$		$7,915 \cdot 10^{-4}$	8,9
1,68	$1,276 \cdot 10^{-3}$		$1,283 \cdot 10^{-3}$	
1,73	$1,485 \cdot 10^{-3}$	3,7	$1,480 \cdot 10^{-3}$	
1,78	$5,575 \cdot 10^{-4}$		$5,565 \cdot 10^{-4}$	
1,83	$7,950 \cdot 10^{-4}$		$7,950 \cdot 10^{-4}$	16,0
1,88	$1,296 \cdot 10^{-3}$		$1,302 \cdot 10^{-3}$	
1,93	$1,725 \cdot 10^{-3}$	10,5	$1,715 \cdot 10^{-3}$	
1,98	$2,800 \cdot 10^{-3}$		$2,796 \cdot 10^{-3}$	
2,03	$4,700 \cdot 10^{-3}$	17,8	$4,702 \cdot 10^{-3}$	
2,08	$7,645 \cdot 10^{-3}$		$7,700 \cdot 10^{-3}$	
2,13	$1,223 \cdot 10^{-2}$	16,1	$1,218 \cdot 10^{-2}$	
2,23	$2,553 \cdot 10^{-2}$	18,9	$2,550 \cdot 10^{-2}$	16,1
2,28	$3,354 \cdot 10^{-2}$		$3,344 \cdot 10^{-2}$	
2,33	$2,750 \cdot 10^{-2}$		$2,725 \cdot 10^{-2}$	
2,43	$2,750 \cdot 10^{-2}$	23,6	$2,725 \cdot 10^{-2}$	16,2
2,63	$2,750 \cdot 10^{-2}$	29,7	$2,725 \cdot 10^{-2}$	23,6
2,83	$2,750 \cdot 10^{-2}$	42,8	$2,725 \cdot 10^{-2}$	
2,93	$2,750 \cdot 10^{-2}$		$2,725 \cdot 10^{-2}$	43,2

to the initial contents of the compound oxidized in the investigated solution m_o within the range 4,6 mg - 100 mg. The investigations were carried out at three different temperatures: 313 K, 323 K, 333 K. The obtained results are presented in Table 2. Fig. 2 presents the results of the experiments carried out at temperature 323 K. From the experimental data it follows that the degree of oxidation of AAS increases with the growth of electric charge value and depends on the initial contents of the compound investigated in the

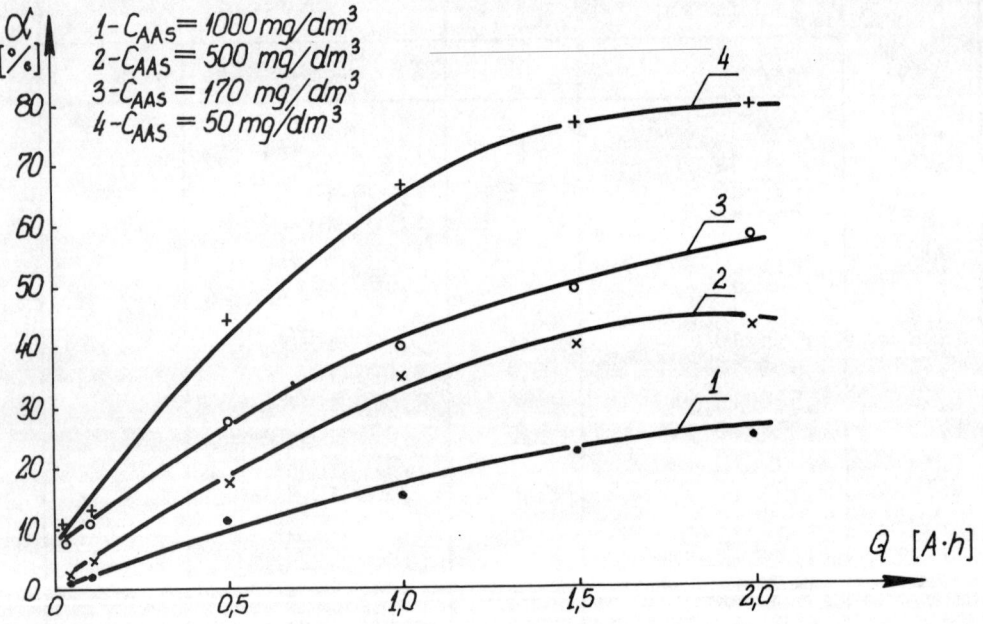

Fig.2 Dependence of the degree of oxidation of sodium salt of cetyl sulphate on electrode charge.

electrolyte solution. At the same value of electric charge a higher initial content of AAS corresponds to a lower degree of its oxidation. Increase in temperature has an essential influence on the efficiency of electrochemical oxidation process as it lowers anodic overvoltage, increase the rate of oxidation reaction and accelerates diffusion of the oxidized compound to anode. The effect of temperature on the degree of AAS oxidation was investigated within the temperature range 303 - 353 K for different sizes of anode

Table II Amount of the oxidized compound depending on electric charge and initial contents of AAS for different temperatures / experimental conditions: NaCl concentration 30 g/dm^3, anode surface area 1 cm^2.

n	Q Ah	m_o mg	m mg			n	Q Ah	m_o mg	m mg		
			TK 313	TK 323	TK 333				TK 313	TK 323	TK 333
2	0,10	100	2,2	2,0	3,6	14	0,10	17	1,0	2,0	2,4
3	0,50	100	6,5	9,5	11,0	15	0,50	17	2,6	4,7	4,5
4	1,0	100	10,6	15,0	18,5	16	1,0	17	4,2	6,2	7,1
5	1,5	100	14,0	22,5	24	17	1,5	17	5,6	8,3	9,5
6	2,0	100	17,9	24,6	30	18	2,0	17	7,0	9,9	11,8
8	0,10	50	1,5	2,5	2,5	19	0,05	4,6	0,25	0,4	0,4
9	0,50	50	4,5	14,0	17,6	20	0,10	4,6	0,44	0,55	0,7
10	1,0	50	7,5	17,3	18,0	21	0,50	4,6	1,0	2,0	2,2
11	1,5	50	10,0	19,5	17,0	22	1,0	4,6	2,0	3,2	3,6
12	2,0	50	12,4	19,9	20,6	23	1,5	4,6	2,9	3,5	4,8
13	0,05	17	0,5	1,6	1,8	24	2,0	4,6	3,6	3,62	5,9

surface with the following constant parameters : charge 1,5 Ah, initial concentration of AAS 170 mg/dm^3, and sodium chloride concentration 30 g/dm^3.

The obtained results of the experiments are presented in Fig.3 and 4.

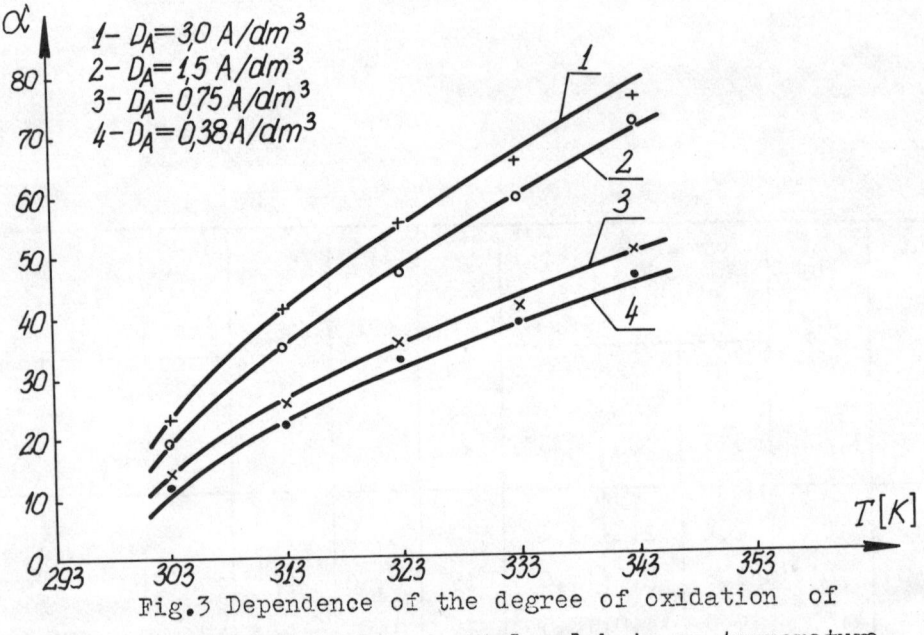

Fig.3 Dependence of the degree of oxidation of sodium salt of cetyl sulphate on temperature

From the figures it follows that with the growth of temperature the degree of oxidation of the investigated compound is increased. It is higher the denser the current D_A. The effect of NaCl concentration on the efficiency of the process was investigated within the concentration range 5 - 30 g/dm^3. Concentrations lower than 5 g/dm^3 of NaCl were not considered because of too small conductivity of the electrolyte solution. Increased NaCl concentration above 30 g/dm^3 seemed aimless because of too a high degree of waste waters salinity. Within the investigated NaCl concentration range no essential influence of it on the degree of AAS oxidation was observed.

For the performed experiments electric energy consumption was calculated. Its value for the parameters investigated ranged from about 40 to about 500 KWh/kg of SAS.

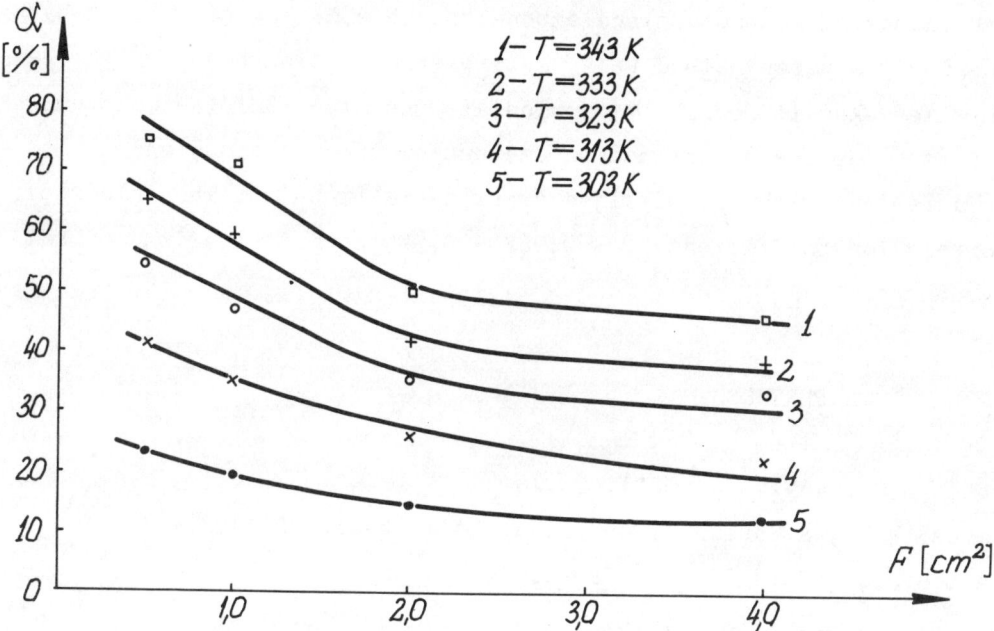

Fig.4 Dependence of the degree of oxidation of sodium salt of cetyl sulphate on electrode surface area.

The experiments proved a complete utility of the electrochemical oxidation method for oxidation of sodium salt of cetyl sulphate in solutions containing sodium chloride. The degree of oxidation of the compound investigated was different depending on the parameters used and its maximum value was about 85 %.

The obtained results of experiments indicated that the amount of the oxidized compound m expressed in mg during the electrochemical oxidation reaction can be calculated from the following formula:

$$M = A \ Q^a \ m_o^b \ T^c \ F^d$$

where: Q - the value of electric charge passing through the investigated solution Ah, m_o - initial contents of the compound oxidized in the investigated solution mg , T - reaction temperature K F - anode surface area cm^2 , A - constant, a,b,c,d - relevant exponents.

The values of A constant and exponents a,b,c,d were calculated by the least squares method using multiple regression equation. The formula makes it possible to calculate the amount of the compound oxidized for the determined parameters of the reaction, and it is in agreement with the obtained experimental results. The values of correlation coefficients r = 0,962 - 0,999 seem to confirm that agreement.

REFERENCES

Abbot,D.C., Analyst. (1962). 80,167
Bames, W.V., S.Sobson (1969).J.Wat.Poll.Abstr. 42,1428.
Easton,J., (1967).J.Water. Poll. Centr.Feder. 39, 1621.
Gorzka,Z., A.Jóźwiak (1976).Metody fizykochemiczne oczyszczania wód i ścieków. Referaty z Konferencji Naukowo Technicznej Lublin.
Kaczmarek,T., R.Dylewski (1975). Chemik 1975 1,3.
Lieber,M., Water and Sewage Works (1969).116,28.
Longwell,J., W.D. Maniece (1955). Analyst. 80, 167.
Mieluch,J., A.Sodowski, P.Złotowski (1975).Przemysł Chemiczny 54,513.
Miller,H.C., W.Kuipe (1965).U.S. Public Health Serv. Publ. AWTR-13, 58.
Pohl,B. (1964). Saponaty ve vodnim hospodarstvi Praha
Sturm,B., (1973). J.Am.Oil Chemist Soc. 50,159.

ELECTROCHEMICAL OXIDATION OF "ROKAPHENOL N-6"

Zbigniew Gorzka, Krystyna Jasinska and
Adam Socha

*Institute of General Chemistry, Technical University of Łódź,
Poland*

ABSTRACT

The commonly known and applied methods of purification of waste waters prove useless for neutralization of concentrated solutions of nonionic surface-active substances (NSAS) . Because of their high resistance to biochemical decomposition the substances are very dangerous for water environment and their neutralization in the available waste water treatment plants is a very difficult task. Experiments have shown that an electrochemical method is one of the most effective methods of decomposition of the substances. Investigations of electrochemical oxidation of Rokaphenol N-6 basic component of a commonly used detergent Roksol IT allowed to establish a dependence of the degree of oxidation of the investigated compound on the below mentioned parameters: current value $0,01 - 0,3A$, current density $0,01 - 0,1 A/cm^2$, temperature $293 - 333 K$, initial concentration of NSAS $100 - 800 mg/dm^3$, concentration of NaCl $2,5 - 35 g/dm^3$.
It has been found that with the determined parameters of the process: current density $0,025 A/cm^2$, temperature $333 K$, concentration of NaCl $20 g/dm^3$ and initial concentration of the solutions of surface-active substances $500 mg/dm^3$ a high degree of oxidation of "Rokaphenol N-6" about 90 % can be obtained. Consumption of electric energy for

the above mentioned parameters has been relatively low of the order of 10 KWh/kg of oxidized Rokaphenol N-6. It has been found that the amount of oxidized Rokaphenol N-6 is proportional to its initial concentration in an electrolyte, whereas consumption of electric energy decreases with the growth of initial concentration of NSAS and the increase in NaCl concentration.

KEYWORDS

Nonionic surface-active substances ; electrochemical oxidation ; electric energy consumption.

INTRODUCTION

Surface active substances (SAS) contained in all kinds of detergents commonly used in industry pollute water reservoirs more and more. The commonly known and applied methods for purification of waste waters prove useless for neutralization of concentrated solutions of non-ionic surface-active substances (NSAS) of e.g., Rokaphenol type because of their high resistance to biochemical decomposition (Albanese,1974; Wirth,1975; Sturm,1973) . Rokaphenols are products of condensation of nonylphenol with ethyleneoxide.
Investigations of electrochemical oxidation of SAS carried out in our Institute proved that Rokaphenol N-6, a component of commonly used detergents called Roksols IT, undergoes a considerable oxidation at potentials exceeding 0,9 V except for the areas of potentials 1,1 ; 1,4 - 1,6 and 2 V at which the electrode shows some passivity resulting from irreaversible character of oxygen adsorption an anode. The highest values of the degree of oxidation were obtained above the potential 2,1 V.
The aim of the present paper was to obtain data about the effect of some parameters on the degree of oxidation of Rokaphenol N-6 within

the potential range $\Delta\varphi \geqslant$ 2,1 V by a galvanostatic method.
The investigations included the influence of a number of basic parameters such as current intensity I 0,01 - 0,3 A , current density 0,01 - 0,1 A/cm^2 , temperature T 293 - 333 K , concentration of sodium chloride C_{NaCl} 2,5 - 35 g/cm^3 on the course and efficiency of the process.

We have also determined the dependence of the amount of oxidized Rokaphenol N-6 and the consumption of electric energy on the initial concentration of Rokaphenol N-6 C_o , and on the concentration of sodium chloride C_{NaCl} .

METHODS

Galvanostatic measurements were carried out in a laboratory apparatus, the main part of which was a glass electrolyzer of a capacity about 250 cm^3. Anodic space of a capacity about 200 cm^3 was separated from cathodic one by a ceramic membrane target anode . Electrodes were made by sheet platinum. Cathode surface area was constant in all the experiments whereas anode surface area was different and depending on the type of experiments equal 0,5,1,2 or 4 cm^2. Before measurements the electrode was properly prepared. Electrolyte of 100 cm^3 was intensively stirred with a magnetic mixer. The electrolyzer was placed in a thermostat, securing temperature stability. The required value of the current was maintained by means of a stabilized feeder. In the course of experiments, the potential of the electrode investigated in relation to SCE was registered using a recorder and millivoltmeter. The experiments were carried out periodically. The time of all the experiments was constant and equal one hour. The procedure of the experiments was as follows: To the electrolyzer electrolyte was poured and mixed in a thermostat. After the required temperature had been reached a voltmeter,recorder and feeder with a proper current value ware switched on. During the experiments stability of the assumed parameters was maintained and the voltage on the electrolyzer clips was recorded. After completing the experiment the concentration of Rokaphenol N-6 in the electrolyte was determined by the colorimetric method in accordance with the Polish Standard, and from the difference in the contents of N.S.A.S before and after the reaction the degree of oxidation of the inve - stigated substance was found (Hermanowicz, 1976).

RESULTS AND DISCUSSION

The results of the investigations of the effect of the current value on the degree of oxidation of Rokaphenol N-6 are presented in Fig.1 During all experiments the below mentioned parameters were constant; time-1 hour, initial concentration of NSAS - 363 mg/dm^3, initial concentration of NaCl - 30 g/dm^3, temperature 313 K; electrolyte volume 0,1 dm^3. From fig. 1 it follows, that with the increase in

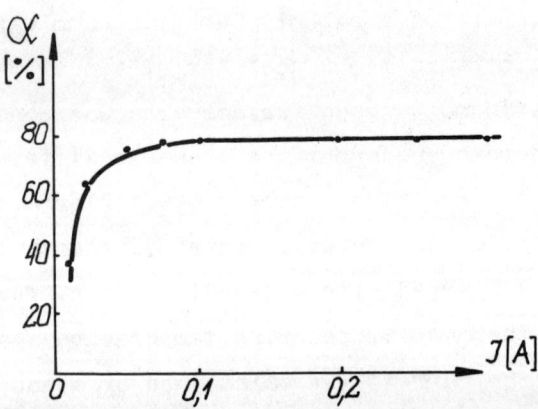

Fig.1 Dependence of the degree of Rokaphenol N-6 oxidation on the current intensity.

current value within the range 0.01 -ca 0.05 A the value of the degree of oxidation grows rapidly. Further increase in current above 0.05 A does not result in any essential changes in the degree of NSAS oxidation. Experiments in which solutions of different initial concentrations of NSAS were used did not indicate any considerable influence of Rokaphenol N-6 concentration on the degree of conversion. The dependence of the degree of oxidation α % on the current value can be presented by the following equation: for $I \geq 0,02$ A

$$\alpha = \frac{I}{AI + B} \qquad 1$$

where

α - degree of oxidation of the investigated compound %

I - current value A

A and B- constants found graphically from the dependence

$$I/\alpha = f(I) \quad (A=0,01259, B=0,00004)$$

Figure 2 presents the results of investigations of the effect of current density on the degree of NSAS oxidation. In all the experiments, current value equal to 0.05 A was used as optimum basing on the previous series of investigations.

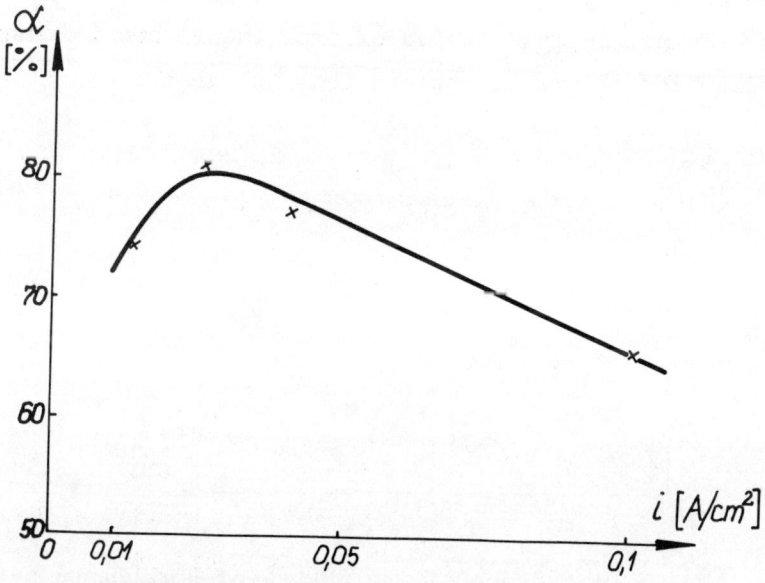

Fig.2 Dependence of the degree of Rokaphenol N-6 oxidation on the current density.

The remaining parameters of the process were identical with those in the previous experiment. The change of current density on anode was obtained by changing its surface area over the range $0,5 \text{ cm}^2 - 4 \text{ cm}^2$. From the dependence presented in fig. 2 it follows that at current density about 0,025 A/cm^2 the degree of oxidation reaches the extremal value of 80 %. Optimum current density corresponding to this extremum, results probably from adsorption of the compound investigated on the electrode and its catalytic effect on the oxidation process. In electro-oxidation reactions the value of anodic current is, as a rule, dependent on the degree of coverage of the electrode with the substance being oxidized. Some general rules governing electrocatalysis and its connections with other chemical properties, especially for the metals from the platinum group follow from Bockris , Suinkels and Breiter's investigations(Kuhn,1963; Conway 1957; Bockris 1964,1966). For the optimum current density 0,025 A/cm^2 the dependence of the degree of Rokaphenol N-6 oxidation on temperature was investigated. All the remaining parameters of the process were analogous to those in the previous experiment. The results obtained in this series of experiments are illustrated in Fig.3. From the figure it appears that the degree of oxidation

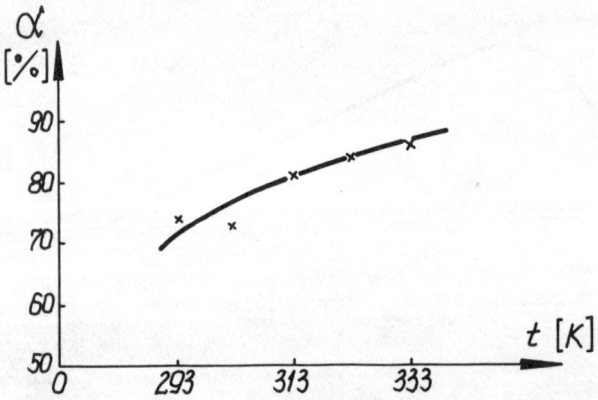

Fig.3 Dependence of the degree of Rokaphenol N-6 oxidation on temperature.

changes insignificantly with the increase in temperature. At the increase from 293 to 333 K the degree of conversion grows from about 73 % to about 87 %. Thus the use of temperatures higher than 313 K seems aimless. Figure 4 presents the results of investigations on the dependence of the degree of NSAS oxidation on NaCl concentration

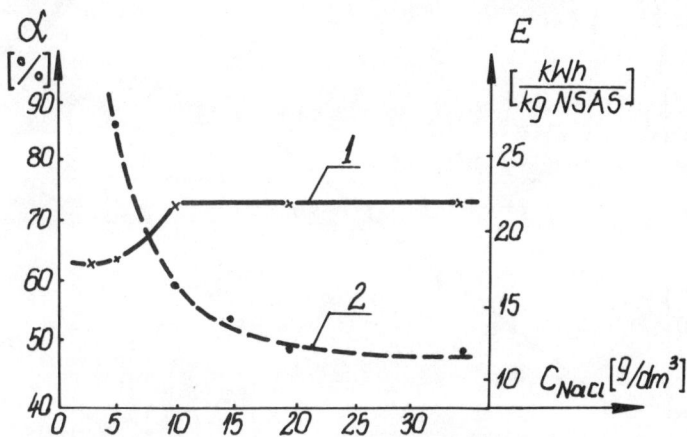

Fig. 4 Dependence of the degree of oxidation /curve) 1 and consumption of electric energy (curve 2) on the NaCl concentration

curve 1. As it follows from the course of the curve 1 the degree of Rokaphenol N-6 oxidation does not depend on the concentration of NaCl above its value equal to 10 g/dm^3. Proper choice of NaCl concentration is also influenced by the consumption of electric energy drop from about 21 KWh/kg$_{NSAS}$ to 12 KWh/kg$_{NSAS}$. A similar influence of NaCl concentration was observed during oxidation of phenol in the solution of NaCl (Mieluch,1975). Results of the investigations of the effect of NSAS initial concentration on the amount of oxidized Rokaphenol N-6 m_o are presented in fig. 5. All the measurements were carried out under conditions following from the previous experiments namely: current density 0,025 A/cm^2; tempertaure 313 K, NaCl concentration 20 g/dm^3, electrolyte volume 0,1 dm^3. From the figure it appears that within the investigated range of NSAS concentrations 100-700 mg/dm^3 the amount of oxidized Rokaphenol N-6 is in proportion to its initial concentration in electrolyte solution. For all concentrations the degree of NSAS oxidation was about 76 %. When the degree of oxidation and initial concentration C_o of Rokaphenol N-6 are known the amount of oxidized substance in the investigated sample can be

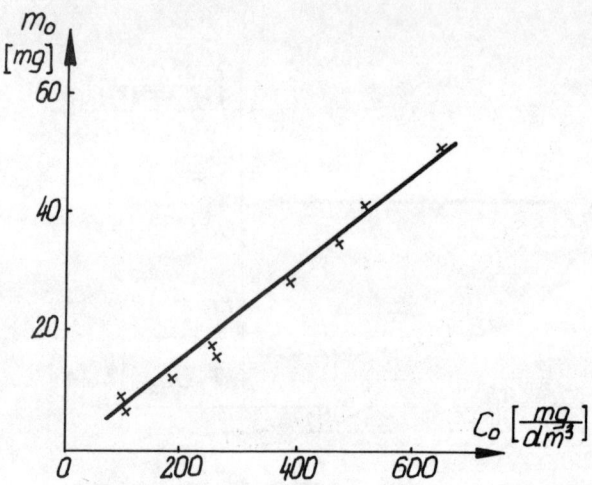

Fig.5 Dependence of the amount of oxidized Rokaphenol N-6 on its initial concentration in
electrolyte solution.

expressed by the following equation :

$$m_o = \frac{I \, C_o \, V \, 10^{-2}}{AI + B} \quad mg \qquad 2$$

where:

m_o - amount of oxidized Rokaphenol N-6   mg
I - current intensity A
C_o - initial concentration of Rokaphenol N-6 mg/dm^3
V - volume of the investigated sample dm^3
A and B - constants as in 1.

The dependence of the consumption of electric energy in KWh/kg of the oxidized Rokaphenol N-6 on its initial concentration is illustrated in Fig. 6 curve 1 . All data for the dependence presented in figure curve 1 were obtained from the experiments in which the initial concentration of Rokaphenol N-6 varied within the range 100-700 mg/dm^3 and the remaining parameters of the process were as follows: current density 0,025 A/dm^2, temperature 313 K, NaCl concentration 20 g/dm^3, electrolyte volume 0,1 dm^3. The figure indicates

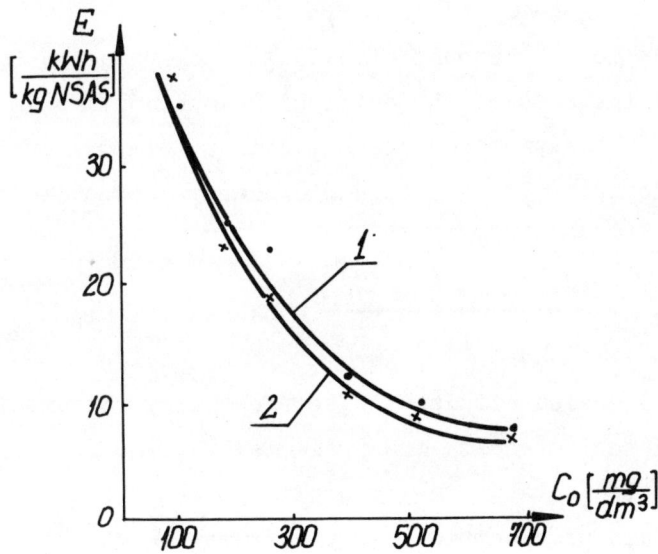

Fig.6 Dependence of the electric energy consumption calculated from formulae 4 (curve 1) and 5 (curve 2) on the initial concentration of Rokaphenol N-6 in electrolyte solution.

that with the increase in initial concentration of NSAS in electrolyte solution, consumption of electric energy is reduced. The observed reduction of electric energy consumption at the increase in initial concentration of NSAS is a consequence of a nearly constant degree of oxidation of Rokaphenol N-6 (α = about 76 %) within the investigated concentration range. Such a course of the dependence of energy consumption is confirmed by the analysis of equations 1 and 2. If in equation 1 the value of product $AI \gg B$ then $\alpha = 1/A$ = const. Substituting expression $I/AI + B$ with the value $1/A$ in equation 2 we obtain its new form

$$m_o = 1/A \; C_o \; V \; 10^{-2} \qquad \qquad 3$$

Unit consumption of electric energy can be calculated form the formula

$$E = \frac{UIt}{m_o} \; \frac{10^{-3}}{10^{-6}} \; \text{KWh/kg}_{NSAS} \qquad 4$$

where:

m_o - stands for the amount of oxidized Rokaphenol N-6 mg.

If in formula 4 the mass of the oxidized Rokaphenol N-6 is replaced by the value of the expression presented in a form of equation 3, a new form of formula 5 for electric energy consumption can be obtained:

$$E = \frac{U \, I \, t \, A \, 10^5}{C_o \, V} \quad KWh/kg_{NSAS} \qquad 5$$

Using formula 5 calculations of electric energy consumption were carried out and their results are presented in fig. 6 curve 2. As follows from the data in the figure there is a high agreement between the values of electric energy consumption calculated according to the formulae 4 and 5. This proves a correctness of the derived dependences 1 and 2. The above defined eqs. 1 2 3 and 5 verified for constant times of the process i.e 1 hour. As follows from the preliminary experiments on the determination of the dependence of the oxidation degree on the time of reaction at current intensity 0,05 A, the process can be shortened to about 30 minutes. The shortening could have an essential effect on the reduction of electric energy consumption without lowering the degree of NSAS oxidation.

The results of investigations presented in this paper allow to conclude the following:

- Oxidation of Rokaphenol N-6 occurrs with a good efficiency about 90 % if concentration of NaCl in the solution is contained within the range 10 - 20 g/dm^3.
- It was found that good effects of NSAS oxidation can be obtained at relatively small current densities of the order of 0,025 A/cm^2.
- Electric energy consumption in the process of electrochemical oxidation of Rokaphenol N-6 is small 10-30 KWh/kg_{NSAS}. It was found that electric energy consumption drops with the increase of the initial concentration C_o= 100-700 mg/dm^3 in accordance with the following equation :

$$E = \frac{U I t A}{C_o V} 10^5 \quad \text{KWh/kg}_{NSAS}$$

- The amount of oxidized Rokaphenol N-6 is in proportion to its initial concentration in electrolyte solution
- The amount of oxidized Rokaphenol N-6 under the conditions of the investigations carried out, can be calculated from the following dependence:

$$m_o = \frac{I C_o V}{AI + B} 10^{-2} \quad \text{mg}$$

REFERENCES

Albanese P., Capuci R., La Rivista Italiana delle Sostanze Grasse, LI 70, 70-81.

Bockris J.O'M., S.Surinivasan (1966). J.Electroanalyt.Chem.,11,350.

Bockris J.O'M., H.Wroblowa (1964) J.Electroanalyt.Chem.,7,328.

Conway B.E., J.O'M.Bockris (1957). J.Chem.Phys.,26,532.

Hermanowicz W., (1976) Fizyczno chemiczne badanie wody i ścieków, Warszawa.

Kuhn A., H.Wroblowa, J.Bockris (1963). J.Electroanalyt. Chem., 6,101.

Mieluch J., A.Sadowski, P. Złotowski (1975). Przemysł Chem.,54,513.

Sturm, R., J. Am. Oil Chemist Soc. (1973). 50,159.

Wirth, W., Tenside (1975). 12,245.

NEUTRALIZATION OF INDUSTRIAL WASTES CONTAINING DETERGENTS BY MEANS OF CATALYTIC OXIDATION METHOD

Zbigniew Gorzka and Marek Kaźmierczak

Institute of General Chemistry, Technical University of Łódź, Poland

ABSTRACT

Investigations have been carried out on the neutralization of industrial waste waters containing detergents by means of their oxidation in a gaseous phase in the presence of copper catalyst of Polish production. Non-ionic surface-active substances (NSAS) of "Rokaphenol" type originating from commercial preparations Roksols IT occurred in industrial waste waters being neutralized at high concentrations ranging from 7.3 g/dm^3 to 47.4 g/dm^3. COD values of industrial waste waters ranged from 26000 mg/dm^3 O_2 to 66000 mg/dm^3 O_2, respectively. Under the experimental conditions: catalyst temperature 670 K, loading of a reactor with waste waters 0,6 $\frac{1}{h}$, consumption of the air 0.7 m^3/dm^3, practically complete oxidation of organic substances to carbon dioxide and water vapour was obtained, as indicated by high values of the degree of conversion ca 99,95 % and COD value of a condensate equal ca 30 mg/dm^3 O_2. Basing on the obtained results of experiments performed on an large laboratory scale, a new technology of neutralization of industrial waste waters containing detergents and other toxic, organic substances in high concentrations, was work out.

KEYWORDS

Neutralization of industrial waste waters ; catalytic oxidation of detergents ; complete oxidation of organic substances.

INTRODUCTION

Investigations were carried out the purpose of which was to work out a technological method for neutralization of industrial waste waters containing non-ionic surface active substances (NSAS) of Rokaphenol type - basic components of commonly used detergents Roksols IT. The method consists in oxidation of organic substances in the gaseous phase in the presence of a cupric catalyst produced in Poland. The subjects of the investigations were: aqueous 2 % solution of Roksol IT (NSAS 12600 mg/dm^3, COD 40000 mg/O_2/dm^3) and industrial waste waters containing this preparation at high concentrations (NSAS 7300 - 47400 mg/dm^3, COD 26500 - 66700 mg O_2/dm^3) . The aim of the investigations presented in this paper was to determine the effect of the basic parameters such as: flow intensity of waste waters and unit loading of a catalyst, air flow intensity, shape of a catalyst layer on the degree of oxidation of NSAS and other organic substances contained in industrial waste waters.

EXPERIMENTAL INVESTIGATIONS

The experiments were carried out in tubular flow reactors of diameters: 19,28,44 and 80 mm containing a stationary layer of a catalyst weighing from 20 to 2000 g. In a reactor of a diameter 80 mm there was a catalyst in the form of tablets 6 x 8 mm, whereas in reactors of smaller diameters the same catalyst of a sieve fraction 1,5-2,0 mm was used consisting of: Cu 49 %, Zn 20 %, Al 3 %. The specific surface area of a catalyst was 60 m^2/g. The reactor was heated electrically. Model solution or waste waters (200 - 1000 cm^3/h) were evaporated and introduced to the reactor in a mixture with air

(200 - 700 dm^3/h). Catalyst temperature in all experiments was 670 K and its unit loading with solution varied within the range 0,2 -17 h^{-1}. The coefficient of air excess ranged from 0,95 to 6,5. The gas mixture leaving the reactor was cooled and the condensed condensate was analysed for NSAS contents by the plarographic method and in some samples by the colorimetric method according to Polish standard. COD was also determined.

RESULTS

The results of the investigations carried out in reactors of different diameters for different masses of a catalyst and its loading with waste waters can be presented with some approximation in the form of one curve in the system of co-ordinates: degree of NSAS oxidation (α), the ratio of the depth of the catalyst layer to its diameter (H/D). The dependence presented in Fig.1 includes the

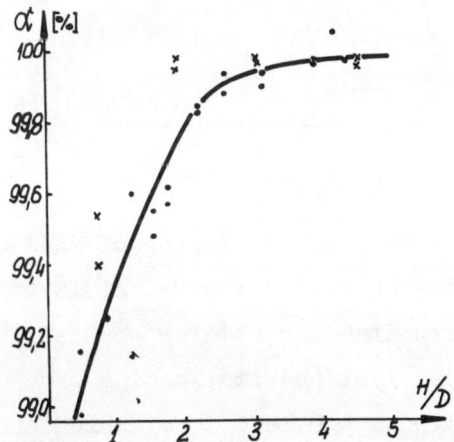

Fig.1 Dependence of the degree of NSAS oxidation (α) on the ratio of catalyst layer height (H) to its diameter (D).

results of experiments obtained under the following conditions: reactor diameters 28,44, 80 mm, catalyst masses 20 - 170 g, 50-380 g, 600 - 2000 g, intensities of waste waters delivery 220,500,700 cm^3/h.

From the data presented in Fig.1 two conclusions can be drawn: the ratio of the depth of the catalyst layer to its diameter in a reactor (H/D) should be higher than 2,5 and the degree of conversion (α) is to a small extent dependent on unit loading of a catalyst with waste waters. An increase in intensity of waste waters delivery to a reactor reduces the time of contact of reagents with a catalyst and simultaneously intensifies the course of the process by the increase in linear flow rate of air mixture, water vapour and organic substances being oxidized. The conclusions were confirmed by subsequent experiments, the results of which are presented in Fig. 2

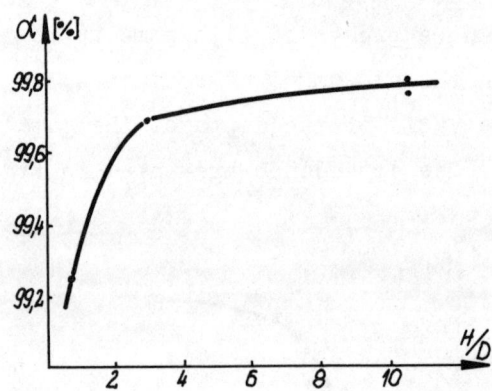

Fig.2 Dependence of the degree of NSAS oxidation (α) on the ratio of catalyst layer height (H) to its diameter (D) for constant amount of a catalyst and its loading with wastes.

and 3. In the experiments the results of which are presented in Fig.2 60 g of the catalyst was used in reactors of different diameters: 19, 28 and 44 mm. Unit loading of a catalyst with waste waters was constant and amounted to 9 h^{-1}. From the figure it follows that the increase in H/D ratio to value 3 results in a considerable increase in the degree of conversion. Further growth of the ratio of the depth of the catalyst layer to its diameter above the value H/D = 3 only slightly affected changes of NSAS concentration in the condensate. High increase in unit loading of a catalyst with waste waters within the range 0,22 - 4 h^{-1} results in an small change of the degree of NSAS oxidation (from 99,96 to 99,85 %) . The dependence presented in Fig. 3 is based on the results of experiments carried out in the following conditions: reactor diameter

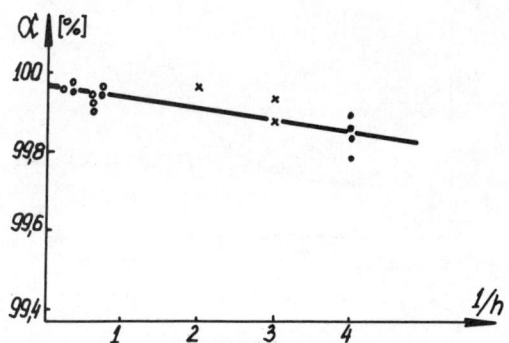

Fig.3 Dependence of the degree of NSAS oxidation (α) on a unit loading of a catalyst with wastes.

28,44,80 mm, unit loading of the reactor with waste waters 2-4, 2-3, 0,22-0,73 h^{-1} respectively. H/D ratio in those experiments was higher than 2,5 and had practically no influence on the obtained degree of conversion. The ratio did not influence the results of the experiments presented in Fig.4 and those obtained at a constant depth of the catalyst layer of 180 mm in the reactors of different diameters and at a constant rate of waste waters delivery 500 cm^3/h. From the data presented in this figure it follows that the increase of catalyst loading above the value 4 h^{-1} results in a considerable increase in NSAS contentration in a condenstate.

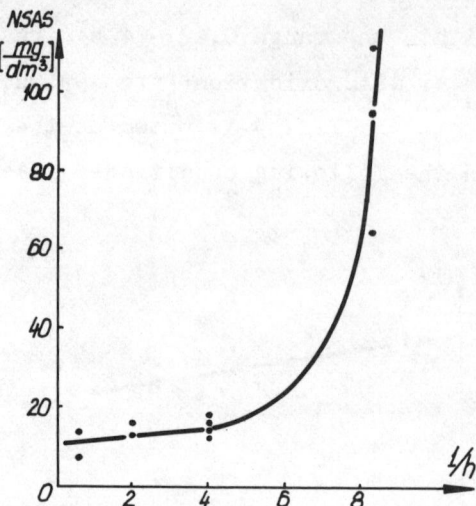

Fig.4 Dependence of the degree of NSAS concentration in a condensate on a unit loading of a catalyst at constant height of a layer.

Figure 5 presents dependence of NSAS concentration and COD of a condensate on the volume of the delivered air. Catalyst mass was

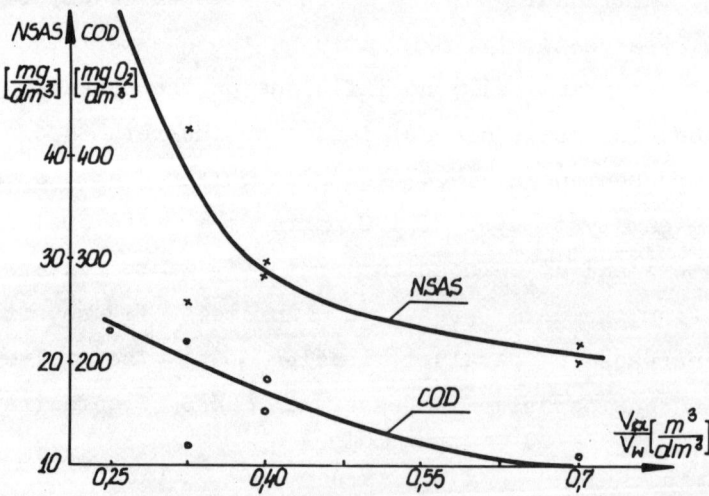

Fig.5 Dependence of NSAS concentration and COD of a condensate on the ratio of the delivered air (V_a) to the amount of wastes (V_w).

1400 g and unit loading of a reactor with wastes 0,8 h^{-1}. Two hundred to 500 dm^3/h of air was delivered to the reactor. It resulted in a change of air excess coefficient from the value 1 to about 3,5. The figure indicates that the increase in air amount above 0,7 m^3/dm^3 of wastes does not effect in the growth of reaction efficiency and is disadvantageous from the energetic point of view. The experiments, results of which are presented above indicate that the following conditions of the process: H/D = 3, catalyst loading with wastes ranging from 0,5 - 1 h^{-1} air amount 0,7 m^3/dm^3 and temperature 670 K should secure oxidation of NSAS and other organic substances contained in the wastes. The conclusions are confirmed by the re - sults of experiment presented in Fig.6. The stable conditions of the process secured very high degree of oxidation above 99,95 %.

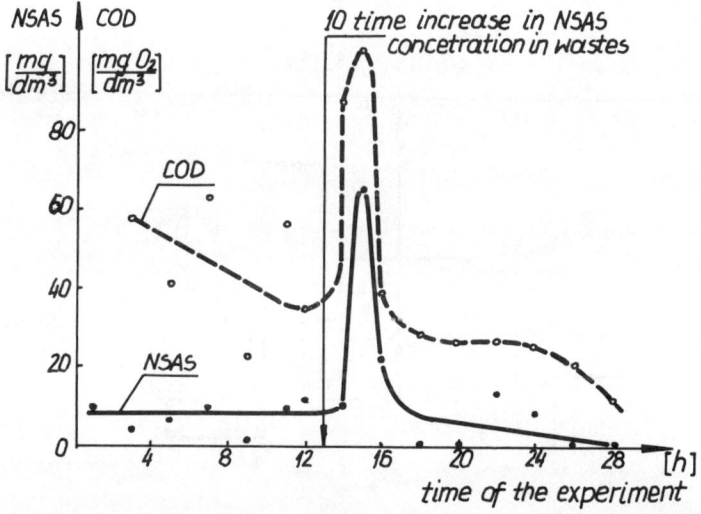

Fig. 6 Dependence of NSAS concentration and COD of a condensate on the time of the experiment.

In the obtained condensate no NSAS determined by colorimetric method according to Polish standard were found and COD of a condensate was about 30 mg O_2/dm^3. NSAS concentration in the condensate by polarographic method was of 10 mg/dm^3 order. Periodical, 10-time increase in NSAS concentration to the value about 170000 mg/dm^3 in the wastes delivered to the reactor effected in a periodical growth of

NSAS concentration in the condensate to the value 65 mg/dm^3. After about 1 hour NSAS concentration in the condensate and COD values were again reduced to minimal ones.

DESCRIPTION OF THE ELABORATED METHOD

Based on the investigations of catalyst oxidation of NSAS and other organic substances a new method of neutralization of detergents occurring in industrial wastes in high concentration was worked out. In order to apply this method at a technical scale a design of installation was made, a simplified of which is presented in Fig. 7. In the installation three cycles can be distinguished: of air, wastes and waste gases.

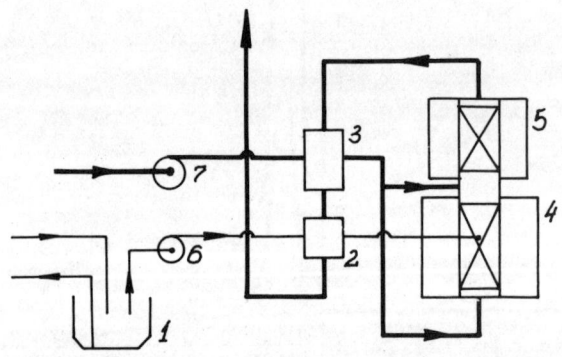

Fig. 7 A scheme of installation to catalytic oxidation of NSAS in industrial wastes.
1 - sedimentation tank, 2-wastes heater,
3 - air heat, 4 - evaporator, 5 - reactor,
6 - pump, 7 - fan.

Air cycle - fan 7 presses the air through air heater 3 to evaporator 4. The evaporator is a pipe made of acid-resisting steel filled with Raschig's rings. Temperature 770 K is maintained inside the evaporator.

Wastes cycle - wastes from a sedimentation tank are pressed to heater 2 by pump 6, from there after heating in waste gases the wastes are introduced to the upper part of evaporator in counter-current to the air. A mixture of vapours and air passes to a reactor 5 in which a constant temperature 670 K is maintained.

Waste gases cycle - waste gases leaving the reactor are directed through air heater 3 and wastes heater 2 to the atmosphere through a chimney conduit.

REMOVAL OF ORGANIC SOLVENTS FROM WASTEWATERS BY MEANS OF PHYSICO-CHEMICAL METHODS

Tymoteusz Jaroszyński

Institute of Environment Protection Engineering, Technical University of Wrocław, Wybrzeże Wyspiańskiego 27, 50-370 Wrocław, Poland

ABSTRACT

Experimental results obtained for the removals of acetonitrile, chlorobenzene, dioxane and N-methylpyrrolidone in physical-chemical treatment are presented. The experiments were conducted on a laboratory scale and employed the following processes: aeration, adsorption on activated-carbon /granular and powdered/ and coagulation. It was found that powdered activated-carbon was the best adsorbent in acetonitrile, dioxane and N-methylpyrrolidone removal. These compounds were removed from the wastewater in 53, 40 and 92 percent, respectively. The aeration of the wastewater with chlorobenzene decreased it is concentration in 97 percent. The application of a two stage process is recommended for removing the solvents in question. In the first stage the wastewater is subjected to air-stripping. In the second stage the air in which the solvents are contained is passed through an activated-carbon bed.

KEYWORDS

Wastewaters, physico-chemical methods, acetonitrile, chlorobenzene, 1,4-dioxane, N-methylpyrrolidone.

INTRODUCTION

Solvents are known to be of great importance to industrial practice. These chamical compounds exhibit a variety of physical and chemical properties among which the ability of dissolving many substances, chiefly organics /such as fats, oils, lubricants, varnish and paints/ is of special interest. In engineering practice, dissolution is frequently used to facilitate, accelerate or initiate a technological process. However, organic solvents have the disadvantage of being toxic to humans, animals and aquatic organisms, especially when discharged into a watercourse. Organic solvents contained in industrial wastewaters cause serious damage to the natural environment of man and should therefore be removed and reasonably disposed of.

The studies reported in this paper were aimed at determining the removals of acetonitrile, chlorobenzene, 1.4-dioxane and N-methylpyrrolidone /NMP/ in some physical-chemical treatment processes.

SOME SOLVENTS PRESENT IN INDUSTRIAL WASTEWATERS

Acetonitrile occurs in the wastewaters discharged by chemical industries. Its concentration is especially high in the wastewaters generated during acrylonitrile, acetophenol, 1-naphthaleneacetic acid and Vitamin B_1 production. In organic chemistry acetonitrile is used to extract fatty acids from petroleum, as well as to remove tar and phenols from petroleum hydrocarbons, whereas in food industries acetonitrile is employed to remove fatty acids from vegetable and animal fats. It follows that acetonitrile occurs in wastewaters discharged by either of these industries.

Chlorobenzene is discharged along with the wastewaters generated during the manufacture of phenols, plant pesticides, solvents and dyes.

Wastewaters from the synthesis of isoprene-base rubber /conducted with the use of Furberov's method/ carry isoprene together with some semi-products, and among them dioxane /Klimkina, 1959/, which is also present in the wastewaters produced by varnish, paint, shoe polish and synthetic resin manufacturing plants /Urbański, 1977/.

NMP occurs in emergency discharges of wastewater from oil industries where this organic solvent is used to separate gaseous mixtures /obtained in hydrocarbon cracking/ and hydrocarbon mixtures /e.g., for butadiene extraction from C_4 hydrocarbons/, /Urbański, 1977/, to extract aromatic compounds after completion of the Arosolvan process and to remove acidic constituents /hydrogen sulphide and carbon dioxide /from natural gas after completion of the Purisol process /BASF, 1965/.

PHYSICAL-CHEMICAL PROPERTIES AND TOXICITY OF THE ORGANIC SOLVENTS UNDER STUDY

The physical properties of acetonitrile, chlorobenzene, dioxane and NMP are listed in Table 1.

TABLE 1 Physical Properties of Acetonitrile, Chlorobenzene, Dioxane and NMP

Compound	Molecular weight	Density at 293 K /g/m^3/	Melting point /K/	Boiling point /K/	Water-solubility /g/m^3/
Acetonitrile	41.05	0.788	228	354.5	Unlimited
Chlorobenzene	112.56	1.106	228	405	500/303K/
Dioxane	88.10	1.034	285	374-375	Unlimited
NMP	99.14	1.031	256	475	Unlimited

Note: These data are available in handbooks.

Acetonitrile is a colourless and almost odourless liquid classified among dipolar aprotic intermediate solvents. Aqueous solutions of acetonitrile at near-boiling temperatures exert a corrosive action on steel-made apparatus. Acetonitrile is a toxic substance which may cause extensive or chronic skin diseases. Any exposure to the action of toxic nitriles leads to death from suffocation in a relatively short period of time. Repeated exposure of the organism to a sublethal dose can damage the central nervous system by causing aphasia, paralysis, anaemia and heart diseases. Rats exposed to the atmosphere containing from 2,800 to 25,000 g of acetonitrile in one cubic meter of air undergo acute intoxication /Meinck, Stooff, Kohlschütter, 1975/.

Chlorobenzene is a colourless liquid characterized by an aromatic odour and volatilizes in steam. The physical-chemical properties of chlorobenzene are similar to those of benzene. Chlorobenzene is an aprotic solvent. It does not react either, with chlorobenzene-soluble acids and bases which do not undergo dissociation. The toxic effect of chlorobenzene is the same as that of halogen derivatives. Chlorobenzene is a narcotic whose influence on blood-generating organs is somewhat weaker than that of benzene.

1.4-dioxane is a colourless flammable liquid characterized by an agreeable odour and belongs to the group of inert /aprotic/ solvents which do not cause dissociation. In ether form dioxane is a stable combination and does not readily participate in a chemical reaction. This industrial solvent is highly toxic to the human organism and a number of lethal intoxication from dioxane are reported in the literature /Browing, 1965/. The highest admissible concentration to which man can be exposed during an eight-hour working period without observable changes in his organism equals 10 mg/m^3, whereas the highest admissible dioxane concentration for man is 300 mg/m^3 /Dutkiewicz, 1974/. Lethal doses by ingestion to mice, rats and guinea pigs are 5.66 g/kg, 5.17 g/kg and 3.90 g/kg, respectively /Browing,1965/.

NMP is a hygroscopic, weakly basic, transparent, light-yellow liquid easily dissolving many of those substances that are almost insoluble in other solvents. NMP does not readily participate in chemical reactions. This property of NMP, together with its high ability to dissolve various substances, permits its application as a reaction medium. NMP has an irritating effect on the skin and mucous membrane /BASF, 1965/. White rats exposed to NMP inhalation died after 6 hours, NMP intake through ingestion is negligible.

The solvents under study are also toxic to aquatic organisms. Acetonitrile penetrating into natural waters kills fishes life after 96 hr exposure to lethal concentrations which vary from 1000 to 1850 g/m^3 /Meinck, Stooff, Kohlschütter, 1975/. However, microorganisms /and particulary nitrogen-fixing bacteria/ are able to degrade acetonitrile /Kelly, Postgate, Richards, 1967/.

Chlorobenzene is less susceptible to biochemical degradation. The admissible chlorobenzene concentration in biological treatment processes equals 10 g/m^3 /Stasiak, 1975/.

The author's own experiments with these solvents /carried out in respirometers described by Gomółka, Szypowski, 1971/, show that chlorobenzene is characterized by the highest toxicity /Fig.1, curve 2/ while the toxicity of dioxane /Fig.1, curve 3/ is lower and that of acetonitrile or NMP the lowest /Fig.1, curves 1 and 4/.

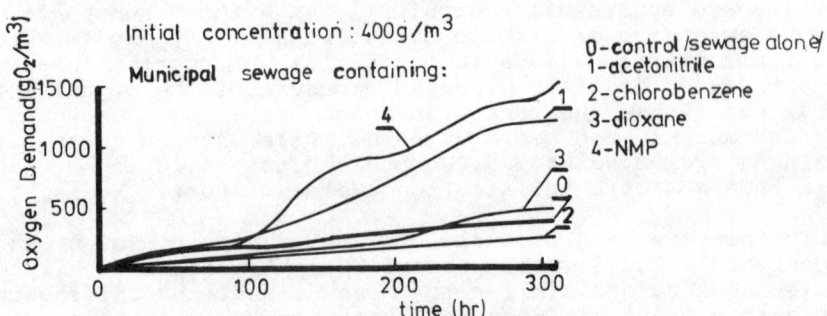

Fig. 1. Oxygen uptake versus sewage aeration time for wastewaters containing selected organic solvents.

Chlorobenzene and dioxane in concentrations of up to 50 g/m^3 only slightly inhibit biological treatment processes. This inhibition becomes evident when the concentration of these substances exceeds 300 g/m^3. Acetonitrile is highly toxic in concentrations up to 1000 g/m^3, and NMP in concentrations up to 2000 g/m^3.

EXPERIMENTAL

Scope of Investigations

The experiments were run on a laboratory scale. The organic solvents under study were removed in alum coagulation, during sorption on pulverized /powdered/ activated carbon and in the aeration process.

Alum Coagulation

The process parameters employed were as follows. Rapid mix: 180 s /100 rev/min/; slow mix: 1200 s /25 rev/min/, sedimentation: 18,000 s; temperature of aqueous solution: 293 K; aluminium sulphate doses /Al$_2$/SO$_4$/$_3$. 18H$_2$O/: from 50 to 200 g/m^3. The initial concentration of each solvent was constant and equaled 100 g/m^3. The results are shown in Fig. 2.

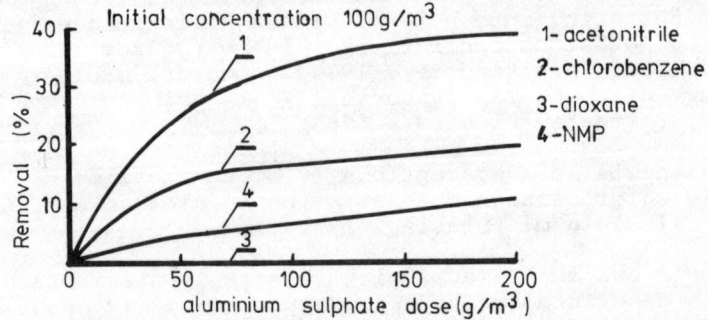

Fig.2. Efficiency of alum coagulation in the removal of solvents.

As can be seen from this figure, the removals of acetonitrile and NMP tend to increase slightly when the coagulant dose is increased /curves 1 and 4/. The increase of the coagulant dose has no effect on the removal of dioxane /curve 3/ and chlorobenzene /curve 2/. A decrease of about 20 % in the chlorobenzene content /which has been observed in the course of the experiments/ is due to the escape of chlorobenzene vapours in the processes of rapid and slow mixing rather than to the effectiveness of the coagulation process. With an aluminium sulphate dose of 200 g/m^3 about 40 percent of acetonitrile was removed, whereas the NMP removal was as low as 10 percent. It can generally be concluded that coagulation is insufficient to remove these solvents from industrial wastewaters.

Sorption on Powdered Activated Carbon

The sorption process was conducted under static conditions with the use of a domestic powdered activated carbon which is sold under the brand-name Carbopol Z-4. The initial concentration of each solvent tested was 100 g/m^3. After having added various carbon doses, the samples were shaken for 1800 s. Following carbon separation /by centrifugation/ the concentrations of the solvents in the effluent were measured. The tests were also conducted without activated carbon. The results obtained are given in Fig. 3.

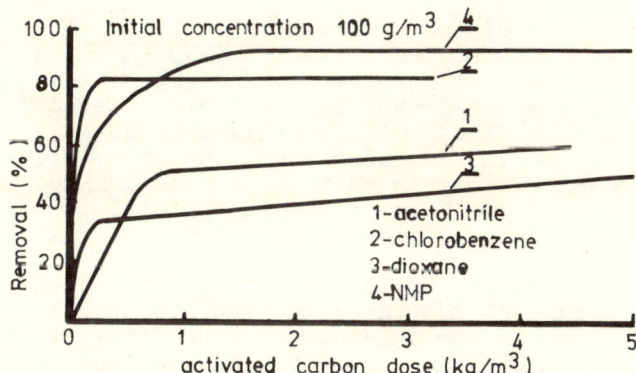

Fig. 3. Efficiency of sorption in the removal of solvents /contact time: 1800 s/.

From these curves it is evident that the best results are achieved for NMP, of which 92.5 percent has been removed with the optimum carbon-dose of 1.4 kg/m^3 /curve 4/. Chlorobenzene removal was 82 percent and had been achieved with a 0.35 kg/m^3 carbon-dose /curve 2/, acetonitrile removal was 53 percent due to a carbon dose of 1 kg/m^3 /curve 1/, and the removal of dioxane equated 40 percent, which had been obtained with the use of a 2 kg/m^3 carbon-dose/ curve 3/.

Aeration Process

In these experiments a combination of aeration and sorption on granulated active carbon was employed. For this purpose a laboratory set was used, which is shown in Fig. 4.

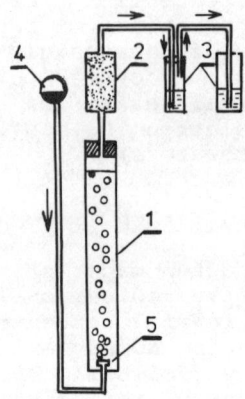

Fig. 4. Laboratory set used in the experiments.
/Details in the following text./

The procedure consists of the following steps. Aqueous solutions of the solvents studied are successively supplied to the glass pipe indicated as /1/, which has an effective height of 1 m and an internal diameter of 0.04 m. Then the solutions are aerated for 2 hours with an intensity of 0.72 m^3/m^3h. For this purpose an air compressor /4/ and an aerating screen /5/ were used. The mixture of air and the corresponding solvent vapour is next passed through the adsorber /denoted as /2// which has been packed with granulated activated carbon /Carbopol Z-4/. The bed is 0.2 m high and 0.04 m in diameter. /A system like the one presented here prevents air contamination/. The solvent portion present in the residue is passed to two separators connected in series /3/. The temperature of the aqueous solutions was 293 K, whereas the initial concentrations of the solvent were various and equaled 400 g/m^3, 500 g/m^3 and 1000 g/m^3 for chlorobenzene, acetonitrile and NMP, respectively. The initial concentration of dioxane was also 1000 g/m^3.
The results are listed in Fig. 5.

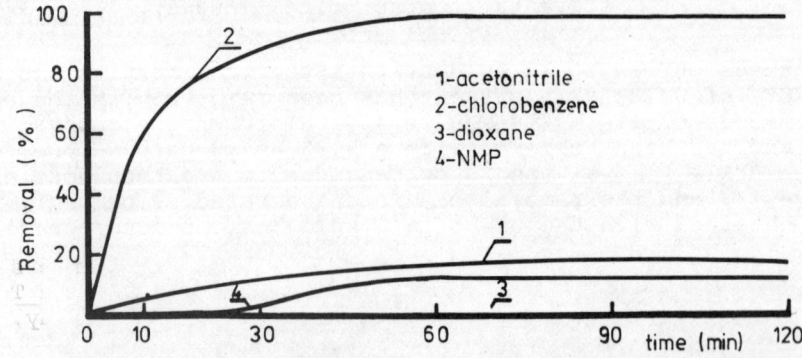

Fig.5. Efficiency of aeration in the removal of solvents.

It is easily seen that chlorobenzene volatalizes within the shortest period of time /curve 2/. After 1 hour of aeration chlorobenzene removal was as high as 97 percent, whereas the removals of acetonitrile, dioxane and NMP were as low as 17 percent, 0 percent and 12 percent, respectively. Chlorobenzene removed from water by air stripping was adsorbed in activated carbon column in 96 percent. Acetonitrile and NMP were adsorbed in 14 percent and 12 percent, respectively. It means, that only 1 percent of initial amount of chlorobenzene, and 3 percent of acetonitrile was discharged to the atmosphere. NMP was totaly removed from gaseous phase by activated carbon adsorption.

CONCLUSIONS

On the basis of the experimental results it can generally be concluded that:
1. Adsorption on pulverized activated carbon is the most effective method of removing acetonitrile, dioxane and NMP. The results achieved via this route are: a 1 kg/m^3 carbon dose yields 53 percent removal of acetonitrile; with a carbon dose of 2 kg/m^3 some 40 percent of dioxane is removed; the removal of NMP is 92.5 percent and has been obtained with a 1.4 kg/m^3 carbon dose.

2. Aeration is most suited to remove chlorobenzene whose concentration in the solution after an 1-hour aeration is as low as 3 percent. However, it is advisable to combine aeration with sorption on granulated activated carbon in order to prevent air contamination. The air-chlorobenzene vapour mixture generated during the aeration process should be passed through the adsorber packed with granulated carbon.

3. To achieve satisfactory removals of all the solvents studied, a procedure is proposed which consists of the two following successive steps: aeration of the wastewater and passing the air /in which the solvents removed are contained/ through a granulated activated-carbon bed or any other medium capable of sorbing organic solvents.

REFERENCES

BASF /1965/ Booklet of Badische Anilin und Soda Fabrik.
Dutkiewicz, T. /1974/. Toxicological chemistry. PZWL, Warszawa.
Gomółka, E., Szypowski W., and others /1971/. On the Application of a Respirometer for the biodegradation of detergents. Prace Naukowe Inżynierii Sanitarnej i Wodnej PWr., Nb.11, Studia i Materiały Nb. 11.
Kelly, M., J. R. Postgate, and R. L. Richards /1967/. Reduction of cyanide and isocyanide by nitrogenase at Azotobacter chroococcum. Biochem.J., 102 /1/.
Klimkina, N. V. /1959/. Grenzwärte für schädliche Abfallprodukte aus Werken zur Erzeugung von synthetischen Kautschuk auf Isopren-Basis. Gigiena i Sanitaria. 24, fasc 6,8.
Meinck, F., M. Stooff and H. Kohlschütter /1975/. Industrial wastewaters. Arkady, Warszawa /Polish Translation/.
Stasiak, M. /1977/. The Influence of mineral and organic substances on biological processes in wastewater treatment. Nowa Technika w Inżynierii Sanitarnej. Wodociągi i Kanalizacja, Arkady, Warszawa. Nb. 5.
Urbański, J. /1977/ Solvents used in the seperation of butadiene from C_4 hydrocarbon mixtures by extraction distillation. Przemysł chemiczny, 56, 3,129-132.

POLYURETHANE RESINS AS OIL SORBENTS IN WASTEWATER TREATMENT

Edward Gomółka and Janusz Rybka

Institute of Environment Protection Engineering, Technical University of Wrocław, Wybrzeże Wyspiańskiego 27, 50-370 Wrocław, Poland

ABSTRACT

The usability of elastic polyurethane resins for the sorption of oils from a non-homogeneous oil-water mixture was studied in laboratory-scale experiments on Polish polyurethanes and SD-10 oil. The sorption ability of the resins in question was tested under static conditions and in a turbulent flow. The results obtained are promising and very interesting.

INTRODUCTION

Oils penetrating into wastewaters and natural water streams are especially troublesome in the management of water resources. These substances exert an undesirable effect on: wastewater treatment, the self-purification process, the biocenosis of natural water coerses and the physical-chemical properties of water. Oils contained in wastewater occur either as solutes or insoluble substances. In engineering practice a number of different devices for oil and fat separation have been used so far /Chojnacki, 1966; Meinck and others, 1975; Cywiński and others, 1972/. To treat the non-homogeneous oil-water mixture, it is conventional to employ two-liquid tanks, centrifugal separators as well as coalescence filters involving coke-, activated-carbon or sand beds /Koziorowski, 1975; Koziorowski, Kucharski, 1964/.

The experiments carried out in our Institute show that elastic polyurethane resins /made of polyethers or polyestehrs, such as the ones described by Olczyk, 1973/ are charakterized by a high capability of sorbing oils from the surface of a liquid or separationg non-homogenous oil-water mixtures which move in a turbulent flow /Gomółka, Gomółka, 1979/.

Considering the physical-chemical and mechanical properties of the polyurethane resins and, additionally, the high sorption rate, a device was designed and patented, which permits the separation of oils from the surface of a fluid.

In the present paper the capability of the polyurethane resins to sorb oil from the surface of a liquid and from a non-homogeneous oil-water mixture moving in a turbulent flow, as well as the desorption of the oil from the resins under turbulent flow conditions, are tested and the results obtained are discussed.

OIL SORPTION BY DRY AND IMPREGNATED POLYURETHANE RESINS FROM WATER SURFACE

The experiments were run tith an elastic polyurethane resin dry or wetted of T-25 type. Wetted resins were obtained by saturation in water /or SD-10 oil/ and wringing in a press. Thus, a small amount of water or oil was retained on the resin surface to from a thin film /impregnation/. The averages from three experimental runs for a T-25 polyrethane resin /cylindrical in shape, 8 mm in fiameter and 20 mm long, i.e. 1.005 cm^3 in volume/ are listed in Table 1.

The oil layer employed in the experiments was 1,2,5,7,10 and 12 mm thick.

As can seen from Table 1, the sorption of oil by a fresh dry resin proceeds slower. After 600 s the volume of the oil sorbed was as low as some 15-19 % of the initial resin-volume. With the use of a water-impregnated resin the volume of the oil absorbed during 600 s was between 20 and 40 % of the initial resin-bolume. The volume of the oil absorbed also depends on the oil layer thickness.

Table 1. Oil Sorption from the Surface of a Liquid

Parameter	Unit	Thickness of the Oil Layer					
		1	2	5	7	10	12
		Fresh Dry Resin					
Time of Contact	s	600	600	600	600	600	600
Volume of Oil Sorbed in Relation to Reson Volume	%	15	17	17	16	18	19
Time of Contact		Fresh Resin Wetted with Tap Water					
	s	600	600	600	600	600	600
Volume of Oil Sorbed in Relation to Resin Volume	%	20	26	30	40	40	40
Time of Contact	s	Fresh Resin Sturated in Oil					
		240	135	60	45	40	30
Volume of Oil Sorbed in Relation to Resin Volume	%	107	111	112	108	116	114

The best results were obtained with an oil-impregnated resin, which was completely saturated after a 240 seconds' time of contact with a 1 mm thick oil layer. For thicker oil layers the times of resin saturation are shorter /Table 1/.

The effect of the oil layer thickness on the saturation time for a 8mm dia. resin is shown in Fig.1, whereas the oil layer thickness effect for 4,6,10 and 12 mm dia. resin is illustrated in Fig.2.

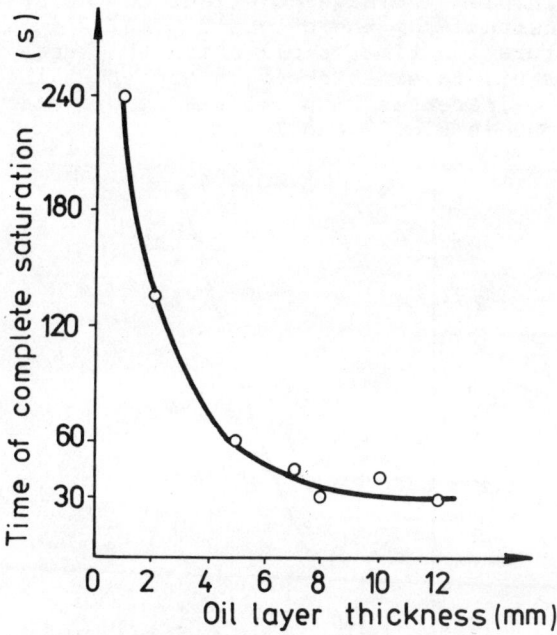

Fig. 1. Time of saturation versus oil layer thickness

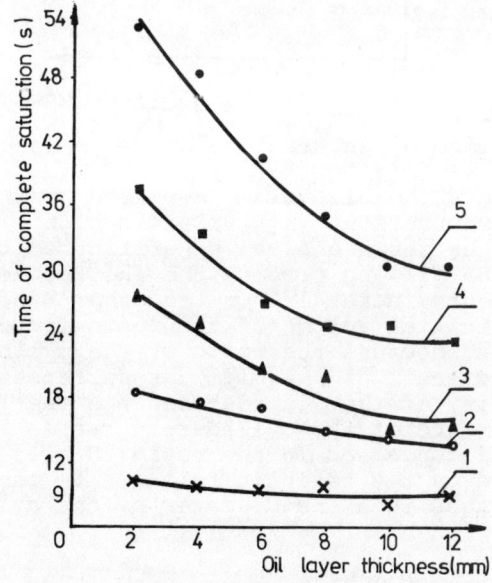

Fig.2. Time of saturation versus oil layer thickness for different resin diameters:1=4 mm; 3=8 mm; 4=10 mm; 5=12 mm

From Fig. 1 it is evident that the oil layer thickness exerts a noticeable effects on the time of resin saturation until the oil layer and the resin diameter become equal /8 mm/. Oil layer thickness higher than the resin diameter have no effent on the saturation time, which remains constant. From the curves of Fig. 2 it is noted that the dependence of saturation time on oil layer thickness behaves in a similar manner. For oil layer thickness higher than the resin diameters the saturaion time increases with increasing diameter. This relationship is also shown in Fig. 3 and Table 2.

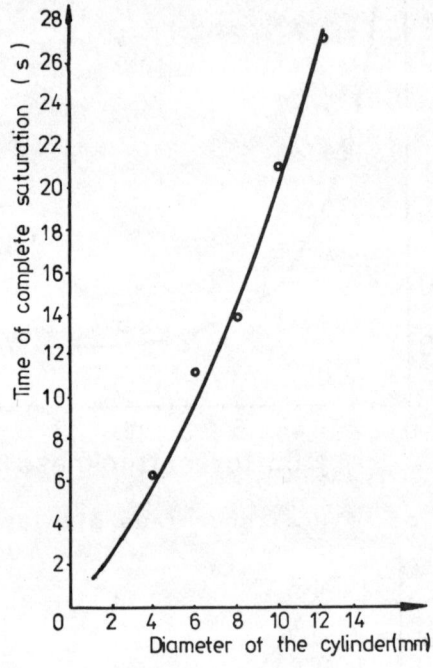

Fig. 3. Time of saturation versus resin diameter

Table 2 also includes the oil volume absorbed by the resin up to the state of complete saturation. This value is higher than 100 % of the initial volume of the resin and can be attributed to the fact that in the course of the filling process the initial volume of the elastic polyurethane resin increases. On the other hand the ratio of the oil volume to the initial volume of the resin increases with decreasing resin diameter, because the ratio of the cylinder surface to the cylinder volume increases as the diameter decreases. It follows that the relative quantity of the oil adsorbed /in the form of a thin film/ on the external surface of the cylinder increases in relation to the quantity of the oil contained in the resin. Hence, the total volume of the oil absorbed by the resin in relation to the initial bolume of the resin increases then the diameter of the cylinder decreases.

Table 2. Results of Static Investigation for a Cylindrical Sample

Resin Diameter /mm/	Time of Complete Saturation /s/	Oil Volume in the State of Resin Saturation /cm^3/*	Resin Volume /cm^3/	Volume of Oil Absorbed in Relation to Resin Volume /%/
4	6,3	0,35	0,251	139,4
6	11,2	0,75	0,565	132,4
8	13,9	1,23	1,005	122,4
10	21,1	1,85	1,570	117,8
12	27,3	2,42	2,261	107,0

* Average from 5 experimental runs

The experiments were also carried out with cube-form 20x20x20 mm resins of a volume of 8 cm^3. The results are given in Fig.4 and Table 3.

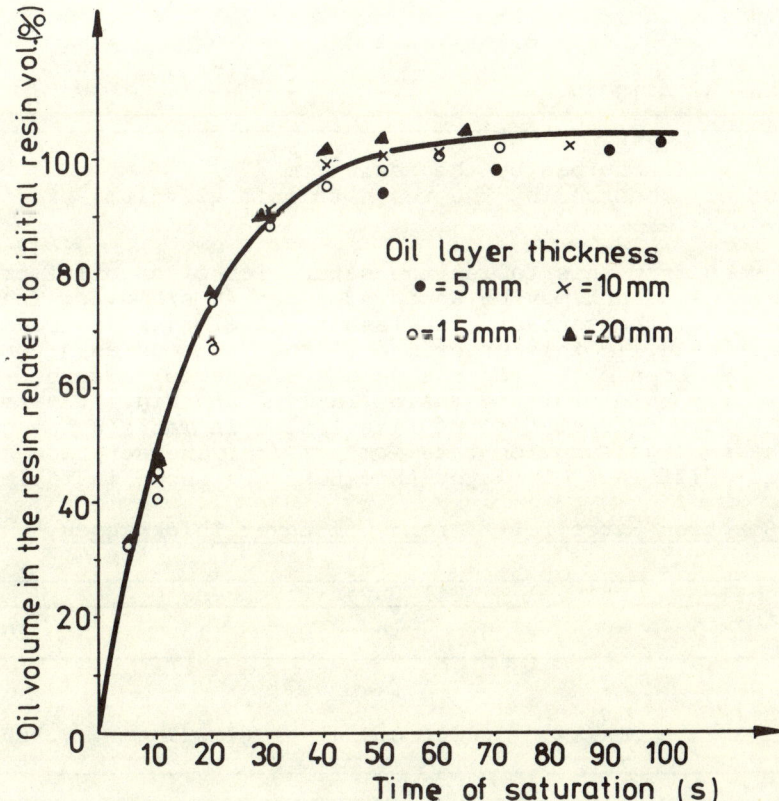

Fig.4. Volume of oil sorbed versus saturation time

Table 3. Results of Static Investigation ofr Cube-Form Samples

No.	Time of Contact	Oil Layer Thickness /mm/							
		5		10		15		20	
		V_o^*	η^{**}	V_o	η	V_o	η	V_o	η
	s	cm^3	%	cm^3	%	cm^3	%	cm^3	%
=1	2	3	4	5	6	7	8	9	10
1	5	-	-	-	-	2,6	32,5	2.65	33.1
2	10	3.26	40.7	3.46	43.2	3.63	45.4	3.83	47.0
3	20	5.35	66.9	5.46	68.3	6.03	75.4	6.15	76.9
4	30	7.23	90.4	7.26	90.7	6.96	87.0	7.23	90.4
5	40	-	-	7.96	99.5	7.63	95.4	8.16	102.0
6	50	7.56	94.5	8.03	100.4	7.83	97.9	8.36	104.5
7	60	-	-	8.1	101.3	8.06	100.7	-	-
8	64.3	-	-	-	-	-	-	8.36	104.5
9	70	7.86	98.3	-	-	-	-		
10	70.8	-	-	-	-	8.18	102.2		
11	83.4	-	-	8.16	102.2	-	-		
12	90	8.1	101.3						
13	98.8	8.24	103.0						

* Volume of oil absorbed by the resin /cm^3/
** Ratio of oil absorbed by the resin to initial volume of the resin /%/

As can be seen from this table, the saturation of an oil-impregnated resin is very fast and may be achieved after 40 s/Fig.4/. The further extension of the contact time results in a marked decrease of the sorption rate. In this experimental run /as soon as a complete saturation had been obtained/ the volume of the iol adsorbed was higher than the volume of the resin /Table 3 and Fig.4/. Based on the V_o value and contact time of 30 s listed in Table 3, we can calculate the average sorption rate /cm^3/cm^3/s/. The sorption rates obtained for different oil layer thickness are shown in Table 4.

Table 4. Sorption Rates for fferent Oil Layer Thicknesses

Oil-Impregnated Resin	Unit	Oil Layer Thickness			
		5	10	15	20
Average Sorption Rate	cm^3/cm^3/s	0.030	0.030	0.029	0.030

It is easily seen that the average sorption rate for 5 - 20 mm thick oil layers in the initial stage of the process is constant and equals about 0.03 cm^3 of oil per 1 cm^3 of resin during 1 second.

With these all in mind, a device for the mechanical separation of oils from fluids in two-liquid flow through tanks was designed and patented. The device employs polyurethane resins of an arbitrary width and a thickness of 20 mm.

SELECTIVITY OF THE RESIN FOR OIL SORPTION FROM A NONHOMOGENEUS OIL--WATER MIXTURE

Elastic polyurethane resins impregnated in oil prior to the sorption process are charakterized by a very high sorption capacity. They are also capable of separating larger oil drops from a nonhomogeneous oil-water mixture moving in a turbulent flow. These abilities have recently been confirmed in our experiments. In the first experimental run the volumetric ratio of SD-10 oil to water in a non-homogeneous mixture equaled 1:5; 1:10; and 1:20. The sorbing resin was cylindrical in shape, had a diameter of 8 mm and was 20 mm long. In the second experimental run the volumetric oil: water ratios were 1:5; 1:10; 1:15; 1:25 and 1:50. The sorbing resin was a 20x20x20 mm cube. The experimental temperature was kept constant /293K/ and so was the mixing speed. The results of the first run are shown in Fig.5, whereas those of the second run are given in Fig.6 and Table 5.

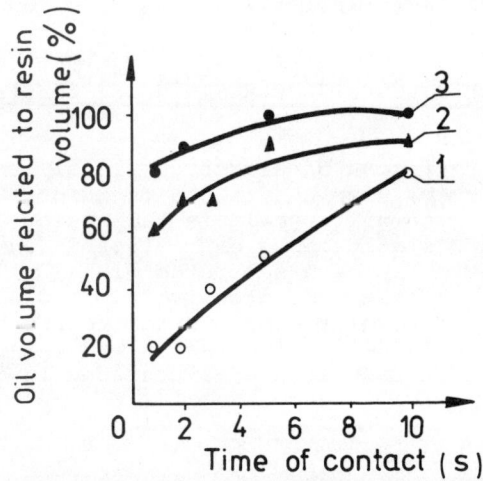

Fig.5. Volume of oil sorbed versus contact time for various oil-water ratios: 1=20; 2=1:10; 3=1:5

From the curves of Figs.5 and 6 it is easily seen that, at a constant mixing rate, the volume of the oil sorbed from the non-homogeneous mixture depends on the following factors: oil content in the mixture, dispersion and contact time. The increase of the oil content in the resin with increasing time of contact is an indication that a selective separation of oil droplets from a nonhomogeneous mixture moving in a turbulent flow takes place.

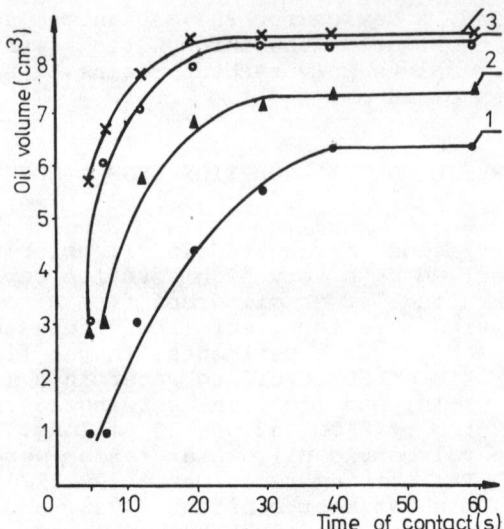

Fig.6. Volume of oil sorbed versus contact time for various oil-water ratios: 1=1:50; 2=1:25; 3=1:10; 4=1:5

DESORPTION OF OIL FROM POLYURETHANE RESIN UNDER CONDITIONS OF TURBULENT WATER FLOW

Oil contained in the resin can be almost completely removed with the use of the press. However, uder conditions of turbulent flow some portion of the oil content will remain in the resin. This has been confirmed lately in two simultaneous expreminetal runs. Polyurethane resins /which had previously been saturated with SD-10 oil /were transported into water /which was intensively mixed with a stirrer at a constant number of revolutions/. In the course of the experiments the oil volume contained in the resins was measured after various time of mixing. The results are illustrated in Figs.7 and 8.

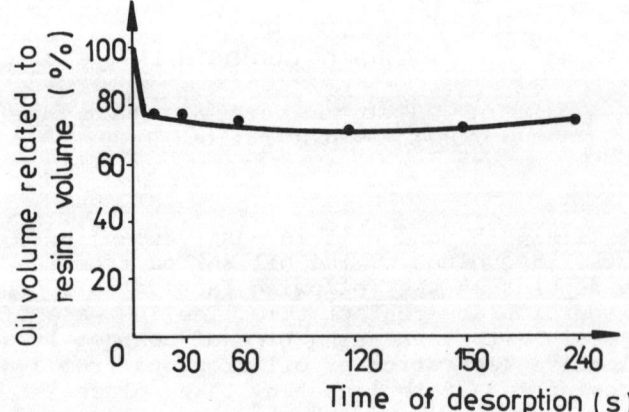

Fig. 7. The influence of desorption time on the degree of oil desorption under conditions of turbulent flow

Table 5. Experimental Results for a 20x20x20 mm Cube-Form Resin

Parameter	Unit	Time of contact /s/						
		5	7	12	20	30	40	60
							1:50	
V_o*	cm^3	1.96	1.96	3.06	4.4	5.56	6.3	6.36
V_w**	cm^3	0.8	0.7	0.96	0.83	0.8	0.96	1.1
V_o/V_w	-	2.45	2.8	3.18	5.3	6.95	6.56	5.78
							1:25	
V_o	cm^3	3.26	3.36	5.7	6.73	7.2	7.33	7.36
V_w	cm^3	0.76	0.9	0.8	8.86	0.96	0.83	0.73
V_o/V_w	-	4.28	3.73	7.12	7.86	7.5	8.83	10
							1:10	
V_o	cm^3	3.36	6.1	7.13	7.8	8.36	8.23	8.23
V_w	cm^3	0.83	0.76	0.66	0.66	0.8	0.73	0.8
V_o/V_w	-	4.0	8.0	10.8	11.8	10.4	11.2	10.2
							1:5	
V_o	cm^3	5.7	6.63	7.7	8.36	8.43	8.43	8.43
V_w	cm^3	0.73	0.83	0.8	0.73	0.73	0.8	0.9
V_o/V_w	-	7.8	7.98	9.62	11.4	11.5	10.5	9.36

*   Volume of oil absorbed by the resin
** Volume of water absorbed by the resin

As can be seen from these figures, there is a rapid decrease in the quantity of oil contained in the resins tested only in the initial storge of the mixing process. The further mixing has no effect on oil removal, which remains almost constant. This is an indication that the decrease of the oil content in the resin is due to the hydraulic removal of the oil film coverint the external surface of the resin, which is also confirmed by the constant water content observed during desorption /curve 3 of Fig.8/. It follows that the quantity of oil contained in the pores and ducts of the resin will not be desorbed during intensive mixing in a water medium.

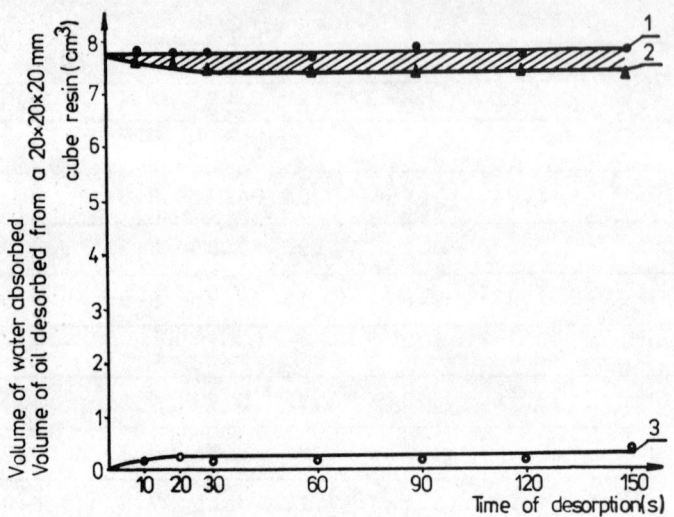

Fig. 8. The influence of desorption time on the defree of oil desorption under conditions of turbulent flow: 1=oil content before desorption, 2=oil content after desorption, 3=water absorbed by the resin

CONCLUSIONS

1. Elastic polyurethane resins of T-25 type have a high sorption capacity after impregnation in oil.
2. An impregnated polyurethane resin is able to separate oil from an oil-water mixture moving in a turbulent flow.
3. The total sorption capacity of the resin is about 100 % of the initial resin volume.
4. The sorption rate for a 20x20x20 mm cube-from resin and an oil layer thickness of 5-20 mm is 0.03 cm^3 of oil per 1 cm^3 of resin during 1 second.
5. Oil contained in the ducts and pores of the polyurethane resin cannot be removed under conditions of turbulent flow.

REFERENCES

Chojnacki,A./1966/.Technology of Industrial Wastewater treatment. PWN, Warszawa. /In Polish/
Meinck,F., H.Stooff, and H.Kohschuter /1975/. Industrial Wastewaters. Arkady, Warszawa./Polish Translation/
Cywiński,B. /1972/ Treatment of Municipal Sewage. WNT,Warszawa. /In Polish/
Koziorowski, B. /1975/ Treatment of Industrial Wastewater. WNT, Warszawa /In Polish/
Koziorowski,B. , and J.Kucharski /1964/ Industrial Wastewaters. WNT, Warszawa. /In Polish/

Olczyk, N. /1973/ <u>Polyurethanes.</u> WNT, Warszawa. /In Polish/
Gomółka, E., and B. Gomółka /1979/ Oil removal from Wastewaters with the use of polyurethane resins. <u>Gaz woda i Technika Sanitarna.,</u> 1, 27

THE EFFECTS OF THE THICKENING OF VARIOUS SEWAGE DEPOSITS SUBJECTED TO THE INFLUENCE OF ULTRASONIC FIELD

J. Bień, E. Kowalska and E. Zielewicz

Environmental Protection Institute, Silesian Technical University, Poland

ABSTRACT

The paper discusses the influence of ultrasound and synthetic polymers on the sedimenting properties of the sludge from the waste water of the metal, cellulose and paper industries and the municipal wastes from the sewage treatment plants. The tests were carried out on sludges unconditioned and conditioned with flocculation agents Zetag 63 and Zetag 92 and subjected sound irradiation by means of ultrasonic waves at frequency of 20 kHz. The optimal sound irradiation time has been determined experimentally.

KEYWORDS

Gravitational thickening; flocculation of sludge; sound irradiation of sludge.

INTRODUCTION

One of the chief aims of the treatment of sewage deposits is to reduce thier volume. Even a comparatively slight reduction of their volume results in a considerable reduction of economical costs, if we take into account the enormous amounts of sewage in all sewage treatment plants. Thus, for instance, the thickening activated sludge with a water content of 98-99 % to 95-97 % brings about a two- to three-fold reduction of their volume /Zakrzewski, 1963/.
Sewage deposits can be thickened by means of coagulating agents or without them. The thickening process is affected not only by the constructional parameters of the installations, but also by the kind of the treated deposit, the concentration of the suspension, the surface tension, the compressibility coefficient and the viscosity of the liquid.
All these parameters draw attention to the effect of the ultrasonic field as a factor which can change the structure of the deposit and cause evident changes in its properties.It might be generally said that an ultrasonic field with strictly defined parameters affects the change of such properties of the liquis as its surface tension, viscosity, the wettability of the surfaces of solid bodies, etc. /Elpiner, 1960^a; 1960^b Kowalska, 1978/.

The present paper deals with the influence of ultrasonic waves with a frequency of 20 kHz upon the change of the properties of sludge discharged by metal industry, municipal sewage treatment, cellulose and paper industry.
To aid the process of thickening, two polyelectrolytes have been chosen, i.e. Zetag 63 and Zetag 92, /produced by the Allied Colloid Manufactoring Company Ltd., England/.
Gravitational thickening, usually applied as initial treatment, leads to a reduction of the volume of sedimented sludge. Sewage deposits containing heavy macromolecules display a small stability and a tendency to settle down at a definite rate. At the same time diffusion occurs in the opposite direction, and after some time dynamical equilibrium is reached between the sedimental and the diffusional flow.
All these phenomena affect the complexity of the thickening process in various degress. Depending on the accompanying conditions and parameters, the ultrasonic field may affect the size and shape of the grains of the thickened substance, the rate of its settling and crystallization, and even the electrostatic phenomena occurring at the interphase. Thaus, ultrasounds may have a bearing upon the final effect of thickening.

EXPERIMENTAL

Investigations were carried out in Imhoff funnels after insonification of the sludge by means of an ultrasonic disintegrator produced in this country. The experimentally determined optimal time of sound irradiation was 2 min in all three kinds of sludge. The optimal dosage of flocculating agents was found of means of a CST-test /time of capillary suction/.
The thickening of metal industry sludge, unconditioned and conditioned with both these flocculating agents took a similar course. Within two hours the volume of the sludge was uniformly reduced, so that after two hours unconditioned sludge had a volume of 621 cm^3, whereas the volume of the sludge conditioned with Zetag 63 or Zetag 92 amounted to 640 cm^3 and 620 cm^3, respectively Fig. 1.
More advantageous effects have been achieved in the case of municipal sewage sludge. Unconditioned sludge, thickened for about two hours, reduced its volume to 911 cm^3; when Zetag 63 was applied, the final volume of the thickened sludge amounted to 400 cm^3, and with Zetag 92 it was 412 cm^3. The thickening of unconditioned sludge took a regular course for about two hours, whereas the volume of conditioned sludge decreased quickly and after 80 minutes did not practically undergo any change Fig. 2.
Similar effects have been obtained in the case of sludge from cellulose and paper-mills. Unconditioned sludge, thickened for two hours, reduced its volume to 810 cm^3; when Zetag 63 or Zetag 92 was used, a volume reduction to 740 cm^3 and 750 cm^3 respectively was observed Fig. 3.
Fig. 1, to 3 show also the effects of the insonification of the particular deposits.
The results of the investigations on the effect of sound irradiation upon the susceptibility of the sludge to thickening indicate a distinct, though varying effect of ultrasonic waves upon the sludge.
In sludge from metal industry it has been found that the insonification of unconditioned sludge resulted in other changes of the thickening process, and that after two hours of the gravitational thickening the volume of sound irradiated sludge amounted to 720 cm^3,

while in the case of sludge not subjeced to sonification it was 621 cm³.

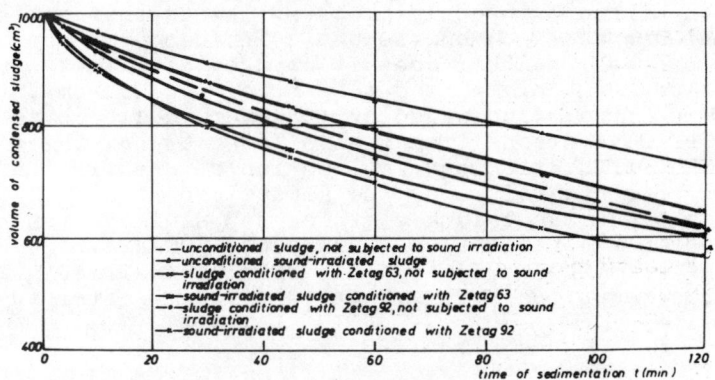

Fig. 1. Thickening curves of sludge from metal plants in an Imhoff funnel.

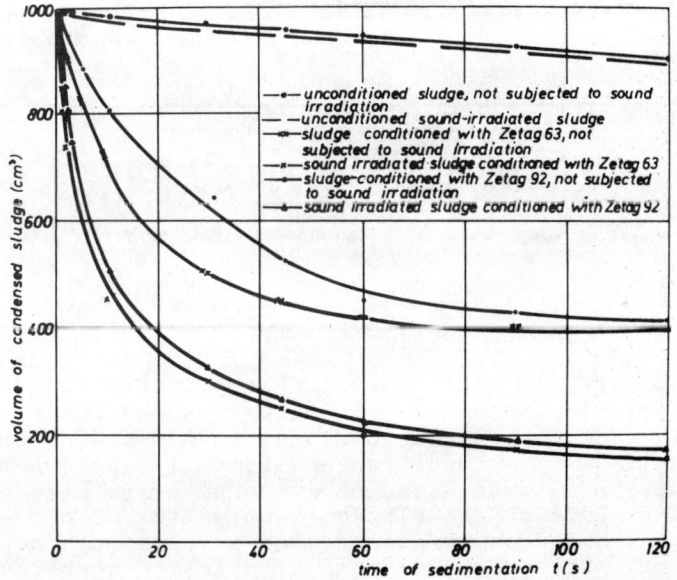

Fig. 2. Thickening curves of sludge from municipal sewage treatment plants in an Imhoff funnel.

Larger volume of sludge thickened for two hours in identical conditions were observed in the case of sludge conditioned both with Zetag 63 and Zetag 92. These volumes amouned to 610 cm³ and 580 cm³, respectively.
In the presence of the above agents the thickening of metal industry sludge, both sound irradiated and not subjected to insoification, took a similar course, as may be concluded from the similar

shape of the curves in Fig. 1.
The conditioning of municipal sludge with Zetag 63 and Zetag 92 considerably reduces the volume of the thickened sludge as well as the duration of the thickening process. The insonification of unconditioned sludge does not on the whole influence the course of the thickening process, neither does it change the final volume of the condensed sludge which amounted to 906 cm^3 and 911 cm^3, respectively. The sound irradiation affected, however, quite obviously the thickening of sludge conditioned with Zetag 63 and Zetag 92. In the case of sound irradiation and application of the first of these flocculating agents the volume of the sludge amounted to 160 cm^3 after two hours, whereas the volume of thickened sludge which had not been subjected to insonification amounted to 400 cm^3. The conditioning of sludge with Zetag 92, together with sound irradiation, resulted in volume reduction from 412 cm^3 to 170 cm^3 after two hours of thickening.

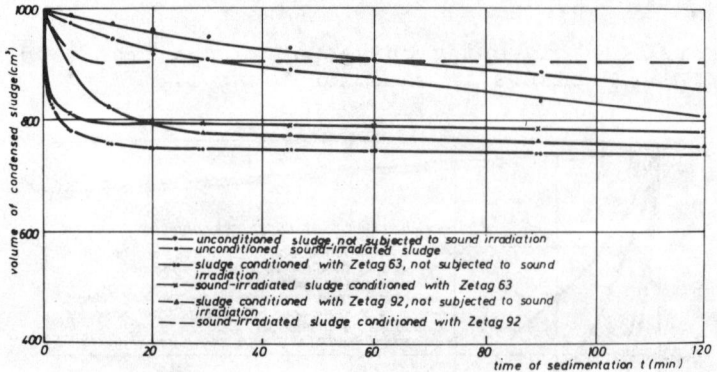

Fig. 3. Thickening curves of sludge from the cellulose and paper- industry in an Imhoff funnel.

In sewage from cellulose and paper industry it has been found that the insonification of unconditioned sludge causes an increase in the final volume of the thickened sludge to 860 cm^3 after two hours, whereas in the case of unconditioned sludge it was 810 cm^3.
If sludge conditioned with Zetag 63 is sound irradiated and then thickened, the volume of the sludge amounted to 780 cm^3 after two hours, whereas sludge non irradiated had a volume of 740 cm^3. Conditioning with Zetag 92 and sound irradiation resulted also in an increase of the sludge volume to 890 cm^3, after two hours thickening whereas sludge was not sound irradiated displayed a volume of 750 cm^3.

DISCUSSION OF RESULTS

The tested kinds of sewage sludge constitute polydispersive systems in which the addition of a polyelectrolyte leads to a series of processes resulting in the improvement of the structure of these sludges, to a decrease of their basicity etc. As to the sludge from municipal sewage treatment, it has been found that ultrasound affects them positively, decreasing the volume of the thickened sludge in the case of applying Zetag 63 or Zetag 92.

Preliminary insonification of sludge from the metal industry conditioned with either Zetag 63 or Zetag 92 before its thickening was of little influence upon the volume of the sludge, and in the case of unconditioned sludge its volume increased as a result of insonification. The insonification of sludge from cellulose and paper industry did not give the expected effects. A considerable increase of the effectiveness of the thickening process, however, was obtained in the case of the sludge from municipal sewages. The application of Zetag 63 and Zetag 92 has made it possible to reduce the volume of the thickened sludge within two hours by more than 50 % in comparison with unconditioned sludge. It should also be noted that if these flocculating agents were used, the actual effect was reached already after 90 minutes. The insonification of unconditioned sludge did not improve the effects of the thickening process, whereas the insonification of conditioned sludge resulted in a further decrease of its volume, which was reduced to about 40 % of the volume of conditioned sludge not subjected to sound irradiation, and below 20 % of the volume of unconditioned sludge that had been thickened in identical conditions. The obtained results indicate a technically and economically important possibility to intensify the thickening process by means of conditioning the sludge from municipal sewages as well as the possibility of improving considerably the effects of such a process by preliminary sound irradiation before the condensation sets in.

The distinct difference of the obtained between municipal sewage, the sludge of the metal industry and the suldge from cellulose - and paper industry results from the varying character and structure of the three investigated sewage deposits. In the sludge from metal industry crystalline mineral particles of sulphates and carbonates are predominat together with metal hydroxites, whereas the sludge of municipal sewage contains considerable amounts of organic colloids as well as particles of the sludge from cellulose and paper industry, besides lignin, fillers and cellulose fibres as well as large amounts of organic substances are found. Thus, by sound irradiation municipal sewage deposits are much more susceptible to phase changes of the condensation and dilution of the medium and the local percussive and thermal effects which accompany them. The sludge from metal industry, on the other hand, due to sound irradiation udergoes much slighter changes, observed only in the from of an insignificant improvement of the process of gravitational thickening. The structure of cellulose- and paper plants sludge changes as a result of insonification so that it hampers the normal thickening process. When the same polyelectrolytes are applied all these effects confrim the different character of the investigated kinds of sewage sludge, indicating in particular the differing electric charge of the particle surface of the solid phase. The obtained positive results may also be considered to be due to the favourable effect of ultrasounds upon the increase of the activity of the applied reagents /Milam, 1971/. Due to the acoustic energy, changes of the structure and properties of the macromolecules take place in the aqueous medium, which has been stressed in quite a number of publications /Freundlich, 1938; Mostafa, 1958/.

The physico-chemical effect of ultrasonic waves upon long-chained and branched macromolecules can also be observed both in water and in organic solvents, i.e. in a medium of polar and apolar particles. The results of the affect may very considerably, depending on quite a number of conditions accompanying them. The observed changes of the sedimentation rate may, therefore, be also connected with the intermolecular rearrangement of the polyelectrolyte, influenced by the ultrasonic field. The appearance of pulsating cavitation bubbles in the sound irradiated medium may lead, among others, to a degradation of the macromolecules, which - as several suthors /Gaertner, 1954; Gooberman, 1960/ have stated, increases with the growth of the intensity of the ultrasonic field. For instance the higher sedimentation rate depends on the presence of long-chained macromolecules /of larger molecular mass/. The obtained effects of the joint activity of ultrasounds and flocculating agents the process of gravitational thickening may, therefore, as results of changes in the activity of polyelectrolytes in sound-irradiated media, and also of the direct influnce of ultrasonic waves on the systems investigated.

REFERENCES

Zakrzewski, I., and J. Zaborowski /1963/. Sewage Utylization., /Warsav/, Arkady.
Elpiner, K. J. /1960a/. Vodosnabzhenie i Sanitarnaia Tekhnika.8,27.
Elpiner, K. J. /1960b/. Gigiena i Sanitriya., 11, 8.
Kowalska, E., J. Bień and E. Zielewicz /1978/. Acustica, 40, 99.
Milam, Q. P., and J. Pop /1971/. Investigation on the influence of ultrasound waves on the desorption of xanthate from sulphurcontaining minerales surface. Proceedings 7-th Intern. Cong. Acoust., Budapest.
Freundlich, H. and D. W. Gillings /1938/ Trans. Faraday Soc., 34, 649.
Mostafa, M. A. K. /1958/ J. Polymer Set., 33, 126
Gaertner, W. /1954/. J. Acoust. Soc. Amer., 26, 977
Gooberman, G. /1960/. J. Polymer Set., 42, 25

PROBLEMS OF CHOICE OR SORPTIVE FILTERS ON THE BASIS OF BIOT NUMBER QUANTITIES

A. Grossman, B. Kucharski and W. Kusznik

Environmental Protection Institute, Silesian Technical University, Poland

Abstract

In this paper it has been shown that the BIOT Number determined in static experiments is not reliable for choosing and dimensioning the sorptive filters, since the hydraulic conditions existing in these devices are different. A method of determining the BIOT Number in dynamic tests, where hydraulic conditions existing in sorptive filters were maintained has been worked out and tested. Measurements carried out by this methods have given real quantities of BIOT Numbers, confirmed by technological tests. This method allows to evaluate properly the mechanism of the process in sorptive filters.

KEYWORDS

Water conditioning; sewage treatment; activated carbon; sorpting; sorptive filters; BIOT number.

INTRODUCTION

Application of sorption on activated carbon in water conditioning processes, as well as the results obtained with the prototype and pilot systems have revealed that sorption together with chemical clarification creates an encouraging alternative for conventional methods of purification. The basic method for this process on a production scale may be primarily sorption on sorptive filters /Grossman, 1975/.
Parameters of the design for operating the sorptive filters and choice of the filter construction have to comprise /Johansson, 1974/:
- prolonged work cycle for a filter bed to reduce the costs of regeneration and consumption of carbon,
- use of suitable sorptive filter systems and proper hydraulic conditions to reduce the design and energy consumption costs.
Contact of activated carbon with sorbent may be proceeded:
- on filters that operate individually or in parallel and in series arrangements under the pressure or gravity flow conditions,
- on stationary or fluid filters with up-the-line flow.
The sorption may be conducted counter- or cocurrently /Johansson, 1974; Westermark, 1975/.

To establish the working parameters and select a proper filter construction, the most important is to determine a mean sorbing capacity of carbon under particular hydraulic conditions, at a high initial concentration of a sorbate and the extent required for removal of the sorbate from the liquid phase.

The mechanism of sorption is not sufficiently taken into consideration by the selection and dimensioning criteria used for the sorptive filters. It can be assumed /Neretnieks, 1976; Sontheimer, 1976/ that transfer of sorbate from the liquid to the solid phase proceeds in two stages:
- by diffusion through the interfacial film of liquid /film diffusion, membrane diffusion/,
- by diffusion through grains, which, for some simplification, may be regarded as a homogeneous phenomenon.

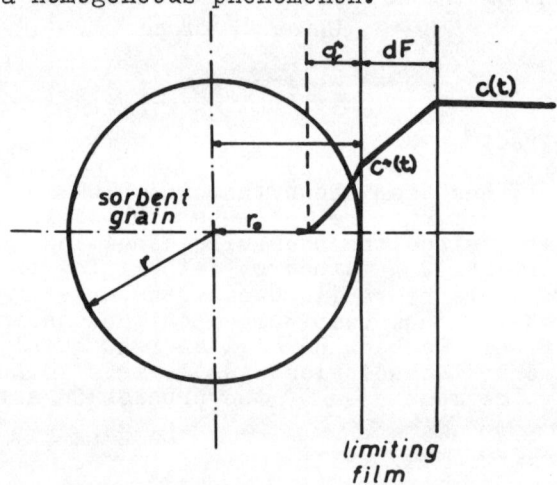

Fig. 1. Simplified diagram of sorption mechanism in a single spherical sorbent grain.

Transport of sorbate is given by the formulae:
in film diffusion

$$\dot{n} = 4\pi r^2 \beta_F /c/t/ - c^* /t/ / \qquad /1/$$

in grain diffusion

$$\dot{n} = \left(\frac{1}{r_0/t/} - \frac{1}{r} \right) = 4\pi D \cdot c^* /t/ \qquad /2/$$

The coefficients, β_F and D, and radius, r, of sorbent grain are interrelated by a dimensionless BIOT Number /Sontheimer, 1976/:

$$BIOT = \frac{\beta_F \cdot r}{D} \qquad /3/$$

It has been found /Brauch, 1975; Sontheimer, 1976/ that with the BIOT Number smaller than 5 the sorption process is limited by the diffusion through the interfacial film, whereas kinetics of the sorption process is determined by the grain diffusion when the BIOT Number is greater than 50. Depending on the BIOT numerical quantity, i.e. on the sorption mechanism characterized by this number, an adsorptive capacity of carbon can be utilized in the best possible manner in filters with fixed or fluid beds. Methodology used for evaluation of the BIOT Number and factors, β_F and D consists in establishing the kinetics of the sorption process and determining the sorption isotherms for a selected sorbing agent and sorbate in bath tests. It has been considered that thus obtained values are not reliable in processes involving sorptive filters, although similar Reynolds numbers are maintained. When measuring the kinetics and sorption isotherm, it is indispensable that hydraulic conditions identical to those in the designed processing on a sorptive filter be created to determine correctly the characteristic quantities for sorption filters.

Object and Scope of the Study

The object of this study was to develop and test a modified methodology for estimation of the BIOT Number and the β_F and D coefficients. The range of the study, which was determined by its purpose, included:
- developing a modified methodology for estimation of BIOT Number and coefficients β_F, D,
- evaluating these quantities for sorption of phenol on carbon N in static trials,
- determining the required values in accordance with a method modified at Instytut Inżynierii Ochrony Środowiska Politechniki Śląskiej Poland,
- examination of the obtained quantities during manufacturing processes.

EXPERIMENTALS

The following substrates were used for testing:
- activated carbon N manufactured at Hajnowskie Zakłady Suchej Destylacji Drewna, Poland
- waste-waters obtained by dissolving phenol in tap water.

Method of Analysis.

The sorption process was controlled by determination of phenol concentration. This value was estimated with a spectrophotometric method using SP 850 B UNICAM spectrophotometer. The thickness of the liquid layer and wavelength used were 1,0 cm and 273 nm, respectively. The spectrophotometric determination of phenol was performed on the basis of the analytical curve plotted previously /Westermark, 1975/.

Methodology of Technological Testing.

The technological tests were conducted with a laboratory model of sorptive filter, in which the diameter and depth of the bed were 37 mm and 500 mm, respectively. 335 A UNIPAN metering pump was used for directing the wastewater onto the filter, whereupon a gravity flow proceeded. The process was followed until a full breakthrough curve was obtained.

Method of Estimating BIOT Number and β_F and D Coefficients.

The BIOT Number and factors β_F, D were evaluated from Eqs. 1 and 2 equation of kinetics of sorption, and a sorbate balance in the periodic test /Sontheimer, 1976/

$$\frac{dq}{dt} = \frac{N}{S} \qquad /4/$$

$$S = \frac{dq}{dt} = -L\frac{dc}{dt} \qquad /5/$$

Under boundary conditions:

for $t=0$ $c^*(t) = 0$ and $c = c_o$

$$\beta_F = \frac{1}{a_s \cdot c_o} \frac{dc}{dt}\bigg|_{t_o} \qquad /6/$$

After determining successive dimensionless quantities:

- time parameter, τ,

$$\tau = \frac{c_o \cdot D \cdot t}{r^2 \cdot q \cdot q^*} \qquad /7/$$

- capacity parameter, C,

$$C = \frac{S \cdot q^*}{L \cdot c_o} \qquad /8/$$

the Biot Number and grain diffusivity, D, can be evaluated.
Directions of the kinetics and sorption isotherm curves are shown in Fig. 2 and 3.
A model set, represented in Fig. 4. was proposed to determine the correct kinetics and sorption isotherm curves.
In this model the periodic character of testing was maintained; however, contacting the active carbon with a liquid phase proceeded under dynamic conditions that occurred in sorptive filters.

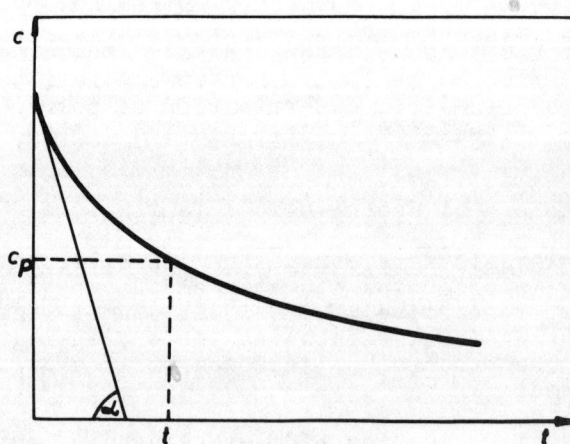

Fig. 2. Diagram for utalization of kinetics of sorption to determine BIOT Number.

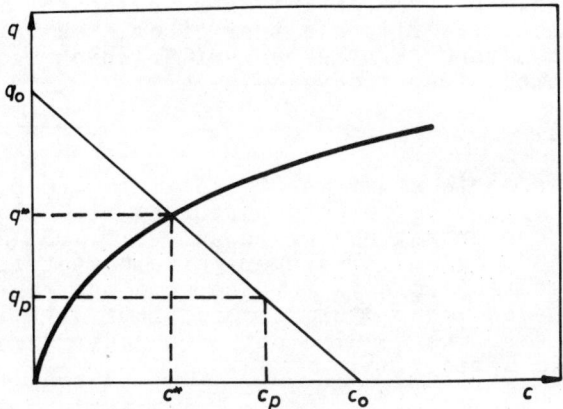

Fig. 3. Diagram of utilizing a sorption isotherm for determination of BIOT Number.

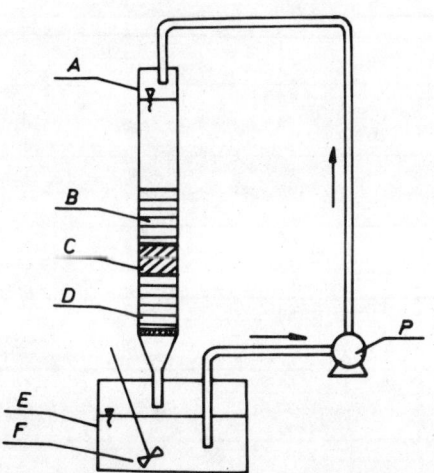

A - Filter with inside diameter of 37 mm
B - Covering layer, 4 cm
C - Sample of carbon
D - Supporting layer, 4 cm
E - Tank
F - Mixer
P - Metering pump

Fig. 4. Scheme of the device for estimation of BIOT Number under dynamic conditions.

Supporting and covering layers for the de-aerated weighed sample of activated carbon consisted of elements made of organic glass, with shapes and dimensions similar to those of the tested carbons. To make measurements, a filtrate that circulated in a closed cycle was sampled from a tank /E/ the content of which was continuously agitated. The samples were recycled after taking measurements.

Results.

The first stage of this study was to estimate the BIOT Number under static conditions. The tests were performed with prepared wastewater, in which concentrations of phenol were 100, 200, 300, 400, 500, 1000, and 1500 mg/dm^3. The average velues of the tests carried out are summarized in Table 1. For the same wastewater, the BIOT Number was estimated under dynamic conditions with the method modified by the authors. The results obtained under dynamic conditions are also given in Table 1.

TABLE 1. Quantities of BIOT Number obtained during the Sorption of Phenol on formed N Carbon under different conditions

Concentration of phenol in prepared wastewater mg/dm^3	Measurement under static condition		Measurement under dynamic conditions	
	Rotational speed of mixer $\frac{1}{min}$	BIOT Number	Filtration velocity m/h	BIOT Number
100	60	less than 1	7,5	less than 1
	90	less than 1	15,5	2,0
	120	less than 1	25,0	3,5
200	60	less than 1	7,5	1,5
	90	less than 1	15,0	2,5
	120	1,2	25,0	5,0
300	60	less than 1	7,5	4,0
	90	less than 1	15,0	6,0
	120	1,4	25,0	7,5
400	60	1,8	7,5	8,0
	90	4,0	15,0	10,0
	120	4,0	25,0	16,5
500	60	4,0	7,5	10,0
	90	4,2	15,0	12,0
	120	11,5	25,0	39,0
1000	60	5,0	7,5	15,0
	90	6,0	15,0	26,0
	120	12,5	25,0	73,0
1500	60	12,0	7,5	16,0
	90	14,0	15,0	48,0
	120	14,0	25,0	120,0

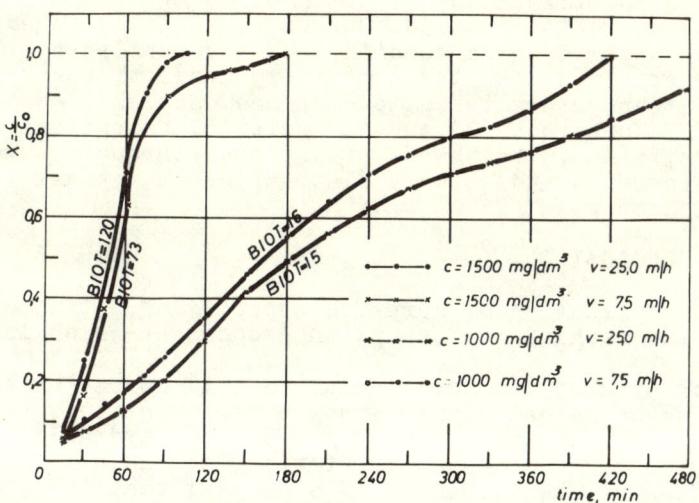

Fig. 5. Breakthrough curve for laboratory sorptive filter with various ranges of BIOT Number.

Using the laboratory model of a sorptive filter, technological tests were conducted to check the values obtained for the BIOT Number. For prepared aqueous wastes with concentration of 1000 and 1500 mg/dm^3, two exchange isoplanes were plotted at two filtration velocities of 7,5 and 25,0 m/h. The shape of the breakthrough curves obtained is presented in Fig. 5.

Discussion

Utilization of sorption capacity of activated carbon is directly connected with the sorption mechanism and ir depends on the values of transfer coefficient through an interfacial film and grain diffusion factor. The correlation of these two factors and the radius of sorbent grain is characterized by BIOT Number. When the BIOT Number is smaller than 5 the sorption process is dependent on diffusion through interfacial film and can be controlled with the variation of hydraulic conditions. These possibilities may be utilized in filters with fixed bed. If the BIOT Number is greater than 50, it is the grain diffusion that determines the sorption effects. This makes it possible to utilize the sorptive filters with a stationary as well as fluidized bed. Thus, the BIOT Number is the parameter that influences the choice of the from for sorptive filters.
A correct determination of the BIOT Number under hydraulic conditions that correspond with the work of a sorptive filter is of great importance in designing the filters. It was shown that the possibilities of measuring the BIOT Number under static conditions using a tank with mechanical stirrer were limited. The measuring range of the mixer rotations from 60 to 120 r.p.m. was imposed by the density and mechanical properties of activated carbon N. The values

of BIOT Number /Table 1/ ranged to 14. Measurements taken with the modified method under dynamic conditions and similar concentrations of phenol have shown that the actual values of BIOT Number were much greater and reached 120 /Table 1/. At the concentration of 1000 and 1500 mg/dm^3, and filtration velocity of 25 m/h the BIOT Numbers obtained were greater than 50. Correctness of the measurement of the BIOT Number under dynamic conditions was supported by the breakthrough curves obtained with the laboratory model of a sorptive filter /Fig. 5/. Steep, characteristic exchange isoplanes were received for the BIOT Numbers of 73 and 120 however. Using wastewater at the same concentration, the shapes of the breakthrough curves with the BIOT Numbers of 15 and 16 were flat, which is typical for this range of BIOT Number.

CONCLUSIONS

1. Optimal utilization of sorption capacity of activated carbon requires the mechanism of sorption process be taken into consideration.
2. The mechanism of the sorption process is characterized by a BIOT Number which is dependent, among others, on the kind of sorbent, the type and concentration of sorbate and hydraulic conditions.
3. The BIOT Number estimated in static tests is not reliable for evaluation of the mechanism of a process that is proceeding on sorptive filters.
4. Determination of the BIOT Number with dynamic tests, besed on the actual hydraulic conditions, makes it possible to obtain the final quantities for selection and dimensioning the sorptive filters.

NOTATION

a_s - characteristic surface, m^{-1}
β_F - treansport coefficient during diffusion trough interfacial film, m/s
c_0 - initial concentration of sorbate in liquid phase, mole/m^3
$c^0/t/$ - concentration of sorbate in liquid phase, mole/m^3
$c^*/t/$ - cencentration of sorbate at interface, mole/m^3
D - grain diffusion factor, m^2/s
L - volume of liquid phase, m^3
$\dot{N}$ - flux of sorbate, mole/s
$\dot{n}$ - stream of sorbate absorbed by a spherical grain of the sorbing agent, mole/s
r - radius of sorbent grain, m
r_0 - radius of inner sphere, the surface of which was reached by sorption - front, m,
S - mass of sorbent, kg
t - working time of sorption, s
$q^*/t/$ - load equivalent to concentration of $c^*/t/$, mole/kg
ρ - density of dry sorbent, kg/m3
$\frac{dq}{dt}$ - change of sorbate load in solid phase, mol.kg^{-1}.s^{-1}
$\frac{dc}{dt}$ - change of sorbate concentration in liquid phase, mol.m^{-3}.s^{-1}

REFERENCES

Brauch, V., and E. U. Schluender /1975/. Chem. Eng. Sci. 30. 539-548.

Grossman, A. /1975/ Zeszyty Nauk. Politechn. w Krakowie 23, 4, 35-46.
Johansson, R., and J. Neretnieks /1974/. Vatten 1, 54-69.
Neretnieks, J. /1976/. Chem. Eng. Sci. 31, 465-471
Hoell, W., and others /1975/. Verfahrenstechnische Grundlagen von Adsorption und Jonenaustausch, Engler Bunte Institut 8, Karlsruhe.
Swietoslawska, J. /1962/. Absorption spectrophotometry, PWN, Varsaw
Westmark, M. /1975/. JWPCF 47, 704-719.

THE CHEMICO-PHYSICAL METHOD OF WASTEWATER TREATMENT FROM GLULAM WOOD STRUCTURES

Z. Niewęgłowska and B. Bartkiewicz

*The Institute of Water Supply and Water Engineering,
Warsaw Technical University, Poland*

Abstract - The wastewaters contain phenol-formal
dehyde resins used for glueing the particular
wood elements into large dimensional structures,
were considered to be purified with chemical-phy
sical methods. A two-stage wastewaters purtifica
tion processes was suggested. A coagulation of
pollutants was obtained in the first stage by in
troducing 0,5 - 2,5 cm^3 concentrated sulphuric
acid per dm^3 of wastewater. The suspended matter
which appeared is separated from the wastewaters
by the vacuum filtration using the filtration
materials PT-15 or BT-16. Effluent is purified
in the second stage by means of activated carbon
Carbopol Z-4 in dosage of 20 g/dm^3 of wastewater.
This method quarantees the removal of COD in 97%
and BOD in 97,5%.

INTRODUCTION

In 1975, within the Wood Product Industry, a modern factory was put into operation for the first time in Poland, which, on a West Germany licence, started the production of glulam wood structures. The Factory of Large Dimension Wood Structures at Cierpice near Toruń processes about 25 thousand m^3 of raw wood yearly, producing about 15 thousand m^3 glulam elements in layers and about 125 thousand current meters of lattice elements /1/.
During the production, adequately prepared wood elements are glued into large dimension structures of high strength. They are mainly used for the construction of sport and entertainment halls.
The combining means are the glues of phenol-formaldehyde rasins, imported so far from West Germany.
A further development is planned for other factories in the coming years, which have similar, or the same specialization. It will be possible to apply home made glues according to the research institutes where the investigations in this matter have been completed. The new developments bring in a new research problem connected with

wastewater management in the above mentioned factories.

A new kind of toxic wastewaters produced by the aggregates for glueing appeared, which are difficult to treat.

The wastewaters occur while washing the aggregate and production halls, i.e. after each production shift or before the exchange of the aggregate contents.

According to a rough estimation, about 6-10 m^3 of post-glulam wastewaters per day originate, which gives about 4% of the total waste waters of the factory.

The post-production wastewaters contain the glue "Penacolite G 4400 A" in amount approximately corresponding to 1% of that of glue used daily for the production, i.e. about 1 g of glue for one dm^3 of wastewaters.

A troublesome kind of post-glulam wastewaters analyzed both in the aspect of a harmful effect over the sewerage as their high toxicity made to treat the post-glulam wastewaters separately. The permissible concentration of the glue in a stream is 2 mg/dm^3.

REVIEW OF METHODS FOR WASTEWATERS TREATMENT CONTAINING PHENOL AND FORMALDEHYDE

As the concrete information dealing directly with the post-glulam wastewater treatment is scant, and having assumed that the main pollution is caused by phenol, the paper includes first of all printed information concerning the methods for dephenolyzing the wastewaters.

The conclusions included in literature concerning the usefulness of chemical methods, i.e. oxygenation by means of chlorium, potasium permanganate, ozone and oxygenated water are not univocal. It can be generally assumed that chemical oxygenation leads only to a partial treatment, and costs included are higher than with biological methods /3, 4/.

Out of the physico-chemical methods of phenol removal from wastewaters the most frequently used is extraction and adsorption on the activated carbon /2,4,5,6,7/. By means of extraction, 90-95% of phenol can be removed using the solutants like: anthracene oil, fluorene oil, disopropyl ether or mixtures of butyl acetate and isopropyl alcohol /4/.

The treatment cost of the wastewaters depends on the production conditions.

The adsorption method on activated carbon has some good points. It is not dependent upon toxicity or phenol compounds. There is even a tendency to increase the removal of phenol with the phenol loading of the carbon beds. Even as high as 2000 to 5000 mg/dm^3 phenol concentration will reduce to about 1 mg/dm^3 /4/.

In the method of carbon beds, the problem of wastes does not always occur. Recurrently, carbon regeneration leads at the same time to phenol recuperation, which lowers the treatment costs.

Another pollution of post-glulam wastewaters is formaldehyde.

As the chemical methods used so far for its removal from industrial wastes have proved hardly economic, a new approach is necessary.

The essential thing in technology of post-glulam wastewaters treatment is neutralization of sludges resulting from this treatment.

The basic purpose is volume reduction and obtaining such a stage of neutralization which would allow to store the product or its utilization in a safe way without a harmful effect on the environment.

The decision concerning the choice of the method for dehydration and neutralization of the sludges should be taken after the primary laboratory research works as well as technical and economic analysis. It should be noticed that in some cases the dewatering process of sludges in the devices on a technical scale has smaller efficiency than in the devices on pilot-plant scale. As an instance the comparison research works on dewatering of sludges by means of laboratory and technical vacuum filter with allowed for determination of the following transfer coefficients: /8/
- for typically non-organic sludges   1,0
- for organic sludge treated by means of ferrous
  sulphate and calcium hydroxide   0,7 - 1,0

Sludge dewatering, however, remarkably diminishing the demand for storage land does not fully solve the problem of neutralization. The final stage is incineration of sludge.

TECHNOLOGICAL METHODS

The primary treatment of the post-glulam wastewaters by means of precipitation and separation of the glue was made in half-liter samples of wastewaters. Concentrated sulphuric acid in doses from 0,5 to 5,0 ml/dm^3 was added to wastewaters. The contents was intensely stirred for about 1 minute, until the wastewaters changed the consistency and homogeneous emulsion was formed. The latter further underwent dewatering process in the laboratory centrifuge and in pressure and vacuum filter.

The centrifugation was made at /2000 rotations per minute, for 5 minutes/.

Pressure filtration was carried in the range from 3 to 9 atn using the filtration materials: PT-15, BT-16, ET-95, and combined PT-15 and BT-16. The same materials were used during vacuum filtration carried out at average subpressure vacuum of 0,9 atn and 0,3 atn.

The filtrate obtained was used as the input material for the second stage of treatment, and the separated glulam sludges underwent combustion to define its utilibility as a fuel.

In the second treatment stage the adsorption method was used on activated carbon.

The above method could be successful thanks to the selection of adequate kind of carbon. By means of elimination resulting from technological conditions, the carbons of Carbopol alkali were chosen. The cheapest Carbopol Z-4 granulated and powdered was used for economic reasons.

Filtration over the granulated carbon of the grain d=1-3 mm was done in a model 30 cm high and 2,5 cm in diameter, and filtration velocity from 0,5 to 3,0 m/h.

Powdered carbon was added to half-liter samples of wastewaters in of suspend form in doses of 2,5 to 25 g/dm^3. The contents was stirred for 20 minutes, put off for 2 hours, the effect of treatment was decantated and studied.

EXPERIMENTAL INVESTIGATIONS
CHARACTERISTICS OF THE INVESTIGATED WASTEWATERS

Raw wastewaters were taken from the retention tank in the factory at Cierpice. They were characterized by high turbidity, brown-cherry colour and strong specific odour. They contained a considerable amount of contamination of organic character.

TABLE 1   Indicators of the Quality of Post-glulam Wastewaters

PH		8.3
Permanganate value	$mg/dm^3 O_2$	22.300
BOD_5	$mg/dm^3 O_2$	39.840
COD	$mg/dm^3 O_2$	30.320
Phenols	mg/dm^3	1.116
Dry rest	mg/dm^3	10.230

According to the licence data, the contamination load expressed in BOD_5 falling to 3 kg of wasted glue was 1900 P.E., thus, it was equal to 34,2 kg BOD_5/kg of glue.

In the working conditions, the use of glue was 200 t/year/1, i.e. about 670 kg/d, and average concentration of wastes pollution expressed in BOD_5 was about 30 kg/m^3, hence the contamination load in the wastewater, falling to one kilogram of glue used for production was 0,36 kg BOD_5/kg of glue. It means that about 1% of glue passed to the wastewaters.

Earlier research works over synthetic wastewaters in the pilot plant showed the correlation between the concentration of glue in water solutions and COD values. It has been assumed that 100 mg/dm^3 glue corresponds to 80 mg/dm^3 COD. Similarly, determination of phenols showed that 1000 g of glue contains 16 g of phenols.

PRECIPITATION OF GLULAM SLUDGE

In research works, the results from the primary treatment of post glulam wastewaters by means of coagulation of aluminium sulphate in PH value correction have been utilized.

However, the method guaranteed the separation of about 50% sludge and 50% of water, the quantity of arising post-coagulation sludge was very high. It has been calculated that out of 4 m^3/d of wastewater about 60 t of sludges will be produced monthly.

To decrease this quantity of sludges, attempts to eliminate some reagents have been made. Only after addition of sulphuric acid, it has been noticed the wastewaters change their consistency, out of the liquid ones having brown-cherry colour they change into dense, rust-coloured emulsion, which, within 24 hours has concentrated and separated from the water. Addition of aluminium sulphate has been given up. Using the dosages of concentrated H_2SO_4 from 0,5 to 5,0 cm /dependedly on the wastewaters concentration/ the drop of COD of water 93,6 to 94,6% has been obtained.

TABLE 2 Effect of Post-glulam Wastewater Treatment by Means of Concentrated Sulphuric Acid

H_2SO_4 mg/dm^3	raw wastewaters			treated wastewaters		
	PH	COD mg/dm$^3O_2$	dry rest mg/dm^3	PH	COD mg/dm$^3O_2$	dry rest mg/dm^3
0,5	8,6	28.200	10.820	2,5	4.000	2.500
2,0	8,9	80.000	34.000	1,5	4.320	6.700
2,5	8,9	80.000	34.000	1,2	4.320	5.600
5,0	8,9	80.000	34.000	1,2	4.120	9.300
10	8,9	80.000	34.000	1,0	4.800	16.200
20	8,9	80.000	34.000	1,0	3.360	30.900
25	8,9	80.000	34.000	1,0	3.200	39.600

The increased dosage of acid to 25 ml/dm^3 caused the growth of COD reduction; however the salinity of the wastewater increased considerably. The dry residue reached the value about 40 g/dm^3, while at the dosages considered as optimum ones it ranged 6-9 g/dm^3.
It was also proved by the attempt at the doses 0,5 and 5,0 ml/dm^3 for the wastewaters 3 times less concentrated. The reduction of COD ranging 70% has been obtained, and the absolute values of this indicator were close to those previously obtained.
To accelerate the treatment and sedimentation processes, experiments with flokulants added were carried out. They have not given, however, satisfactory results and stimulated for the application of mechanical dehydration of the emulsion.

DEHYDRATION AND NEUTRALIZATION OF THE GLULAM SLUDGE

Introductory attempts to dehydrate the sludge in the laboratory centrifuge, allowed the sludge volume to be decreased within 56-60% at the change of water contents on average from 98,4 to 93,0%.
The decrease of pollution in the water phase was on average 94,1% reduction of COD and 78,0% reduction of dry rest. In the laboratory pressure filter and laboratory vacuum filter, in the course of the research work, it has been found that the filtration is the most effective through the cotton technical BT-16 and steelon technical PT-15. At the pressure of 3-6 atn and filtration time of 95-110 min the reduction of the sludge volume ranging 54-60% has been obtained as well as the decrease of water content in the sludge on average from 98,8 to 91,9%. The decrease of pollution in the water phase ranged 86,7% reduction of COD and 78% reduction of dry rest.
Within a shorter time, i.e. 70 minutes the sludge was filtrated at the pressure of 6 atn through the combined materials PT-15 and BT-16. The filtration resulted in sludge volume decrease to 84%, and hydration deminished to 85,1%. The same combination of materials was used at low filtration pressures, i.e. 0,9 and 0,3 atn.
The dehydrated sludge and filtrate showed similar pollution indicators to the previous ones; on the other hand the filtration time

was prolonged to 2 and 3 hours. Aiming at shortening the time and improving the effects of dewatering, a series of attempts to filter the sludges mixed up with wooden chips have been made. At the quantity of chips 25 g/dm^3 of sludge and pressure 6 atn, the filtration time was cut short to 35 min., and at the quantity of 50 g/dm^3 to 8 min. The water contents in the sludge was 80,9-83,6%; however, the amount of obtained filtrate decreased, while the sludge volume increased. From 500 ml of wastewaters, 254-154 ml of filtrate could be obtained.

Good results were obtained when Rokrysol WF-2 at a dosage of 2 ml/dm^3 was added to the filtrated sludge. The sludge was filtrated under the pressure of 6 atn through the combined materials PT-15 and BT-16. The filtration lasted 55 min., bringing about reduction of the sludge volume ranging 88%, and the water content in the sludge 83%. A similar effect, but with relatively shorter time, was obtained by heating the sludge undergoing filtration to $60^\circ C$. The raw sludge was filtrated for 12 min., and the one with addition of Rokrysol WF-2, for 2 min.

Both in vacuum and pressure filtrations the regularity proved that the degree of dewatering and quality of filters depend on the applied pressure only to a small extend.

Basically though the diminished pressure caused the prolongation of filtration time. It was particularly evident at the vacuum filtration, where at the pressure 0,9 atn the optimum filtration time was 2 hours, and at the pressure 0,3 atn - 3 hours.

Attempts of combustion were made for dewatered sludges in similar conditions, i.e. at the pressure 6 atn, when the material BT-16 was used. The sludge with addition of chips and without was combusted.

The fuel values of 10000 kcal/kg for sludge with chips and 5000 kcal/kg for sludge without chips allow to conclude that combustion of post-glulam sludge should not be difficult, and the chosen method is the most appropriate in the aspect of environmental protection.

THE TREATMENT OF WASTES FILTRATE

Since the primary treatment of post-glulam wastewaters resulted in obtaining the water phase of a high pollution /COD at the level 3-5 thousand mgO_2/dm^3/, it was justified to continue investigations on methods of further filtrate purification.

One of the methods which seemed to be appropriate as regards the nature of wastewaters was adsorption over activated carbon. The results of the investigations showed that for diminishing wastewaters concentration to COD of level about 100 mgO_2/dm^3 large doses of carbon ranging 20-25 kg/m^3 of wastewaters were necessary.

It has been defined that the sorption capacity of granulated carbon Z-4 is of the same order as powdered carbon /0,1-0,2 g COD/g of carbon/; i.e. the use and costs of treatment would be similar. The use of that method in technical conditions would involve high costs. Besides the cost of regeneration, dumping or combustion of used carbon would have to be added.

TABLE 3 The Effect of Activated Carbon Carbopol Z-4 Dosage Over Changes of COD Filtrate
- COD before adsorption 4480 mgO_2/dm^3

Dosage g/dm^3	COD		Sorption capacity g COD/g carbon
	$mg/dm^3 O_2$	% red.	
2,5	3,660	18,3	0,29
5,0	2.976	33,4	0,28
7,5	2.492	44,2	0,25
10	1.860	58,3	0,25
15	1.320	70,4	0,20
25	912	79,7	1,14

CONCLUSIONS

The research works carried out showed that the post-glulam waste waters have a high concentration of pollution, as well as a conside-rable toxicity for the natural environment, caused mainly by the content of phenol and formaldehyde. The specific nature of wastewa-ters is also seen in their glueing properties, which primarily indicate the treatment method to be chosen.

In conclusion of the analysis of utility of the method for removing phenol and formaldehyde from wastewaters, and on the basis of technological investigations carried out in order to find the optimum variant in post-glulam wastewaters treatment, a two-stage system of the treatment was found to be the appropriate one. The first stage of the treatment includes removal of non-soluable contamination, reduction of colour and diminishing the general concentration. For this purpose coagulation with sulphuric acid is recommended, and separation of the sludge by filtration through the material. The amount of the acid varies from 0,5 to 2,5 ml/dm^3 of wastewaters dependedly on the concentration of raw wastewaters.

Two criteria for defining the acid dosage were proposed: the PH value of wastewaters after acidifying ought to be lower than 2,0, and at the same time the change of wastewaters consistency from the liquid into jelly-like and that of colour from brown-cherry into light brown should take place.

The precipitated sludge after 15 min. reaction is passed for de-watering. For practical purposes /acid resistant installation/ the use of ceramic vacuum filters with vacuum pressure 0,9 atn is recommended. As a result of filtration, the sludge of water content about 90% is obtained and treated wastewaters which have the concen tration 3-5 thousand COD/dm^3 and a very low pH.

The sludge is passed to dewatering, and filtrate to the second stage of treatment.

The attempts of filtrate treatment by the method of adsorption on activated carbon have showed that there is a possibility to dimi-nish the concentration of pollution to the level required at pass-ing the wastewaters to the sewarage system /COD below 1000 mgO_2/dm^3/, however, the doses of activated carbon are high. This does not result only in high costs of treatment but in the appearence of additional solid waste in the form of used activated carbon as well.

In connection with the above, the other variant of further treatment was used, i.e. biological treatment of neutralized filtrate in mixture with the sewage.

BIBLIOGRAPHY

1. Bartkiewicz B., Niewęgłowska Z. "The solution of watersewage management for the factory of big-dimensional wooden structures at Cierpice near Toruń" - not published
2. Meinck F. at al. "Industrial wastewaters" Arkady, Warsaw 1975
3. Joseph P. at al. "Oxidation of phenols by ozone", JWPCF, ol 48, no 1, January 1976
4. Richard E. at al. "Phenols a water pollution control assessment", Water and Sewage Works, March 1976
5. "Colloids on carbon adsorption" Water and Sewage Works, March 1977
6. Giusti D. at al. "Activated carbon adsorption of petrochemicals", JWPCF, ol 46, no 5, May 1974
7. Hamilton C.E. "A new solvent for system improved ekstraction of activated charcoal", Water and Sewage Works, Reference Number 1964
8. Koziorowski B. et al. "Comparative investigations over the dewatering of sludges by means of laboratory and technical vacuum filter for defining the transfer coefficient /in Polish/, Gospodarka Wodno-Ściekowa nr 4, 1970.

METHOD FOR RECOVERY OF WATER AND VANADIUM COMPOUNDS FROM WASTEWATER

I. Zagulski, L. Pawlowski and A. Cichocki

Institute of Chemistry, Maria Curie-Sklodowska University, Pl. M.C. Sklodowskiej 3, 20-031 Lublin, Poland

ABSTRACT

A method for recovery of water and vanadium compounds from wastewater associated with the production of zirconium-vanadium pigment has been presented. This wastewater containing about 1300 mg V/dm^3 is passed through a strongly acid cation exchanger bed where it is decationized. Then it is forced through two columns filled with a weak base anion exchanger bed, where it is deanionized. A mineral acid is used for regeneration of the cation exchanger bed, and the regeneration effluent is discharged into the sewer.
For regeneration of the anion exchanger bed 20% solution of ammonia is used. The regeneration effluent is evaporated to dry and then calcinated at 550°C. During calcination of ammonium metavanadate and ammonium sulphate are decomposed. The resulting product is V_2O_5.

KEYWORDS

Ion exchange, water recovery, vanadium recovery, purification of wastewater, water recycling, chemicals recycling.

INTRODUCTION

Vanadium compounds are strongly toxic, which have a particularly harmful effect on the circulatory system and disturb metabolism; e.g. only a small amount of vanadium pentoxide initially causes an increase of hemoglobin content in blood and then its decomposition. For this reason the admissible content of vanadium compounds in surface water has been limited to 1 mg/dm^3.

The removal of vanadium compounds from wastewaters is complicated by a variety of forms in which vanadium occurs in them. For example in the fifth state of oxidation in relation to pH, vanadium may occur as cation VO_2^+ or anions VO_3^-, VO_4^{-3}, $V_2O_7^{-4}$, $V_3O_9^{-3}$, $V_6O_{17}^{-4}$.

Attempts have been made (Clark 1968) to remove vanadium from wastewaters either by way of precipitation or extraction. However, they have not any practical value.

First attempts to apply ion exchange also encountered great difficulties caused by blocking ion exchange bed with precipitates. Recent studies (Patent USA no 2,937,034) showed that precipitate is not formed when ion exchange is conducted on a fluidized bed at $1.3 < pH < 2$. The latter, however, arouses doubts in regard to the interval of optimal pH, as it is known in vanadium chemistry that at $pH < 2$ in aqueous solutions it occurs mainly in the form of VO_2^+, i.e. cations which cannot be adsorbed in the ion exchange bed. This supposition has been confirmed by the authors of the patent (Patent USA no 3,186,952) who assumed that at $pH < 1.8$ practically the whole amount of pentavalent vanadium occurs in the form of VO_2^+. Their patented method concerns purification of wastewater formed in production of catalysts, which, besides pentavalent vanadium compounds, contain also Fe^{+3}, Cu^{+2} ions. To recover vanadium compounds, they proposed a system of series connection of two columns packed with a strongly acid ion exchanger. In the initial phase of exhaustion all ions are adsorbed in the first column, and after its exhaustion Fe^{+3} ions entering with the solution begin to displace VO_2^+ and Cu^{+2} ions from the bed, which are adsorbed on the second column. In a moment it comes to a state that in the first column almost only Fe^{+3} ions are present, whereas in the second one VO_2^+ and Cu^{+2} ions. Separate regeneration of the columns makes it possible to obtain a concentrate of vanadium and copper compounds in a form enabling their return to production. The studies presented in this paper concern the production of zirconium-vanadium pigment, where wastewater of the following composition is formed:

pH = 7.3

alkalinity = 20 mval/dm^3

V = 1278 mg/dm^3

Cd, Zn, Ni - 0.0

SO_4 - 1106 mg/dm^3

SiO_4 - 30 mg/dm^3

suspended matter - 43 mg/dm^3

total solid - 3244 mg/dm^3

total solid after calcination - 3064 mg/dm^3

Whenever it is possible, there should always be the tendency to recycling both the water and chemicals recovered from wastewater to production. Approaching the problem from this point of view, vanadium compounds should be separated from wastewater in the form ammonium metavanadate (NH_4VO_3), as this compound can be directly recycled into production. In the case ion exchange is applied to purify this kind of wastewater, ammonium metavanadate solution can be obtained by regenerating the anion exchanger by means of solution of ammonia. For this reason the studies have been limited to weakly base anion exchangers, which can be regenerated by means of ammonia solution. Weakly base anion exchanger Amberlite IRA-94 S was chosen for the studies.

EXPERIMENTAL

The studies were carried out in glass columns 0,02 m in diameter, the depth of the bed - 1.0 m and its volume - 0.3 dm^3. Wastewater was taken directly from the industrial installation. After filtrating it on a gravel-sand filter, it was pumped by means of a peristaltic pump through an appropriate system of columns. During exhaustion fractions were collected in which pH and the content of vanadium and sulphates was determined. For regeneration of the cation exchanger 5% sulphuric acid in an amount of 2 bed volumes was used,

Solution of ammonia was used for regeneration of the anion exchanger. The studies were carried out for the column systems:
WBA; CC + WBA;
SAC + WBA; SAC + WBA + WBA
where: WBA - denotes a column packed with weakly base anion exchanger,
CC - carboxylic cation exchanger (weakly acid),
SAC - sulphonic cation exchanger (strongly acid).

RESULTS AND DISCUSSION

System WBA

Attempts to purify wastewater only on one column filled with weakly base anion exchanger Amberlite IRA-94S did not give positive results, because breakthrough occurred as early as 6 bed volumes of wastewater had passed through the column. To improve sorption, wastewater pH had to be decreased, which can be obtained by passing wastewater through the cation exchanger bed in hydrogen form.

System CC + WBA

Weakly acid cation exchangers can be easily regenerated by means of acid solutions, therefore, their use is advantageous, where regeneration is carried by means of an acid solution. Moreover a decationized solution is acidified moderately, which is important in the case of vanadium, because reduced pH below 2 causes formation of VO_2^+ ions, which in turn are adsorbed in a cation exchanger bed. This induced us to examine the possibility to use weakly acid cation exchanger Wofatit Ca 20 for decationization of wastewater. It was found that pH of partially decationized wastewater ranges from 4 to 4.5. Wastewater decationized in this way sent directly onto the bed of a weakly base anion exchanger did not give satisfactory results because of the occurrence of a high leakage of anion compounds of vanadium; only decreased pH to 2.2 - 2.5 by acidification of partially decationized wastewater by means of a mineral acid resulted in decreased leakage to the level of 30 mg/dm^3. A reduction of pH is likely to cause a transition of polyvanadates to metavanadate, the anion of which is smaller and thus it permeates more readily anion exchanger grains. However, additional introduction of an acid increases the content of anions in wastewater, which are adsorbed together with vanadium in the anion exchanger bed, occupying a part of the exchange capacity of the anion exchanger. Moreover they contaminate the regeneration effluent.

The above disadvantages made us also study decationization on a strongly acid cation exchanger.

System SAC + WBA

Decationization of wastewater on the bed of a strongly acid cation exchanger lead to a high decrease of pH (below 2). In this pH range an unfavourable phenomenon of partial sorption of vanadium in the cation exchanger bed occurred (in the form of VO_2^+ ions). In the initial exhaustion phase the concentration of vanadium in the effluent from the cation exchanger column averaged 50 - 60% of the initial value. It means that 40-50% of vanadium was adsorbed in the cation exchanger bed. After some time the concentration of vanadium in the effluent started to increase until it rached the value of 160% of the initial concentration. It can be concluded that after exhaustion of the cation exchanger bed (vanadium concentration in the effluent equalled that in the initial solution) displacement of VO_2^+ cations by other cations contained in wastewater started. However, a certain amount of vanadium remained in the cation exchanger bed

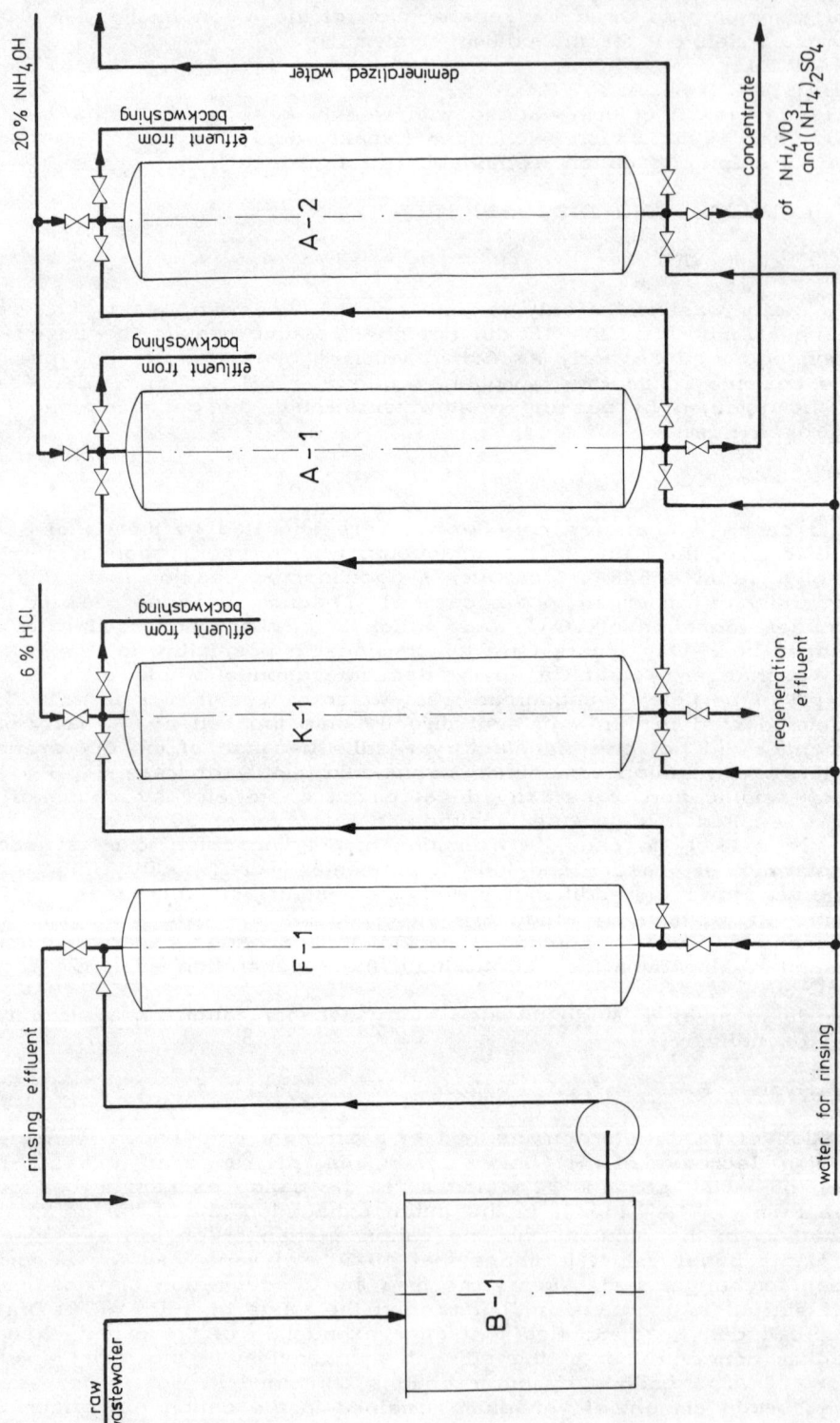

Fig. 1. Schematic presentation of the purification process of wastewater containing vanadium. B-1 — container of raw wastewater; F-1 — gravel-sand filter; K-1 — column packed with a strongly acid cation exchanger; A-1, A-2 — columns packed with a weakly base anion exchanger.

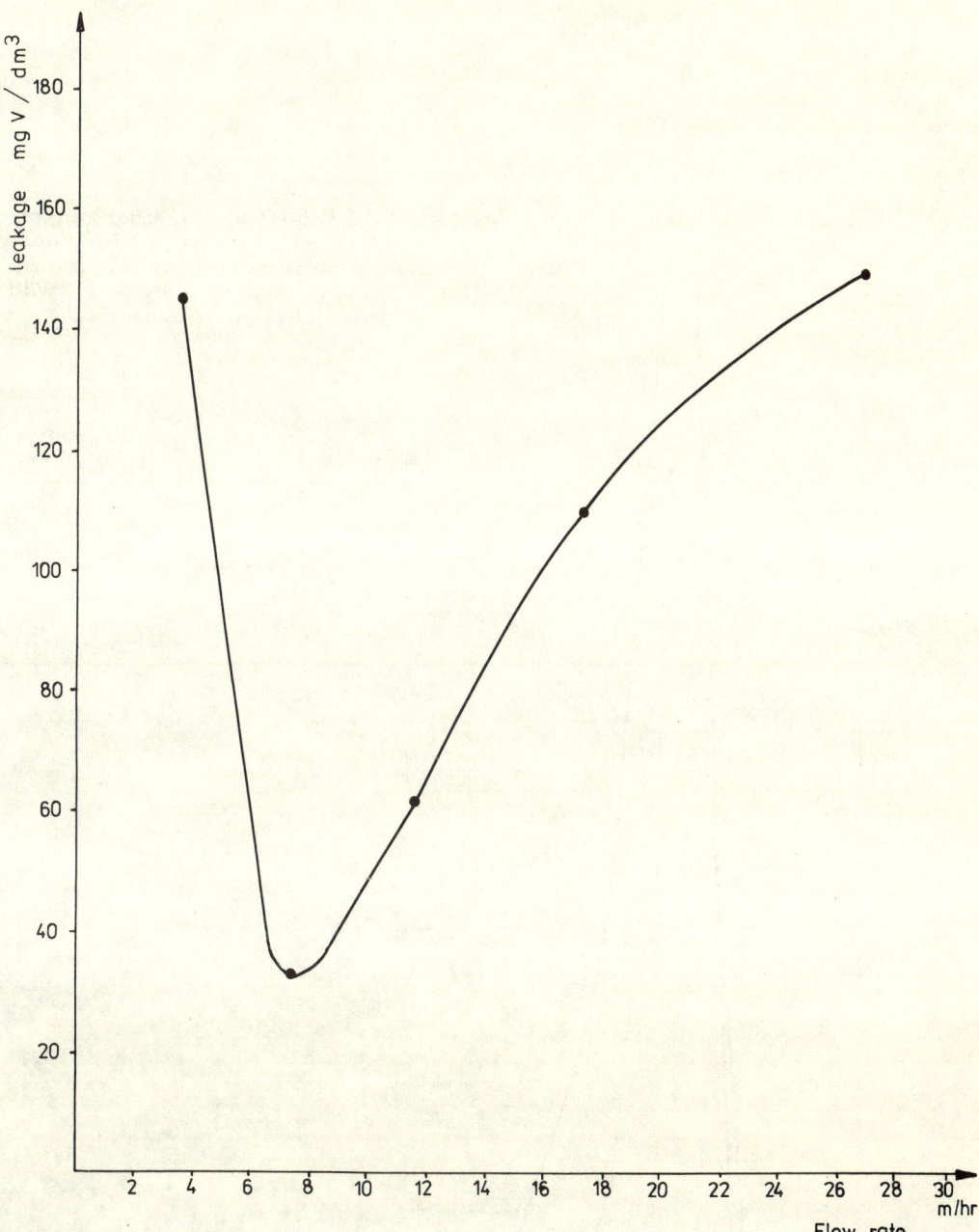

Fig. 2. Characteristics of the effect of lineary velocity of wastewater flow through the bed of anion exchanger (Amberlite IRA-94S) on the leakage of vanadium ions.

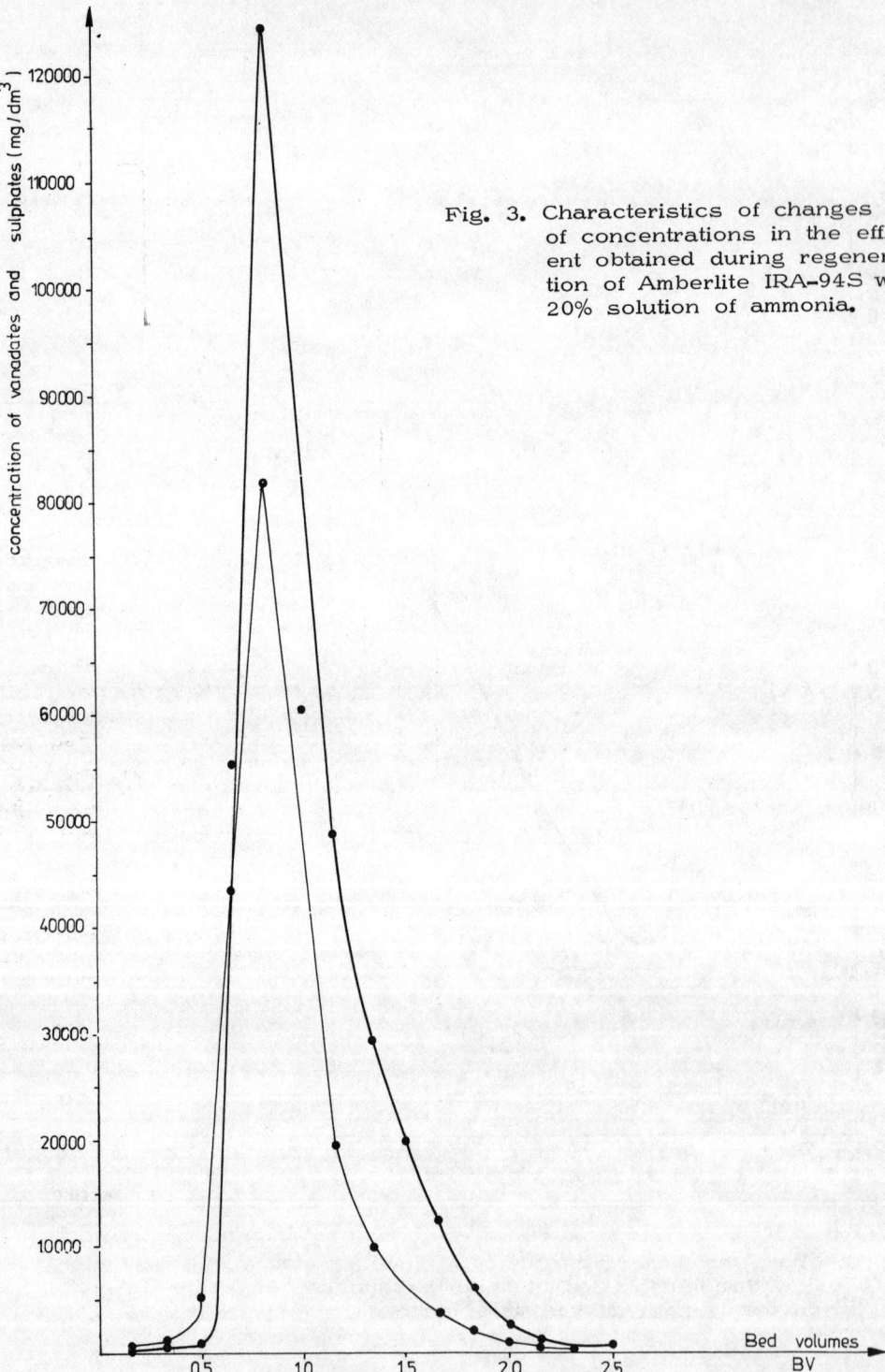

Fig. 3. Characteristics of changes of of concentrations in the effluent obtained during regeneration of Amberlite IRA-94S with 20% solution of ammonia.

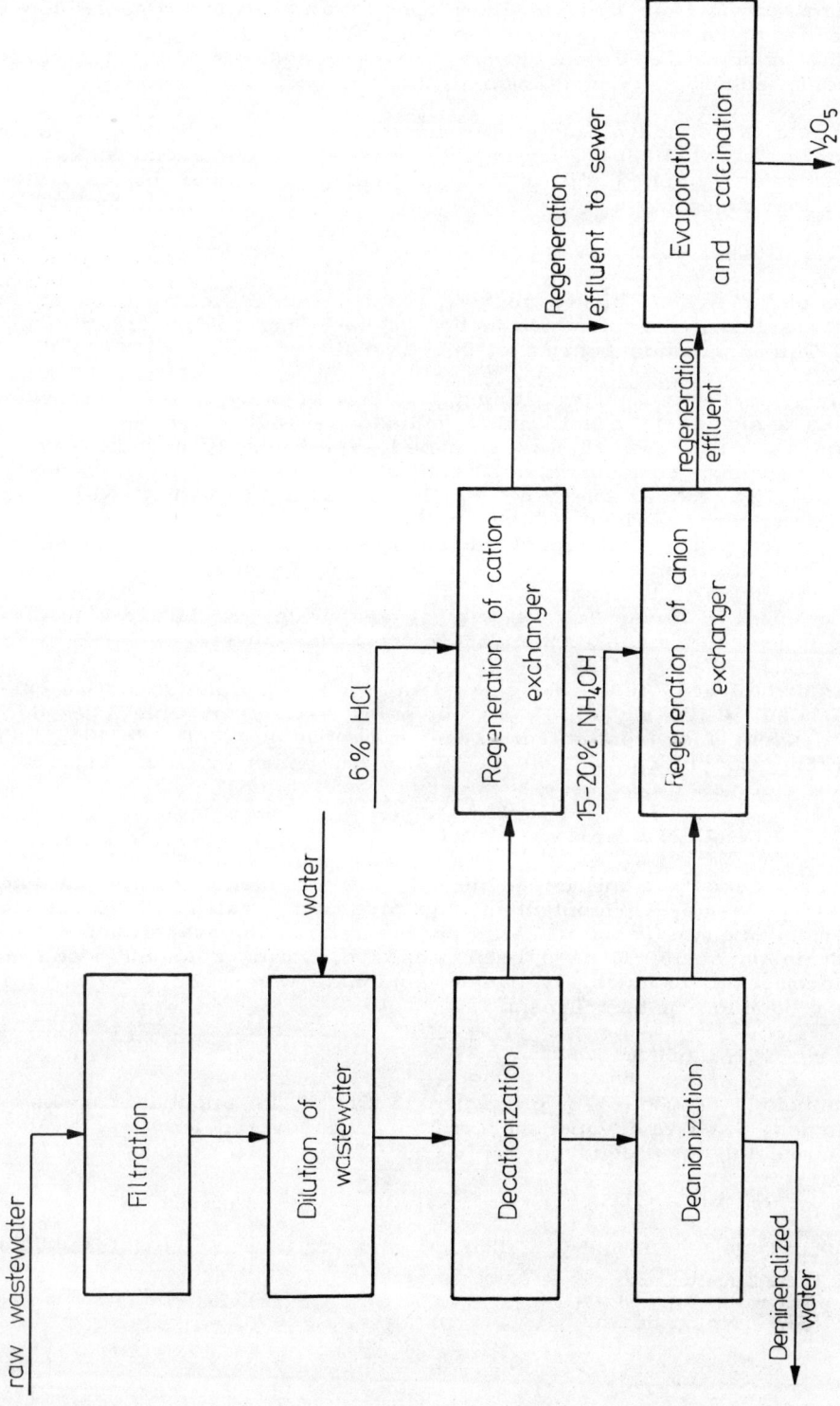

Fig. 4. Chemical flow diagram of a purification plant of vanadium containing wastewater.

in the form of VO_2^+ so that vanadium concentration in the regeneration effluent from the cation exchanger column was 1000 to 1200 mg/dm^3. (The amount of the effluent was 2 bed volumes). This constitutes about 4% in relation to vanadium adsorbed on the anion exchanger bed.

The leakage of vanadium anions through the anion exchanger bed was maintained on the level of 20-50 mg/dm^3. In order to eliminate this leakage an additional column packed with a weakly base anion exchanger, as buffering bed was introduced.

System SAC + WBA + WBA

The use of the system of two columns packed with a weakly base anion exchanger resulted in complete elimination of vanadium compounds from the effluent. The installation scheme of this type is presented in Fig. 1.

On studying this system it was found that flow velocity of decationized wastewater has a significant effect on the value of leakage of vanadium ions trough the anion exchanger bed. It was estimated experimentally that the smallest leakage of vanadium compounds occurs at a lineary flow velocity amounting 7 m/h (see Fig. 2.). At this flow velocity, on average 50 to 60 bed volumes of wastewater were purified, which contained from 800 to 900 mg of vanadium compounds $/dm^3$. The leakage of vanadium compounds on the first anion exchanger bed was maintained on the level of 30 mg/dm^3.

A similar effect of linear flow velocity of wastewater on the work of the buffering bed (second anion exchanger column) was observed.

Regeneration of both beds was carried out with 20% solution of ammonia using a dose of 100 mg of NH_3/dm^3 of anion exchanger. Characteristics of the distribution of concentrations in the effluent is presented in Fig. 3. The regeneration effluent contains about 30% of ammonium sulphate and 70% of ammonium metavanadate.

EFFLUENT MANAGEMENT

The effluent cannot be introduced directly into production of the pigment because of a considerable content of sulphates and a relatively low concentration of metavanadate (about 10%). For this reason its evaporation to dry and calcination at 550°C have been proposed. During calcination salt mixture is decomposed producing vanadium pentoxide which can be easily recycled to production of the pigment.

CONCLUSION

The technology developed is presented in Fig. 4. Its application makes it possible that water and vanadium compounds recovered from wastewater return completely to production.

REFERENCES

Clark, R.J.H., The chemistry of titanium and vanadium, Elsevier Publishing Company, N.Y., (1968).
Patent USA - Nr 2,937,034.
Patent USA - Nr 3,186,952.

ION EXCHANGE METHOD FOR WATER, AMMONIA AND NITRATES RECOVERY FROM NITROGEN INDUSTRY WASTEWATER

J. Barcicki, L. Pawlowski, A. Cichocki and L. Zagulski

*Institute of Chemistry, Maria Curie-Sklodowska University,
Pl. M.C. Sklodowskiej 3, 20-031 Lublin, Poland*

ABSTRACT

An ion exchange method for purification of wastewater formed in the processes of ammonia synthesis is proposed. The process consist of sorption of ammonia ions in the bed of a strong acid cation exchanger and regeneration with concentrated sulphuric acid. It is shown that the ammonia sulphate obtained during regeneration can be used in the production of saltpetre. Addition of regeneration effluent to saltpetre increases the durability of saltpetre granules. It appears this method has considerable economical advantages.

KEYWORDS

Ion exchange; wastewater purification; ammonia recovery, water recovery.

INTRODUCTION

In nitrogen plants considerable amounts of wastewaters polluted mainly with N_{NH_3}, N_{NH_4} and N_{NO_3} can be distinguished. Their discharge is limited by law because of the fact that their excessive content in surface waters causes serious disturbances of hydrobiological processes, which in consequence lead to degeneration of biological life. Moreover, excessive concentration of nitrogen markedly decreases utility qualities of water. For this reason studies on limiting the discharge of wastewaters containing nitrogen compounds have been carried out. One of the most promising methods is ion exchange, which makes it possible to recovery water (Arion 1973; Bingham 1970, 1971, 1972a, 1972b; Pawłowski 1976; Popovici 1974), ammonia and nitrates from wastewaters. The research works carried out in the Institute of Chemistry, Maria Curie-Sklodowska University in Lublin, resulted in the development of a technology permitting complete recovery of water, ammonia and nitrates from wastewaters (Pawlowski 1976). The technical installation has been designed by Organization for Water and Wastewater Treatment in the Polish Chemical Industry „Biprowod". The first factory applying this technology is the Nitrogen Plant "Puławy". The purpose of this paper is to present the experiences gained from operating this technological installation.

CHARACTERISTICS OF WASTEWATERS DIRECTED TO TECHNICAL INSTALLATION

For purification a group of condensates formed in the process of ammonia synthesis was selected. Part of wastewaters comes from expansion installation, adsorption and distillation of ammonia, where it is constituted by the effluent from the distillation column. They contain largely N_{NH_4} in amount from 100 to 3000 mg/dm^3. Other impurities occur in trace amounts and come mainly from corrosion of the apparatuses and piping. Maximum amount of wastewaters of this kind from EAD installation is about 5 m^3/h. Another most important source of wastewaters are installations of copper washing in the particular process stages, which serve for purification of synthesis gas from residual carbon monoxide. These wastewaters contain ammonia in the form of bicarbonate or ammonium carbonate, occasionally with ammonia excess. The concentration of N_{NH_4} in these wastewaters ranges from 300 to 300 mg/dm^3. Moreover they contain metal ions: copper averaging from 0 to 5,0 mg/dm^3, iron averaging to 2 mg/dm^3. Anionic impurities are sulphates, chlorides and silica at the level appropriate for softened water. The total amount of wastewaters from copper washings is about 35 m^3/h. Another wastewaters group is constituted by condensates from potassium washing of the second ammonia plant (Ammonia Plant II), which contain on average 400 mg/dm^3 of N_{NH_4}. This condensate contains methanol in an amount of about 300 mg/dm^3. The average approximate composition of wastewaters directed to the purification installation is (before their dilution) as follows:

pH - 8 ÷ 11
NH_4^+ content - 2000 mg/dm^3
NO_3^- content - 1,0 mg/dm^3
Cl^- content - about 30 mg/dm^3
$SO_4^=$ content - about 50 mg/dm^3
Cu^{+2} content and
    heavy metals - about 5 mg/dm^3
Na^+ content - 1 mg/dm^3
$Ca^{++} + Mg^{++}$ content - 1 mg/dm^3
Carbonates content - about 40 mval/dm^3
Total Fe content - 3 ÷ 4 mg/dm^3
SiO_2 content - about 1,5 mg/dm^3

Other impurities occur in very small amounts and are practically insignificant for the work of the installation.

CHARACTERISTICS OF TECHNOLOGY AND TECHNICAL INSTALLATION FOR PURIFICATION OF WASTEWATERS FROM AMMONIA PRODUCTION

In the process the following stages can be distinguished:
- <u>filtration on gravel - sand bed</u>, the purpose of which is to remove the suspension coming mainly from corrosion of apparatuses;
- <u>decationization on the bed of a strongly acid cation exchanger</u>,
During this process all cations constituted mainly by NH_4 ions are removed from wastewaters. They are replaced by hydrogen ions coming from the ion exchanger;
- <u>desorption of CO_2</u>, the purpose of which is to remove CO_2 formed in the

process of decationization, when carbonates contained in wastewaters undergo decomposition. Desorption is carried out in typical desorbents filled, e.g., with Raschig rings,
- <u>deanionization on the bed of a weakly basic anion exchanger</u>, is carried out in order to remove anions of strong acids;
- <u>further purification on the system of two columns, the first filled with a strongly basic anion exchanger and the second filled with a strongly acid cation exchanger</u>, this operation is performed as polishing unit to obtain demineralized water of high purity.

These are the main elements of the ion exchange method for recovery of water and ammonia from wastewaters coming from ammonia production. In dependence of the kind of the acid used for regeneration (Pawlowski 1976) this process can be realized in two variants. In Fig. 1 the method is presented in which sulphuric acid has been used for regeneration of the cation exchanger (nitric acid can also be used).

Raw wastewaters in amount of about 35 m^3/h flow into an equalizing tank, whence they are pumped onto gravel filter 2. On the effluent pipe from the filter a mixer is placed in which wastewater is diluted with softened or demineralized water fed from the end of the installation. Diluted wastewater flows in amount of 60 m^3/h to tank 3, and then it is pumped through cation exchanger 4. In the installation two identical cation exchangers working alternately have been provided, i.e., while in one of them exhaustion process takes place, in the other the bed is simultaneously backwashed, regenerated and rinsed. An appropriate pipe system is provided for this purpose within the installation of cation exchangers. In order to guarantee a regular flow of the acid during regeneration, special acid distributors of the acid used for regeneration have been designed (bed volume - 9 m^3, depth of the bed in the column - 2 m, cation exchanger - Amberlite 200, linear flow velocity through the bed - about 14 m/h). The end of cation exchanger exhaustion is signalized by a pH-metr probe placed on the pipe carrying decationized wastewater to CO_2 desorber. Backwashing is done by means of raw wastewaters which are returned to tank 1. After backwashing the bed is subjected to regeneration by 40% sulphuric acid (variant I). This solution is prepared from 96% H_2SO_4. The regeneration effluent is divided into two fractions. One fraction is constituted by raw wastewaters filling intergranular spaces of the ion exchanger. This fraction is returned to raw wastewaters, i.e. to tank 1. The other fraction is constituted by about 17% $(NH_4)_2SO_4$ solution in excess of nonreacted sulphuric acid. This fraction is directed to tank 8, where nonreacted sulphuric acid is neutralized with gaseous ammonia the result of which is 23% solution of ammonium sulphate. This solution is pumped to the ammonium sulphate section and submitted to further processing. A small content of Cu^{++} ions in the final product increases the value of ammonium sulphate as fertilizer because it is one of the essential microelements. The flow velocity of 40% sulphuric acid during regeneration is about 4,0 m/h. For single regeneration 1,9 m^3 of 40% H_2SO_4 is used. Directly after regeneration decationized wastewaters are introduced to rinse the bed. The flow velocity is the same as that of acid, i.e. 4 m/h. As mentioned previously, fractionation of the effluent follows: the first fraction 4,5 m^3 in volume is returned to raw wastewater, the second 5,5 m^3 in volume is ammonium sulphate concentrate, the third one 2 m^3 in volume is diluted sulphuric acid which is used for preparing 40% sulphuric acid (dilution of 96% H_2SO_4). After regenerating and rinsing, the bed is suitable for another exhaustion. At a medium content of ammonia in raw wastewater, amounting to about 1000 mg of N_{NH_4}/dm^3 about 170 m^3 of wastewater can be purified during a single exhaustion. After decationization the wastewater is sent to CO_2 desorber and then onto the anion exchange bed, where deanionization takes place. Next it is sent to the polishing system. This part of installation is similar to the classical

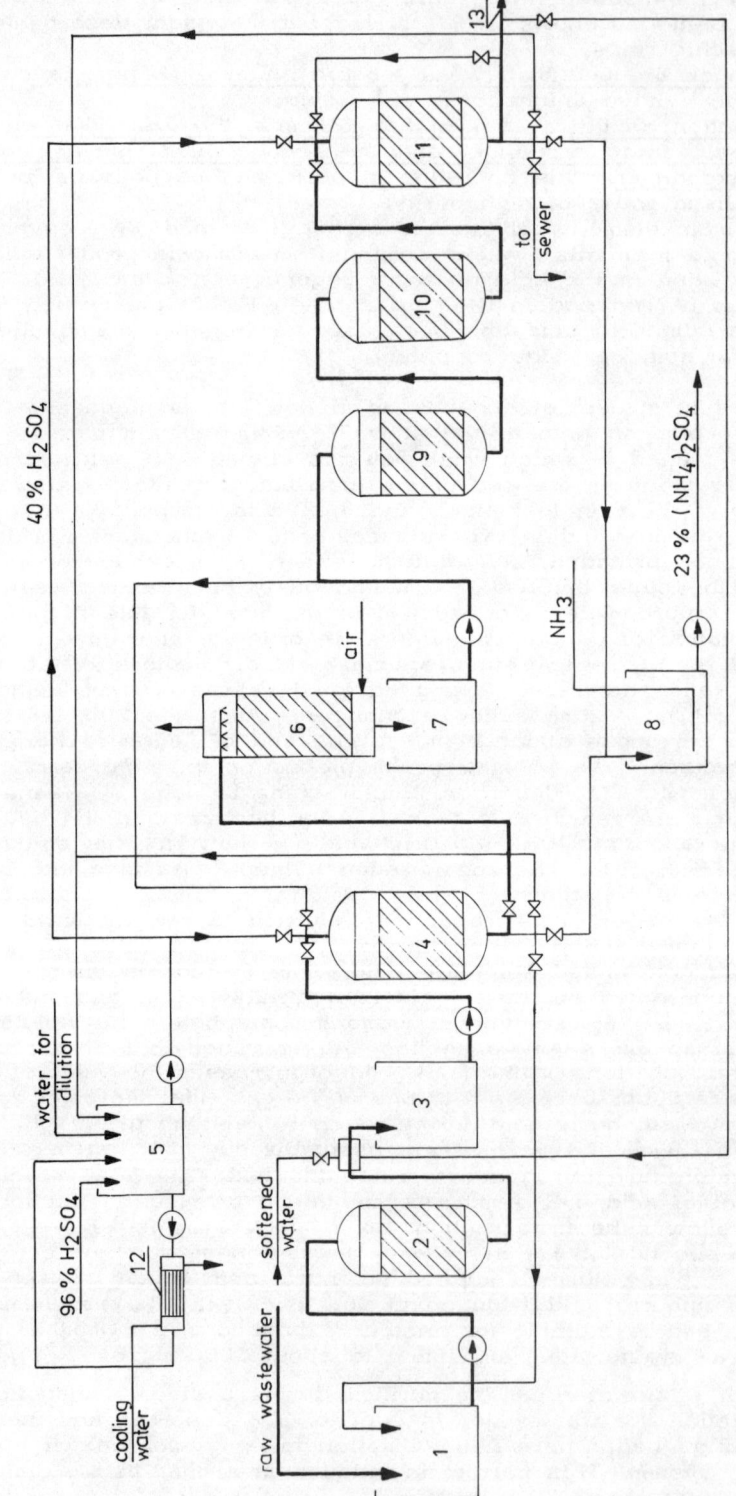

Fig. 1. Flow-sheet of the nitrogen industry effluents purification plant.
1. Raw wastewater tank; 2. Gravel-sand filter; 3. Filtrated effluents tank; 4. Strongly acid cation exchanger bed; 5. 40% H_2SO_4 solution tank; 6. CO_2 desorber; 7. Decationized wastewater tank; 8. $(NH_4)_2SO_4$ solution tank (regeneration effluent); 9. Weakly basic anion exchanger bed; 10. Strongly basic anion exchange bed; 11. Buffer ion exchanger bed (cation exchanger); 12. Cooling of H_2SO_4 solution; 13. Demineralized water.

one for demineralization of water, using well or river water. In this part the following ion exchangers are used: weakly basic anionite Wofatit AD-41, Wofatit SBW - strongly basic anionite and strongly acid buffering ion exchanger with Amberlite 200. Regeneration of both the types of anion exchangers is carried out by means of 4% caustic soda. Regeneration effluent is discharged into the sewer. Weakly basic anion exchanger beds are backwashed with decationized water and rinsed with demineralized water. Strongly basic anion exchanger beds, however, are backwashed with effluents from weakly basic anion exchanger beds and also rinsed with demineralized water. Buffering ion exchanger closing the process of water demineralization is filled with the same cation exchanger which is used in the cationite part of the installation, i.e., Amberlite 200 - a strongly acid cationite. The role of this ion exchanger consists in additional purification of roughly demineralized water. The parameters of the individual operations for the buffering cationite are similar to those for main cation exchanger beds. The process conducted in this way results in detaining demineralized water with conductivity 1 μS.

CHARACTERISTICS OF TECHNOLOGY AND INSTALLATION FOR PURIFICATION OF WASTEWATER FROM PROCESSING OF AMMONIUM NITRATE

In the process of ammonium nitrate production wastewater is formed which is contaminated only with ammonium and nitrate ions. The method of purifying it is approximate to that described previously, except for the anionite part, where an additional weakly basic anion exchanger has been placed, which serves for recovery of nitrates. Organization for Water and Wastewater Treatment in the Polish Chemical Industry "Biprowod", disposes of a technical project of an installation for recovery of water, ammonia and nitrates from wastewaters of ammonium nitrate production. Such an installation will be built in not very distant future.

TECHNICAL-ECONOMICAL EVALUATION OF THE METHOD

The basic, technical advantage of the installation is its simplicity and easiness in operation. Big industrial installations according to the technology described can be automatized to such an extent that the personnel during one shift may be practically limited to two persons. The installation does not practically produce any wastewaters, but a small amount is formed only in the course of classical regeneration of weakly and strongly basic anion exchanger beds by means of 4% NaOH solution. Another advantage of this technology is the fact that concentrated impurities (NH_4^+ and NO_3^-) are returned to production. Moreover, demineralized water of high quality is obtained in amount equivalent to that of purified wastewater, what is significant in the light of water deficit suffered by industry. For big nitrogen plants the amount of such water may reach several dozen cubic meters per hour. The advantage of the technology in economical sense is the fact that its application brings definite and calculable profits with simultaneous environment protection - in this case of surface waters. The balance of profits and costs of one-hour work of the installation presented, in terms of actual prices of raw materials and products, is as follows:
Profits:

At a concentration of N_{NH_4} ions in wastewaters amounting to 1 kg/m^3 and volume flow of wastewaters of 60 m^3/h through the installation about 220 kg/h of $(NH_4)_2SO_4$ is obtained.
- $(NH_4)_2SO_4$ recovery, 220 kg of $(NH_4)_2SO_4 \cdot 0.92$ zł/kg = 202.40 zł,

- recovery of demineralized water,
  Production cost of 1 m^3 of demineralized water is 8.45 zł, which gives:
  $$60 \ m^3 \cdot 8.45 \ zł/m^3 = 507.00 \ zł.$$

- decreased rate-payment for wastewater discharge in terms of COD (owing to purification of the discussed amount of wastewater). Unitary payment of 8.45 zł/kg of COD in 1 m^3 of discharged wastewaters. The amount of COD of discharged wastewaters ranges from 0.15 to 0.40 kg/m^3 which corresponds to the rate-payment from 1.27 zł to 3.38 zł, on average 2.32 zł/m^3, which gives:
  $$60 \ m^3 \cdot 2.32 \ zł/m^3 = 139.20 \ zł.$$

- decreased use of well water for production of demineralized water and therewith decreased payment
  $$60 \ m^3 \cdot 0.45 \ zł/m^3 = 27.00 \ zł$$

  Total profit for one-hour work of the installation:
  $$202.40 \ zł. + 507.0 \ zł. + 139.20 \ zł. + 27.00 \ zł. = 875.60 \ zł.$$

Costs:

- consumption of concentrated sulphuric acid: 155.4 kg/h
  $$155.4 \ kg \cdot 0.85 \ zł/kg = 132.10 \ zł.$$

- consumption of soda lye its average consumption is 21.3 kg/h
  $$21.3 \ kg \cdot 2.60 \ zł/kg = 55.36 \ zł.$$

- consumption of electric energy
  its maximum consumption by electric facilities is about 140 kW
  $$140 \ kW \cdot 1h \cdot 0.47 \ zł/kWh = 65.80 \ zł.$$

  Total cost for one-hour work of the installation:
  $$132.10 \ zł. + 55.38 \ zł. + 65.80 \ zł. = 253.30 \ zł.$$

It appears that annual profit from the installation work, according to the above calculations, is (for 333 days of work in the year):

$$(875.60 \ zł. - 253.30 \ zł.) \cdot 24h \cdot 333 \ days = 4,973,421.60 \ zł.$$

In the calculations such elements as amortization of the facilities, employment and the like have not been taken into account. It can be assumed that the costs connected with employment of the personnel will be relatively low, because in one shift 1-2 persons are planned to be employed. The cost of the installation is about 20 million zł. on the basis of the technical design.

The design of the technical installation and completion of the facilities are carried out by Organization for Water and Wastewater Treatment in the Polish Chemical Industry "Biprowod".

REFERENCES

Arion, N., (1973). Ion Exchanger Treatment of Wastewater from the Fertilizer Industry, <u>UNESCO Seminar on the Chemical Industry and the Environment, Warszawa, Poland.</u>

Bingham, E.C., (1970). Investigations into the Reduction of High Nitrogen Concentrations, <u>5th International Water Pollution Research Conference, San Francisco, Cal., USA.</u>

Bingham, E.C., Chopra R.C. (1971). A Closed Cycle Water System for Ammonium Nitrate Producers, <u>International Water Conference, The Engine-</u>

ers Society of Western Pensylvania 32-nd Annual Meeting, Pittsburgh, Pen., USA.
Bingham, E.C., (1972). Env. Sci. and Techn. 6, no 8, p. 692.
Popovici, N., (1974). Technical Solutions and Technological Advances Made in Romania to Control Environmental Pollution Effects Avising from Fertilizer Plants. ID/WG. 175/18, UNiDO Expert Group Meeting, Helsinki.
Bingham, E.C. (1972). Water Eng., p. F-4.
Pawłowski, L. (1976). Odzyskiwanie związków azotowych ze ścieków na jonitach, (Recovery of Nitrogen Compounds from Wastewater using Ion Exchangers), Ph.D These, Instytut Inżynierii Ochrony Środowiska Politechniki Wrocławskiej, Wrocław, Poland.
Pawłowski, L., and J. Barcicki (1976). Jonitowa metoda odzyskiwania związków azotowych i wody ze ścieków przemysłu azotowego, (Ion Exchange Methods for Recovery of Water and Nitrogen Compounds from Wastewater of Nitrogen Industry). Proceedings of Conference Physicochemical Methods for Water and Wastewater Treatment, 6-7 May, Lublin, Poland.

MEMBRANE SEPARATION OF THE PRODUCT IN THE PROCESS OF BIOLOGICAL TRANSFORMATION

A. Poranek, T. Winnicki and J. Wiśniewski

Institute of Environment Protection Engineering, Technical University of Wroclaw, Wybrzeze Wyspiańskiego 27, 50-370 Wroclaw, Poland

ABSTRACT

A new method of product separation from reaction environment in transformation processes of organic compounds has been described. Fungi Rhodotorula mucilaginosa have been immobilized in the ultrafiltration cell with polysulphone membrane formed on a porous support made of sintered polyvinyl chloride. The membranes have been cast from 15% polysulphone solution in dimethyl-formamide, using different temperatures of casting solutions: 293 K, 313 K and 333 K. The membranes were 20-60 μm thick. It has been found that transport properties of membranes are better in the case of membranes with lower thickness and cast from solution of higher temperature. The obtained rejection coefficient of microorganism cells was 97-99.9%. Decrease of hydraulic permeability of membranes during ultrafiltration of microorganisms connected with microorganism adsorption on membrane surface is inconsiderable and is less than 10%. Adsorption of microorganism does not affect permeability of molecules obtained from microbiological transformation process.

KEYWORDS

Biological transformation ; immobilization ; ultrafiltration cell ; porous material ; polysulphone membrane.

INTRODUCTION

Microbiological transformation consists of a gradual conversion of organic compounds. The interruption of the process at a given moment yields determinable transformation products. If the process involves microorganisms with no ability to assimilate the substrate, only some fragments of the particles are responsible for the detoxication habitat. Owing to their high selectivity and high stereospecificity, these reactions are widely used in pharmacy, medicine, food industry and chemistry to obtain organic compounds whose synthesis raises difficulties and is expensive. The attempts at employing transformation processes in waste water treatment have so far failed to be succesful (Slote, 1970). Nevertheless, there is a possibility of applying some of them to the detoxication of waste waters by transformation of toxic compounds into less toxic forms. This can be illustrated by the example of benzaldehyde to benzyl alcohol transformation carried out with Rhodotorula mucilaginosa (Peczyńska, 1974). Literature reports (Meinck, Stooff and Kohlschütter, 1975) emphasize that benzaldehyde exerts a harmful effect on the organisms at a concentration of 0.40 kg/m^3. Benzyl alcohol inhibits the biological functions of algae at a concentration of 0.64 kg/m^3. Bacteria E. coli are resistant to a concentration of 1 kg/m^3.

Methods of microorganisms immobilization

The transformation process should be carried out with immobilized microbial cells so as to avoid difficulties in the separation of the transformation products from the reaction mixture. The interest in various methods of immobilizing microbiological cells continues to increase. The immobilization of microorganisms is found to be markedly advantageous than that of enzymes, which is both difficult and expensive. On the other hand, immobilization of microorganisms may be divided into four classes :
1. Entrapment in the polymer. This method involves immobilizazion in polyacrylamide gel (Martin and Perlman, 1976) and in a membrane (Saini and Vieth, 1975), as well as the entrapping of the microorganisms into the pores of porous polymer (Mika-Gibała, Siewiński and Winnicki, 1978).
2. Adsorption on the surface of water insoluble carriers as, for example, ion exchangers or cellulose (Kan and Shuler, 1978).
3. Formation of large aggregates of cells by the use of linear polyelectrolytes (Lee and Long, 1976).
4. Chemical (covalent) fixation of cells with the corresponding carrier (Jack and Zajic, 1977).

The methods presented have the disadvantage of being responsible for a marked decrease in the enzyme activities of the immobilized microorganisms as compared to those of the native organisms. It is reported that the enzyme activities may be reduced to 50-70% (Martin and Perlman, 1976) and in some cases even to 15-20% (Saini and Vieth, 1975) of the initial activity. This is the result of diffusion resistance, a factor limiting the inflow of the substrate to the cell which has been included in the polymer carrier. Another disadvantage of the methods in question is the difficult separation of deactivated cellular matter from the carrier. Sometimes, in methods 1 and 4, this separation becomes unfeasible, which necessitates employing new bacterial matter for immobilization. Having this in mind, attempts are made to develop new methods of immobilization. One of them involves membranes which are placed in cells employed in dialysis and ultrafiltration.

At present, dialysis and ultrafiltration (UF) cells are widely used to immobilize enzymes (Zaborsky, 1974). However, to immobilize microorganisms and to separate the product from the reaction environment dialysis cells alone have been used so far (Kan and Shuler, 1978). No experiments are reported on the possibility of employing membrane separation of microorganisms in the UF cells.

This paper is aimed at presenting a novel method which involves polysulphone membrane supported on a sintered polymer.

Immobilization of microorganisms in the UF cells

The application of UF cells have been known for about ten years (Zaborsky, 1974). They were first used for the treatment and concentration of enzyme solutions, but soon appeared to be useful in the immobilization of microorganisms. It is to be noted that immobilization in an UF cell proceeds in a manner different from that employed in other methods. The organisms are placed in a space closed by semipermeable membrane, and there is no attachment to the carrier. The substrate is supplied at a given rate, and the product of microbiological transformation is removed at the same rate. If these requirements are met, the UF cell can be treated as a biological reactor of continuous flow. The membrane should be carefully selected so as to retain high molecular weight emzymes or microorganisms and provide a free passage for the small product particles. If the substrate is a compound with particle size similar to that of the product, the control of the residence time of the substrate in the UF cell wil become indispensible in order to achieve the maximum degree of conversion. These problems do not appear, if we use a high molecular weight substrate e.g. starch or proteins , because the substrate particles will not be able to pass through membrane until after breaking up.

Among the various advantages of immobilization in UF cells, one deserves special attention, namely the direct contact of the microorganisms enzymes with the substrate supplied to the cell. It follows that their enzyme activity is as high as that of the native organisms. There is no toxic effect either, which usually appears when the microorganisms are immobilized in polyacrylamide gel. The removal of deactivated organisms and the supply of active material do not raise any difficulties. The only disadvantage of immobilizing microorganisms via this route is concentration polarization; microorganisms or enzymes aggregates are formed on the surface of the membrane (Hancher and Ryon, 1973). Concentration polarization deteriorates the transport properties of membranes. This deterioration, however, can be partially eliminated by employing rapid mixing or by increasing the flow rate.

Polysulphone membranes

Polysulphone membranes have been used since 1970´s (Majewska, 1979). They are characterized by very good chemical and thermal properties. The chief advantage of using polysulphone membranes is that they may work in a wide range of pH 0.5-13.5 and at a temperature as high as 363 K. The synthesis and properties of the membranes in question are reported in Refs. (Fremount, 1972; Koenst and Mitchell, 1978).

The polysulphone membranes used so far, were synthesized on a non-porous support (Majewska, 1979). A novel technique described by Koenst and Mitchell (1978) involves direct casting inside a porous tubular modulus and yields good results. The tests performed with those membranes have shown that the volume flow might be influenced by the support used. Having this in mind, the authors of the present paper moulded the membranes directly on the surface of a sintered polymer. The theory and technological parameters for sintering pulverized polymers are presented in Refs. (Malczewski, Mika and Steller, 1975).

EXPERIMENTAL

Sintering process

Pulverized polyvinyl chloride was used to prepare a porous support, as this polymer yields a very small grain size (7-10 μm). The sintering process was carried out in steel forms at 443 K for 3,000 s. The apparent density required was achieved by adding a certain quantity of the polymer. The sintered polymer had a density of 0.8 kg/m^3 and 0.9 kg/m^3.

Preparation of polysulphone solution

The polysulphone solution was prepared according to Koenst and Mitchell (1978). Aromatic polysulphone P3500, Union Carbide, was used for membrane synthesis, whereas dimethyl formamide (DMF) was employed as solvent. The quantity of polysulphone required for obtaining a 15% solution was gradually added to a measured quantity of DMF with intensive mixing. Prior to addition the polysulphone was dried at 423 K for 24 h to remove adsorbed water. Mixing was carried out at 288 K for 96 h.

Membrane preparation by direct casting on porous support

The membranes were made from solutions of various initial temperatures 293 K, 313 K and 333 K. Polysulphone was transported to a broad vessel, whereas care was taken that the polysulphone layer thickness did not exceed the thickness of the sinter, i.e. 5 mm. One of the support surfaces was immersed in the solution for several seconds. Then excess solution was removed using the device for casting (Majewska, 1979). Polysulphone together with the sinter were immersed in distilled water for 1800 s to gel the polymer and remove excess solvent. After gelling the membranes cast onto a porous support were placed in a vessel with pure distilled water. The thickness of membranes obtained varied from 20 to 60 μm. Thickness was measured with a micrometer screw, starting from the surface of the sinter.

Transport properties of polysulphone membranes

The experiments were carried out in a autoclave made of organic glass, with a magnetic stirrer (Fig. 1). This apparatus permits a pressure of up to 0.2 MPa. The membrane to be tested was conditioned prior to the experiment run. The conditioning process was accomplished at a pressure of 0.15 MPa until the rate of volume flow through the membrane became constant, which took place after approx. 20 h. In this period of time the volume flow decreased by some 99% as compared to the initial value. The studies of transport properties were carried out at the following pressures: 0.05 MPa, 0.10 MPa and 0.15 MPa. The pressure was applied in two manners - from the lowest to the highest values and vice versa. The rate of the volume flow was measured after a constant value had been achieved.

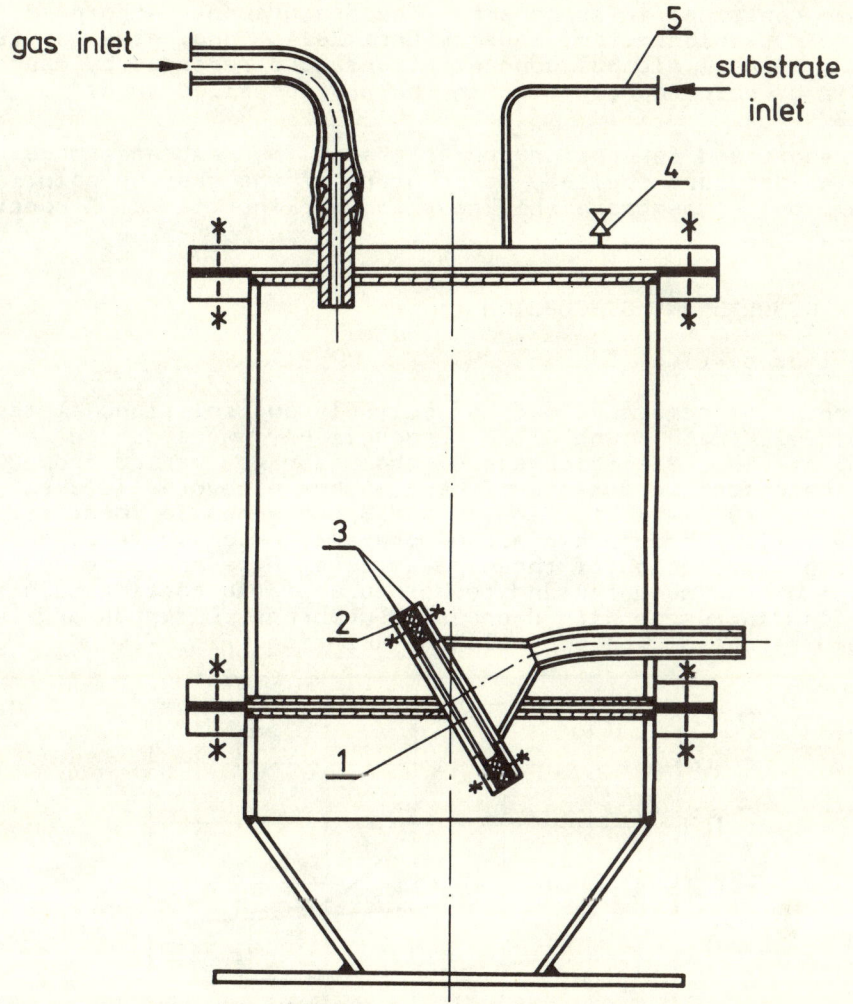

Fig. 1. Testing apparatus: 1 - sinter, 2 - O-Ring, 3 - rubber seals, 4 - valve, 5 - inlet for distilled water.

Separation properties of polysulphone membranes

The studies were conducted in the same autoclave. After having completed the experiments for transport properties, a suspension of <u>Rhodotorula mucilaginosa</u> at a concentration of 0.30 kg/m^3 was placed in the UF cell. The reactor was continuously fed with benzyl alcohol at a concentration of 0.05 kg/m^3 (with methanol admixture). The inflow rate was as high as the rate of the volume flow through the membrane. Microorganisms concentration in the permeate and the product content were controlled during the whole experiment run. A standard curve relating extinction to the dry weight of the microorganisms was plotted with the aim to determine the number of microorganisms contained in suspension. The measurements were performed with a SPEKOL colorimeter, made by Carl Zeiss Jena, at a wavelength of 540 nm. Benzyl alcohol concentrations were recorded by means of a VSU - 2 UV Spectrophotometer in the way described in Refs. (Peczyńska, 1974).

The transport and separation properties of the membranes were related to the pressure applied, temperature of the casting solution (polysulphone), membrane thickness and working time at a constant pressure.

RESULTS AND DISCUSSION

Transport properties

Three groups of membranes made of polysulphone solutions at temperatures of 293 K, 313 K, and 333 K, respectively, were subject of experimental studies. The thickness of the membranes varied from 20 to 60 μm. The effective surface of each membrane covered 10.2 cm^2. The results are listed in Figs. 2 and 3. As shown in these figures, the volume flow through a membrane prepared on a porous support is found to be a function of three parameters - the pressure applied, thickness of the membrane and temperature of the casting solution. Volume flow increases with decreasing membrane thickness and increasing temperature of the polysulphone solution (Fig. 2).

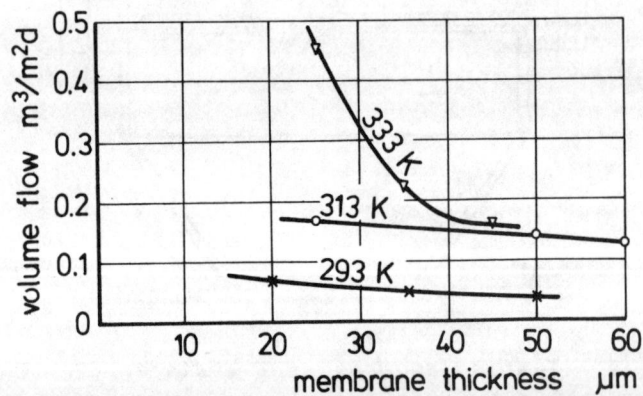

Fig. 2. Volume flow of solution versus membrane thickness at a pressure of 0.15 MPa. Temperatures of casting solutions were: 1 - 333 K, 2 - 313 K, 3 - 293 K.

The rate of volume flow through the membrane, which was determined at pressures varying from lower to higher values, is greater than the one which was determined at pressures varying from higher to lower values (Fig. 3). This is an indication that the membrane undergoes compaction and that the well known phenomenon of pressure hysteresis loop occurs.

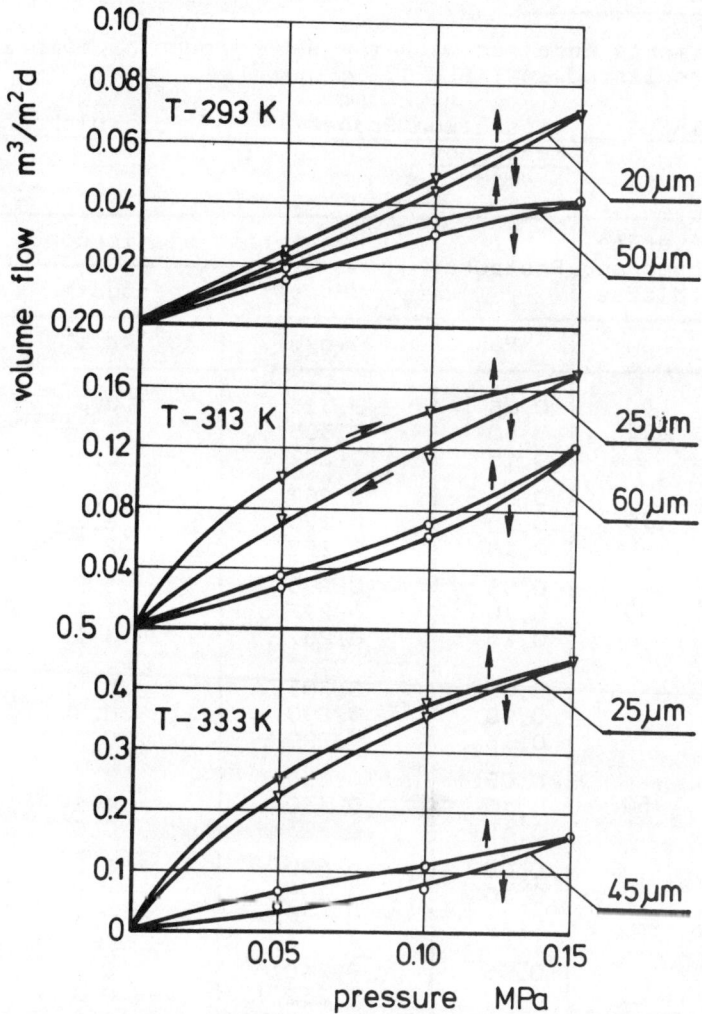

Fig. 3. Volume flow of water versus pressure for membranes of various thicknesses. Temperatures of casting solutions were:
A - 293 K, B - 313 K, C - 333 K,
↑ - increasing, ↓ - decreasing pressures.

A 25 μm thick membrane synthesized from a polymer solution at a temperature of 333 K was characterized by the highest permeability 0.45 $m^3/m^2 d$. This permeability value is ten times as high as that of a 50 μm thick membrane prepared from polysulphone at a temperature of 293 K. The results obtained show that the porosity of membranes increases with increasing temperature of the casting solution.
The high permeability of thin membranes may also be due to the increase in porosity especially in relation to membranes prepared from a casting solution of a temperature of 333 K, however, the flow resistance cannot be neglected, either.

Separation properties

The experiments were run with the same groups of membranes, and the results are listed in TABLE 1.

TABLE 1   Separation Properties of Polysulphone Membranes

Temp. of polysul. solution	Average membrane thickss	Pressure	Concentration of microorg.		Eliminat. coeffic.
			UF cell	permeate	
K	um	MPa	kg/m^3	10^{-3} kg/m^3	%
293	20	0.05	0.315	0.8	99.8
		0.10	0.305	1.0	98.7
		0.15	0.292	0.8	99.7
	35	0.05	0.287	2.4	99.2
		0.10	0.270	2.2	99.2
		0.15	0.267	2.6	99.0
	50	0.05	0.290	2.9	99.0
		0.10	0.277	3.1	98.9
		0.15	0.267	3.1	98.8
313	25	0.05	0.305	0.5	99.8
		0.10	0.290	0.8	99.7
		0.15	0.280	0.6	99.8
	50	0.05	0.353	4.5	98.7
		0.10	0.330	4.1	98.8
		0.15	0.323	5.6	98.3
	60	0.05	0.270	6.0	97.8
		0.10	0.265	7.5	97.2
		0.15	0.258	7.4	97.1
333	25	0.05	0.245	0.2	99.9
		0.10	0.232	0.3	99.9
		0.15	0.220	0.2	99.9
	35	0.05	0.244	3.9	98.4
		0.10	0.230	3.8	98.3
		0.15	0.224	4.5	98.0
	45	0.05	0.253	5.8	97.7
		0.10	0.238	6.2	97.4
		0.15	0.230	6.0	97.5

It was found that the temperature of the polysulphone solution, membrane thickness and the pressure applied did not affect the selectivity of membranes. At the same time, experiments were carried out on the influence of the working time on both transport and separation properties. The pressure applied was 0.15 MPa. Some of the results obtained are represented in Fig. 4. The investigations show that during ultrafiltration of the microorganisms, the volume flow of the solution through the membranes conditioned slightly decreases, which is due to the adsorption of the microbial cells at the membrane surface, despite of the intensive mixing. It was observed that deterioration of the transport properties took place during the first hour of the ultrafiltration process. The decrease of permeability was not high; did not exceed 10%, and did not vary with time elapsed.

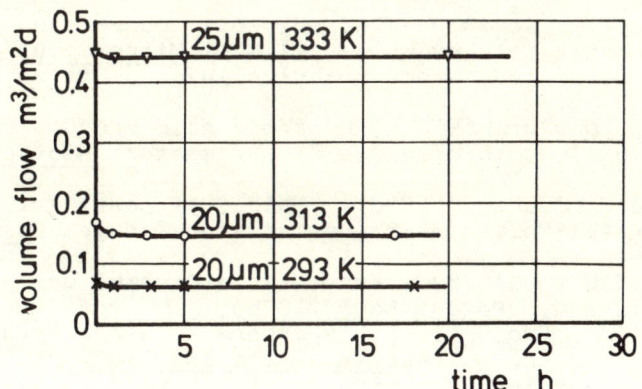

Fig. 4. Volume flow of solution versus working time at a pressure of 0.15 MPa.

The adsorption of microorganisms did not effect the selectivity of membranes throughout the whole working time. It did not limit either the passage of small particles through the membranes. The concentration of benzyl alcohol, which is a product of the microbiological transformation of benzaldehyde, in the reactor, and the concentration of this compound in the permeate, were equal.

SUMMARIZING REMARKS

1. The casting of polysulphone solution on the surface of sintered PVC yields membranes of high hydraulic permeability which increases with increasing temperature of the casting solution and decreasing membrane thickness. Membranes of 25 μm thickness, prepared from polysulphone solution at a temperature of 333 K, are found to have the best transport and separation properties (0.45 m^3/m^2d and 99.9%, respectively).

2. The membranes obtained via the above method may be employed in the UF cell for the immobilization of microorganisms, owing to their high selectivity in relation to microbiological cells. The coefficient of elimination does not depend on the temperature of the casting solution, thickness of the membrane or the pressure applied.

3. During ultrafiltration, a slight decrease of the transport properties of the membranes is observed, which is attributable to adsorption of microorganisms at the membrane surface. The decrease of permeability does not exceed 10%.

4. The adsorption process does not limit the passage of the particles of the transformation product.

REFERENCES

Fremount, H.A. (1972). Verfahren zur Entfernung von Farbkörpern aus Abwassern von der Papier und Zellstoffherstellung. Ger.Offen.Pat. No.2711072.

Hancher, C.W. and A.D.Ryon (1973). Evaluation of ultrafiltration membranes with biological micromolecules. Biotechn. Bioeng., 4, 677-691.

Jack, T.R. and J.E.Zajic (1977). The enzymatic conversion of L - histydyne to urocanic acid by whole cells of Micrococcus luteus on carbodiimide activated carboxymethylcellulose. Biotechn. Bioeng., 5, 631-648.

Kan, J.K. and M.L.Shuler (1978). Urocanic acid production using whole cells immobilized in a hollow fiber reactor. Biotechn. Bioeng., 2, 217-230.

Koenst, J.W. and E.Mitchell (1978). Method of casting tubular polysulphone ultrafiltration membranes in sand modules. U.S.Patent, No.4038351.

Lee, C.K. and M.E.Long (1976). Enzymatic processes using immobilized microbial cells. U.S.Patent, No.3821086.

Majewska K. (1979). Evaluation of membrane transport properties for ultrafiltration process. Reports of the Institute of Environment Protection Engineering, Technical University of Wrocław, No.D-131 (in Polish).

Malczewski, J., A.Mika and R.Steller (1975). Plastics with open pores and high apparent density. Pr.Nauk.Inst.Technol.Org.PWr, No.19, Studia i Materiały, No.15 (in Polish).

Mika-Gibała, A., A.Sławiński and T.Winnicki (1978). Method of microorganism immobilization for the microbiological transformation of chemical compounds. Patent, Poland. Application No.194148.

Martin, C.K. and D.Perlman (1976). Conversion of L - sorbose to L - sorbosone by immobilized cells of Gluconobacter melanogenus IFO - 3293. Biotechn. Bioeng. 2, 217-237.

Meinck, F., M.Stooff and H.Kohlschütter (1975). Industrial wastes. Arkady, Warszawa. Chap.16, pp.835-915 (Polish Translation).

Peczyńska W. (1974). Investigation of the conditions of microbiological reduction of chosen ketones with Rhodotorula fungi. Ph.D.Thesis, Agricultural University of Wrocław (in Polish).

Saini, R. and W.R.Vieth (1975). Reaction kinetics and mass transfer in glucose isomerization with collagen-immobilized whole microbial cells. J. Appl. Chem. Biotechnol., 2, 115-141.

Slote, L. (1970). Development of immobilized enzyme systems for enhancement of biological waste water treatment. U.S.Nat.Techn.Inform. Service, No.203598.

Zaborsky, O. (1974). Immobilized enzymes. CRC Press, Cleveland.

DIALYTIC PROCESSES IN WATER AND WASTE WATER TREATMENT

T. Winnicki, A. Mika-Gibala and G. Blazejewska

Institute of Environmental Protection Engineering, Technical University of Wroclaw, Poland

ABSTRACT

General principles of dialytic separation techniques such as conventional dialysis, dialysis across ion-exchange membranes, electrodialysis and piezodialysis have been described. Development trends of the techniques and examples of their application in water and waste water treatment have been presented.

KEYWORDS

Dialysis; electrodialysis; piezodialysis; charge-mosaic membrane; water renovation.

INTRODUCTION

The notion of dialysis was introduced in 1854 by T. Graham (Graham, 1854) to define selective diffusion through a semipermeable diaphragm, which means "permeable" to dissolved low-molecular-weight substances and "impermeable" to colloids or macromolecules. Today, this notion is used to describe the transport of solute through semipermeable and ion-exchange membranes, the driving force for this process being the chemical or electrochemical potential difference in the solutions on both sides of the membrane. "Dialysis" is also a stem of other terms defining the process of solute transport through the membrane such as electrodialysis and piezodialysis initiated respectively by an electric field or pressure difference. The purpose of this paper is to compare and characterize the individual dialytic techniques. The problem inherent with conventional dialysis and electrodialysis will only be summarized while nonconventional dialysis using ion-exchange membranes will be discussed in detail as well as the concepts and present state-of-the-art of piezodialysis.

CONVENTIONAL DIALYSIS

In this case, conventional dialysis is the process as described by Graham (Fig. 1).

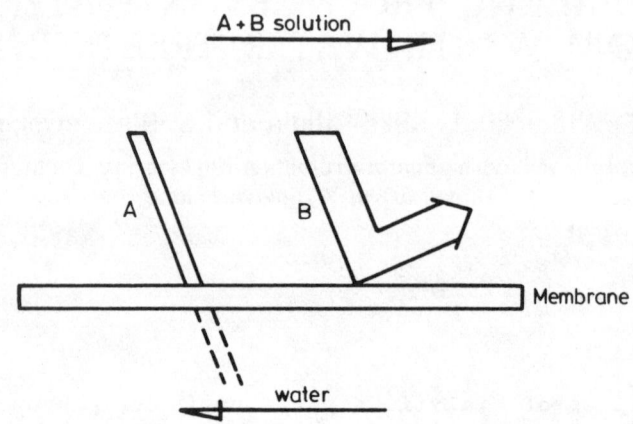

Fig. 1. Dialysis.

The method was first employed in the mid 19th century. However, a rapid development of this technique took place in the 1920ies due to the expansion of the rayon industry using large quantities of sodium hydroxide. In the rayon industry considerable amounts of waste sodium hydroxide polluted with hemicellulose were generated, which raised serious technical and economical problems. The application of dialysis permitted the recovery and reuse of this substance. In the dialysis of sodium hydroxide, porous porcelain, asbestos sheets, parchment paper and cellulose derivatives followed cotton cloth impregnated with magnesium chloride as dialytic diaphragms.
The application range of conventional dialysis is limited. Despite the recovery of sodium hydroxide, it is usually employed in the fields of medicine artificial kidney , pharmaceutics and food industry.

DIALYSIS OF ELECTROLYTE SOLUTIONS THROUGH ION-EXCHANGE MEMBRANES

One of the most difficult to treat waste waters are those carrying mineral acids and heavy metal salts, and concentrated waste acids polluted with organic substances. On the other hand, these waste waters contain valuable components raw materials for, by-products and products of the industrial processes in question , the recovery of which is of great importance to the industry.
Dialytic treatment of acid waste water was first employed to remove nickel sulphate from copper-refining solutions (Tuwiner, 1962).
The dialysis of acid waste waters polluted with metal ions was carried out with commercial synthetic Nalfilm and Hi-Sep semipermeable membranes (Chamberlin and Vromen, 1959 and Vromen, 1962) by taking advantage of the considerable difference between the concen-

trations of sulphuric acid and nickel sulphate, and the significant in their diffusivities. The factor of sulphuric acid from nickel sulphate separation achieved in this process was markedly low and was approximately 15. The separation factor is defined as a ratio of the membrane diffusion coefficients of separated substances. Donnan equilibrium and the resulting co-ion exclusion effect permitted the application of anion-exchange membranes as dialytic diaphragms in the separation of acids and salts. The dialysis of electrolyte solutions using strong-acid and strong-base ion-exchange membranes was studied extensively by Manecke and Heller (Manecke and Heller, 1956) and by Wolf and coworkers (Wolf et al., 1963). They proved that the diffusion coefficients for acids transported through strong-base anion-exchange membranes were ten to one hundred times those of the corresponding metal salts. Bearing these in mind, consideration was also focused on anion-exchange membranes with weak-base ion-exchange groups. The degree of dissociation of these groups depends markedly on the pH of the external solution, which is in equilibrium with the membrane, and increases with decreasing pH. The ion-exchange groups fix the hydrogen ion by means of a free electron pair at the nitrogen atom, thus achieving a positive charge and electrostatically fixing the acid anion. The hydrogen ion does not lose its ability to move from nitrogen to nitrogen so that either kind of ions can diffuse through the membrane. Due to the Donnan mechanism, metal ions are retained outside the membrane phase. The separation factor of acids and metal salts on a weak-base anion-exchange membrane was from several times to an order of magnitude higher than the one obtained on Hi-Sep membrane (Oda et al., 1964).

Anion-exchange membranes also proved to be more effective than non-charged membranes in the dialytic treatment of acids polluted with organic substances (Nishiwaki and Itoi, 1969) and (Hansen and Wheaton, 1966). The application of non-charged membranes in the dialysis of concentrated waste sulphuric acid led to a fourfold diluting of the effluent as compared to the use of anion-exchange membranes. This is the result of the anomalous osmosis phenomenon associated with the dialysis of acids through anion-exchange membranes. The phenomenon of anomalous osmosis both positive and negative had been investigated earlier and was first explained as being due to both the heteroporosity of the membrane and to the generation of currents circulating between the adjacent pores (Sollner, 1930, and Sollner and Grollmann, 1954). Then, Schlögl (Schlögl, 1955) proved the incorrectness of these models and his approach did not require any assumption as to the geometry of the pores. According to Schlögl, anomalous osmosis may occur when the electrolyte solutions, varying in concentration, are separated by an ion-exchange membrane of an arbitrary structure. Considering the transport of the liquid in the pores of the membrane in its hydrodynamical aspect, he has shown that the swelling pressure difference in the membrane and the electric field diffusion potential acting on the charge liquid are the driving forces of the process. Swelling pressure difference yields a positive osmosis, i.e., directed from the solution of a lower concentration to the solution of a higher concentration. The action of the electric field on the charged liquid is either consistent or inconsistent with the action of the pressure difference. The direction of the electric field depends both on the mobility of the ions hydrogen ion and anion and on the permeability of the membrane.

Anion-exchange membrane dialysis of acids is a process in which the mobility of the co-ion (hydrogen ion in relation to the membrane-fixed ion groups) is higher than the mobility of the counter-ion (anion), the diffusion potential being inconsistent with the action of the pressure difference. If the membranes are characterized by a high water permeability, and if the ion mobilities differ significantly e.g. sulphuric acid the osmosis observed is negative, i.e. water is transported from concentrated to diluted solution. In this case, transport of the solute is congruent.

Investigations carried out by the authors on the dialysis of acid solutions containing organic pollutants or heavy metal salts included detailed studies on the transport and separation properties of weak-base anion-exchange membranes. These membranes have been prepared at the Institute of Organic and Polymer Technology, Technical University of Wrocław on polyethylene modified with 4-vinylpyridine and divinylbenzene copolymer. The studies have shown that the separation factor for acids and metals increases with an increase in the degree of membrane crosslinking (Mika-Gibała et al., 1976, and Mika-Gibała, 1977). It has also been shown that in membranes displaying the same molal concentration of ionogenic groups acid transport velocity was a function of one structural parameter alone which had been termed the tortuosity coefficient. This parameter describes the diffusion obstacle created by the polymer chains in membranes.

The application of charge-mosaic membranes as dialytic diaphragms (Mika-Gibała et al., 1978) permitted satisfactory elimination of membrane fouling by organic pollutants which occur in concentrated waste sulphuric acid (Mika-Gibała and Winnicki, 1976).

Weak-acid cation-exchange membranes, whose ionogenic groups are completely dissociated at high pH's, exhibit similar transport and separation properties in the dialytic treatment of alkaline metal hydroxides.

Considering a system in which a porous diaphragm separates two electrolyte solutions, initially one finds the solution on the left-hand side of the diaphragm contains substance AY alone, and on the right-hand side substance BY_2. Substance AY will then diffuse to the right-hand side solution, and substance BY_2 to the left-hand side solution until an equilibrium is reached in which the solutions on both sides of the diaphragm display the same AY and BY_2 concentrations. This process is illustrated in Fig. 2.

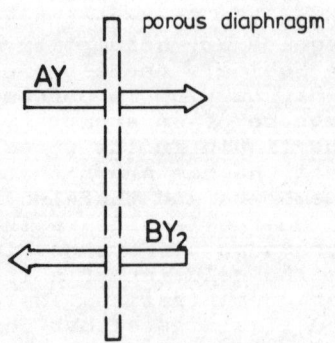

Fig. 2. Diffusion in a system of two electrolytes separated by a porous membrane.

Considering a system in which a porous diaphragm has been replaced by a strong-acid cation-exchange membrane, one finds the membrane is permeable to cations A^+ and B^{++} only, and impermeable to anions Y^- and water. Such a system is shown in Fig. 3.

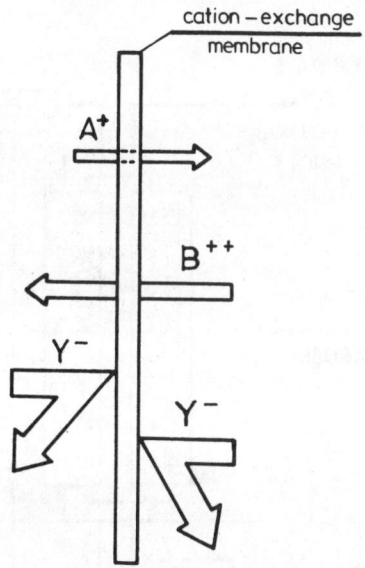

Fig. 3. Donnan dialysis.

Cations A^+ and B^{++} will diffuse in opposite directions until an equilibrium state is reached. According to Donnan (Donnan, 1925) the concentrations of all ions in the equilibrium state are described by the following general expression

$$\left[(c_i)_1 / (c_i)_r \right]^{1/z_i} = \text{const} \tag{1}$$

where C_i denotes the concentration of an arbitrary mobile ion i of a valency z_i, and the subscripts 1 and r refer respectively to the left-hand side and right-hand side of the membrane. Thus, the equilibrium concentrations for ions A^+ and B^{++} satisfy the equation

$$(c_A)_1 / (c_A)_r = \left[(c_B)_1 / (c_B)_r \right]^{1/2} \tag{2}$$

From this relationship it can be seen that if the ratio of the equilibrium concentrations of ion A is, for example, 10 then the corresponding ratio of the ion B concentrations will be 100. Therefore, there is possibility of concentrating one of the cations present in the system. The driving force for the transport of ion B^{++} from the low concentration phase assume it to be on the right-hand side of the membrane to the high concentration phase assume it to be on the left-hand side of the membrane is the concentration potential resulting from the difference in ion A^+ concentrations in the solutions on both sides of the membrane. The process considered here is known as Donnan dialysis and has been introduced by Wallace (Wallace, 1967) who employed dialysis to study the separation and concentration of uranyl ions UO_2^{++}. Later Davies and coworkers (Davies et al., 1971) developed this process on a large scale (Fig. 4).

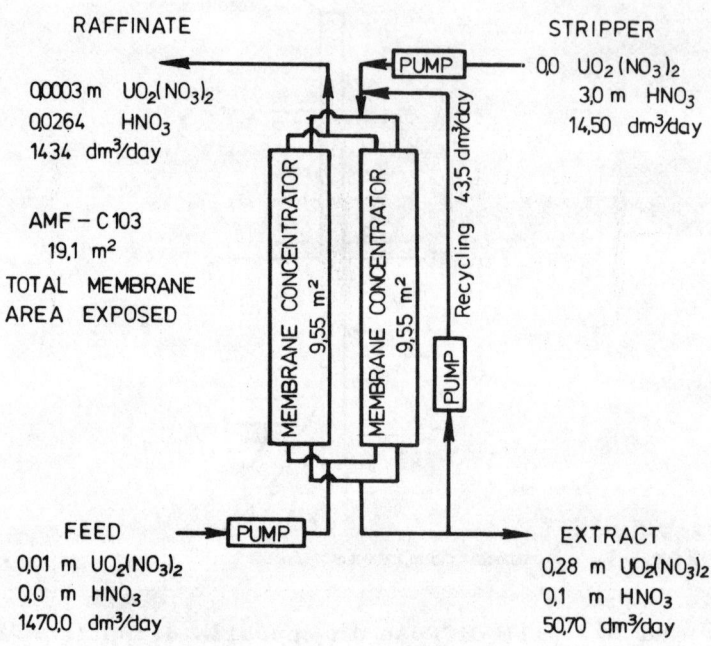

Fig. 4. Uranyl ion concentration in Donnan dialysis.

Donnan dialysis has also been applied to water softening (Smith, 1970 and 1971) as well as to the removal of nitrates, phosphates and ammonium ion present in secondary effluents (Eisenman and Smith, 1973).
Studies conducted by the authors on the application of Donnan dialysis to waste water treatment were focused on two basic problems - deacidification of rinse waters as well as separation and concentration of metal ions.

ELECTRODIALYSIS

The development of electrodialysis is greatly emphasized, since the process can be applied to water desalination. Although the use of reverse osmosis, which had been considered a competetive method, decreases the interests in electrodialytic processes for a period of time, the importance of electrodialysis continues to grow. At present, the development trends in this field tend toward high-temperature electrodialysis 338 K to 348 K which brings about a considerable drop in the resistance of the system and, at the same time, decreases the energy required. In the system used thus far preference is given to alternating arrangements of cation- and anion-exchange membranes, which are again termed conventional system (Fig. 5). Other combinations of ion-exchange membranes or ion-exchange and non-charged membranes (Winnicki and Mańczak, 1978) were studied and applied with various degrees of success.

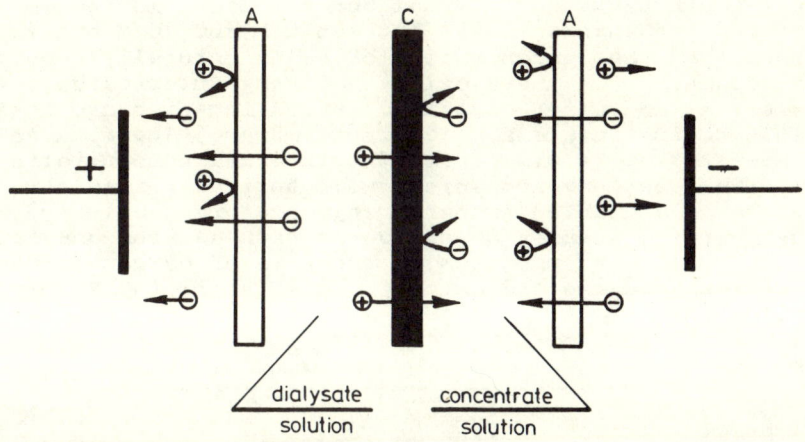

Fig. 5. Electrodialysis.

In addition to desalination of water, especially of water exhibiting a salinity level of several grams of NaCl per liter, the application of electrodialysis to separate valuable ionic components from industrial waste waters is both interesting and promising. Electrodialysis is particularly efficient when combined with other separation techniques such as ion exchange, dialysis or reverse osmosis. A good example of this application is concerned with zinc recovery from industrial waste water rayon industry which has recently been reported (Winnicki et al., 1975).

PIEZODIALYSIS

Piezodialysis was first described in 1932 by Sollner (Sollner, 1932) who predicted that membranes displaying bifunctional ion-exchange properties can also exhibit specific transport properties which are due to the electrical interaction of opposite-charged ion-exchange resins. The idea of mosaic membranes was introduced by Teorell (Teorell, 1935) and later by Meyer and Sievers (Meyer and Sievers, 1936). Neihof and Sollner (Neihof and Sollner, 1950; 1955) found that if mosaic membranes were placed between electrolyte solutions of different concentrations, a circulating electric current would pass between the cation-exchange and anion-exchange resins of the membrane. The concentration effect was first reported by Kollsman (Kollsman, 1961) who carried out a piezodialysis of 0.025 M KCl solution involving charge-mosaic membrane at a pressure of 9.8 MPa.

The fundamental concepts which the transport through mosaic membranes were formulated by Kedem and Katchalsky (Kedem and Katchalsky, 1963 a; 1963b). Later Merten (Merten, 1966), Leitz and coworkers (Leitz et al., 1969) and Weinstein (Weinstein, 1970) developed these concepts and employed them to piezodialytic concentration. The piezodialytic transport of salts proceeds through membranes with a mosaic arrangement of anion- and cation-exchange resins which cover the whole membrane thickness (Kedem and Katchalsky, 1963a). Due to the presence of two kinds of ion-exchange resins, the membranes in question are characterized by high permeability. It is well known that the internal concentration in the most of synthetic ion exchangers ranges between 2 M and 10 M and is noticeably higher than the concentration of salts naturally occurring in brackish waters. Since there exists a strong interaction between the mobile ions and water contained in the ion-exchange regions, it is possible to initiate a forced transmembrane flow of a solution whose concentration is similar to the internal concentration of the membrane. Thus, cations and anions will be transported through cation-exchange and anion-exchange segments, respectively. Each ion-exchange resin assures a continuous pathway from one solution to the other. The flow proceeds in the form of current circulation between cation-exchange and anion-exchange segments (Weinstein, 1971), (Fig. 6).

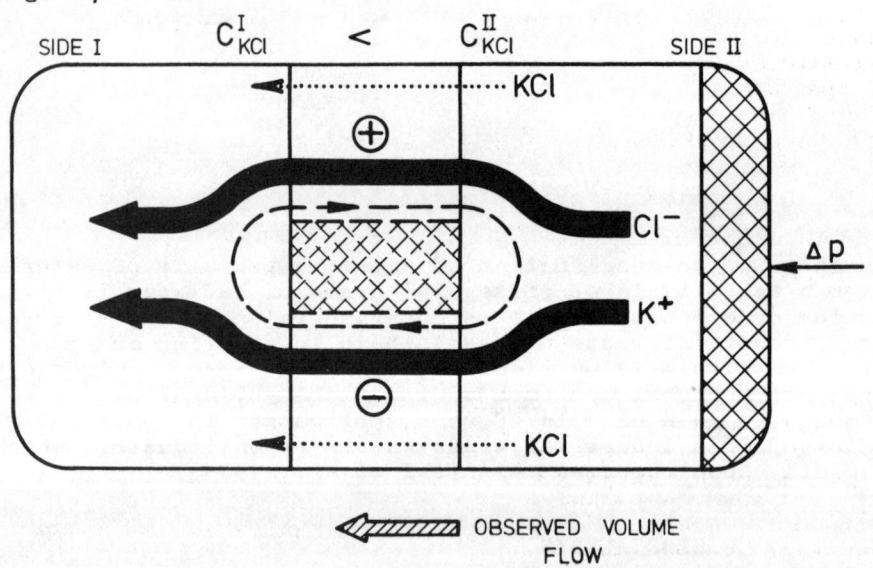

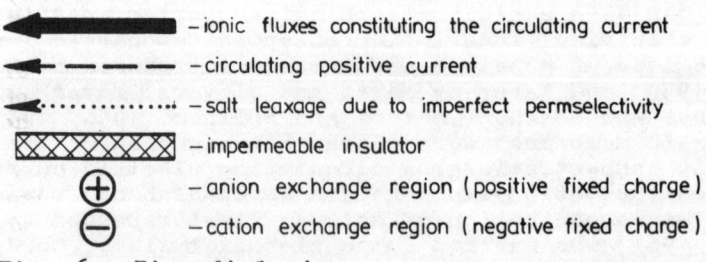

Fig. 6. Piezodialysis.

The transport of electrolytes in the piezodialytic process can be described by the following equations

$$J_v = L_p(\Delta p - \Delta \pi) + c_s L_p(1-\sigma)\Delta\mu_s^c + \beta I \qquad (3)$$

$$J_s = c_s L_p(1-\sigma)(\Delta p - \Delta \pi) + c_s \omega' \Delta\mu_s^c + (\tau_+ z_+ \gamma_+)I \qquad (4)$$

$$E = -\beta(\Delta p - \Delta \pi) - (\gamma_+ z_+ F)\Delta\mu_s^c + (1/K')I \qquad (5)$$

The equation of volume flow (J_v), salt flow (J_s), and potential difference (E) include six phenomenological factors of a practical meaning: filtration coefficient (L_p), reflection coefficient (σ), permeability of salt (ω), electroosmotic coefficient (β), conductivity of the membrane (K'), and transference number (τ). The other symbols have following meaning: (Δp) - the hydrostatic pressure difference, ($\Delta \pi$) - the osmotic pressure difference, (c_s) - the mean concentration of salt, ($\Delta\mu_s^c$) - the chemical potential difference of salt, (I) - the electric current density, (z_+) - the charge on cation, (γ_+) - the fraction of membrane area occupied by cation-exchange region, and (F) - Faraday constant.
These coefficients can be determined experimentally, as shown by Weinstein and coworkers (Weinstein et al., 1973), under conditions of one or two flows equal to zero.
The effectivity of piezodialysis process is described in term of the concentration factor (α)

$$\alpha = c_2/c_0 = J_s/J_v c_2 \qquad (6)$$

where (c_0) and (c_2) denote initial and final concentrations, respectively. This factor depends on initial concentration (c_0), the pressure applied (Δp), and on the membrane properties such as ion-exchange capacity, ion-exchange area, and structure homogenity (Shorr and Leitz, 1974).
The mosaic arrangements of charges permits the application of these membranes to (Weinstein and Bunow, 1972):
- the desalting and concentration of solutions of small non-electrolytes, i.e. charge-mosaic dialysis,
- the determination of binding-constants and activity coefficients of electrolytes in the presence of non-electrolytes by means of equilibrium dialysis,
- the adjustment of ionic constituents in body fluids in an artificial kidney,
- desalination and concentration of sea water by means of piezodialysis.

An artificial mosaic membrane may be used as a model (Weinstein and Bunow, 1972) of biological processes such as secretion of HCl, impulse conduction in nerve and muscle, oxidative phosphorylation, an photophosphorylation.
Space does not permit a complete review of literature on the dialytic processes employed in water and waste water treatment. In conclusion, it may be emphasized that most important work which has been done in this domain has been presented in detail.

REFERENCES

Graham T. (1854). Phil. Trans. Roy. Soc., London, 144, 177
Tuwiner S. B. (1962). Diffusion and Membrane Technology. Reinhold Publ. Corp., New York.
Chamberlin N. S. and Vromen B. H. (1959). Chem. Eng., 60, 117
Vromen B. H. (1962). Ind. Eng. Chem., 54, 20
Manecke G. and Heller H. (1956). Disc. Faraday Soc., 21, 101
Wolf F., Wehland W. and Bachman R. (1963). Z. Chem., 3, 384
Oda Y. et al. (1964). Prod. Res. Dev., 3, 244
Nishiwaki T. and Itoi S. (1969). Jap. Chem. Quart., 5, 36
Hansen R. D. and Wheaton R. M. (1966). U.S. Pat. 3,244,620
Sollner K. (1930). Z. Elektrochem. angew. physik. Chem., 36, 36; 234
Sollner K. and Grollmann A. (1954). Z. Elektrochem. angew. physik. Chem. (Frankfurt), 1, 305
Schlögl R. (1955). Z. physik. Chem. Frankfurt, 3, 73
Mika-Gibała A., Błażejewska G. and Winnicki T. (1976). Proceedings 2nd Congress APLICHEM, Bratislava.
Mika-Gibała A. (1977). Dissertation. Technical University of Wrocław.
Mika-Gibała A. et al. (1978). Pol. Pat. 97,880
Mika-Gibała A. and Winnicki T. (1976). Proceedings Konf. Nauk.-Tech., Lublin.
Donnan F. G. (1925). Chem. Rev., 1, 73
Wallace R. M. (1967). Ind. Eng. Chem. Process Design Develop., 6, 423
Davies T. A., Wu J. S. and Baker B. L. (1971). AIChE J., 17, 1006
Smith J. D. (1970). Office of Saline Water U.S. Dep. Interior Rep., No 506
Smith J. D. (1971). Office of Saline Water U.S. Dep. Interior Rep., No 655
Eisenman J. L. and Smith J. D. (1973). Office of Res. and Develop. U.S. Environmental Protection Agency Rep., No 670/2-73-076
Winnicki T. and Mańczak M. (1978). Proceedings 6th Intern. Symp. Fresh Water from the Sea, Las Palmas.
Winnicki T. et al. (1975). Envir. Protect. Eng., 1, 37
Sollner K. (1932). Biochem. Z.. 244, 370
Teorell T. (1935). Proc. Soc. Appl. Biol. and Med., 33, 282
Meyer K. and Sievers J. F. (1936). Helv. Chim. Acta, 19, 649; 655; 987
Neihof R. and Sollner K. (1950). J. Phys. Colloid Chem., 54, 157
Neihof R. and Sollner K. (1955). J. Gen. Physiol., 38, 613
Kollsman P. (1961). U.S. Pat. 2,987,472
Kedem O. and Katchalsky A. (1963a). Trans. Faraday Soc., 58, 1931
Kedem O. and Katchalsky A. (1963b). Trans. Faraday Soc., 59, 1958
Merten V. (1966). Desalination, 1, 297
Leitz F. B., Alexander S. S. and Douglas A. S. (1969). Office of Saline Water U.S. Dep. Interior Rep., No 452
Weinstein J. N. and Caplan S. R. (1970). Science, 169, 296
Weinstein J. N. (1971). Dissertation. Harvard University.
Weinstein J. N. et al. (1973). Desalination, 12, 1
Shorr J. and Leitz F. B. (1974). Desalination, 14, 11
Weinstein J. N. and Bunow B. (1972). Desalination, 10, 341

CATALYTIC OXIDATION OF INORGANIC SULPHUR COMPOUNDS IN REDUCING SOLUTIONS

A. L. Kowal and E. M. Klocek

Institute of Environment Protection Engineering, Technical University, Wroclaw, Poland

ABSTRACT

The kinetics and stoichiometry of the oxidation process in the mixture of sulphides and sulphites were investigated. This process differs from the oxidation of the individual substrates and yields higher concentrations of thiosulphates and only of a small part of sulphates. The formation of thiosulphates is accompanied by the generation of tetrathionates. In this case tetrathionates are the main reaction product. The distribution of the oxidation products depends on the pH: with decreasing pH the amount of tetrathionates and colloidal sulphur increases and becomes predominant over sulphates. Hopkalit /manganese and iron ore/ and blast-furnace slag were found to be the best catalysts in the oxidation of sulphur compounds and organic matter contained in the pyrometallurgical scrubbing water. In the process on Hopkalit-bed the amount of oxidated sulphides increases from 87 to 97%, that of sulphites from 14 to 95%, that of phenol from 13 to 89%, and COD reduction from 8 to 61% as compared with no catalyst employed.

KEYWORDS

Catalytic oxidation; sulphides; sulphites; pyrometallurgical scrubbing water.

INTRODUCTION

Scrubbing techniques are usually employed to reduce the particulate and gas emissions produced in the smelting process. Nevertheless, the application of this method does not satisfactorily solve the problem of environment pollution, as the scrubbing process produces wastewaters containing settleable solids which are charged with a high pollution load /toxic substances/ and cannot be recirculated without treatment. Due to dissolution of gaseous sulphur compounds in water and some secondary reactions that occur in the liquid phase, sulphates, sulphites, sulphides, thiosulphates and thiocyanates are formed, while colloidal sulphur is formed, which passes into the solids. Table 1 shows the composition of the wastewater generated in the scrubbing of pyrometallurgical gases from a copper smelter.

TABLE 1 Chemical Analysis of Pyrometallurgical Scrubbing Water

Determination	Unit	Contents
pH	pH	4.4
Permanganate value	$mg/dm^3\ O_2$	13800
COD	$mg/dm^3\ O_2$	32800
BOD_5	$mg/dm^3\ O_2$	12840
TOC	$mg/dm^3\ C$	9800
Sulphides	$mg/dm^3\ S^{2-}$	450
Sulphites	$mg/dm^3\ SO_3^{2-}$	850
Sulphates	$mg/dm^3\ SO_4^{2-}$	16200
Total amount of sulphur comp.	$mg/dm^3\ S$	6390
Cyanates	$mg/dm^3\ CN^-$	12
Ammonia	$mg/dm^3\ N$	1750
Nitrite and nitrate	$mg/dm^3\ N$	74
Total phenols	mg/dm^3	980
Thiocyanates	mg/dm^3	5
Chlorides	$mg/dm^3\ Cl^-$	5600
Calcium	$mg/dm^3\ Ca$	500
Magnesium	$mg/dm^3\ Mg$	200
Iron	$mg/dm^3\ Fe$	178
Copper	$mg/dm^3\ Cu$	15
Lead	$mg/dm^3\ Pb$	8
Zinc	$mg/dm^3\ Zn$	18.2
TDS	mg/dm^3	95350
VDS	mg/dm^3	45760

The investigations reported here were aimed at describing the oxidation of sulphides and sulphites which occur jointly in the solution containing sulphates and some substances of a reducing nature. The literature on oxidation of sulphur compounds deals mainly with the kinetics and stoichimetry of the reactions taking place in one-component solution, where sulphides or sulphites are the compounds oxidized; no data, however, are available on the oxidation of sulphur-compound mixtures. Theoretical studies based on the analysis of oxidation-reduction potentials do not sufficiently explain the kinetics of sulphides and sulphites oxidation, as they may be influenced by a reducing solution or by the occurrence of autocatalysis, catalysis or inhibition among the individual sulphur compounds and some other constituents of the solution. It seemed therefore reasonable to carry out experimental studies on oxidation of sulphur compounds in the wastewaters containing reducing agents, with emphasis on the physical-chemical factors that shift the equilibrium of the chemical systems toward the formation of sulphates. The oxidation reaction for sulphur compounds is first of all influenced by pH and by the catalyst used.

EXPERIMENTAL METHODS

Laboratory investigations were conducted at room temperature in a reactor consisting of a glass column /150 cm long and 4.5 cm in diameter/ which was packed with either Raschig rings or a catalyst. Air was supplied to the column through a porous filter, which was placed on the bottom of the column. The aeration process was carried out under conditions of excess air so that an intensive mixing of the fluid could also be arranged. The air flow rate was kept constant /20 dm^3/h / and yielded an air-to-liquid volumetric ratio of 7:1. The wastewaters under study were circulated in a closed system with counter-current flow in relation to the air flow. The wastewaters were circulated with the use of a peristaltic pump 2.5 times per hour. The air leaving the column was passed through the scrubber, where hydrogen sulphide was retained. The aeration time equalled 20 hr. Phenols and COD were determined according to the procedure of Hermanowicz /1976/. Sulphur compounds were determined by titration with modification to allow the determination procedure in a reducing solution /Klocek, 1978/.

STUDIES ON THE OXIDATION OF SULPHUR COMPOUNDS IN
MODEL SOLUTIONS

The model solution contained sulphur compounds at concentrations typical for scrubbing waters. Phenol was selected as the reducing agent, because this compound also occurs in the wastewater studied. The pH varied from 7.0 to 9.5 because of indispensable elimination of desorption of hydrogen sulphide. During aeration at an initial pH 7.0, about 31% of the sulphides were removed in the form of hydrogen sulphide, and at pH 9.5 this procentage was as low as 4.9%. According to Chen and Morris /1970/ the maximum oxidation rate for sulphides is reached at pH 8.0 - 8.5 and 11.0, whereas the maximum oxidation rate for sulphites was found by Roxgurgh /1962/ to occur at pH 7.5 - 9.5. Model solutions of neutral or slightly alkaline pH were found to be most suitable, which is due to the less probable formation of tetrathionates that are more resistant to oxidation /Avrahami, 1968; Charlot, 1976/. Figure 1 shows the oxidation of sulphur compounds at an initial pH of 9.5. The dynamics of the oxidation of sulphur compounds were studied with various catalyst beds. The catalysts were made by the

Chemical Works of Oświęcim /ferrosa,FC-2,Co-Mo/ and by Inowrocławskie Zakłady Sodowe /Hopkalit/.Blast-furnace slag from the copper smelters of interest were also studied as catalysts.A single packing of the column consisted of a catalyst /750 g/ or blast-furnace slag /2500 g/.The catalyst was employed in high amounts to prevent its deactivation.

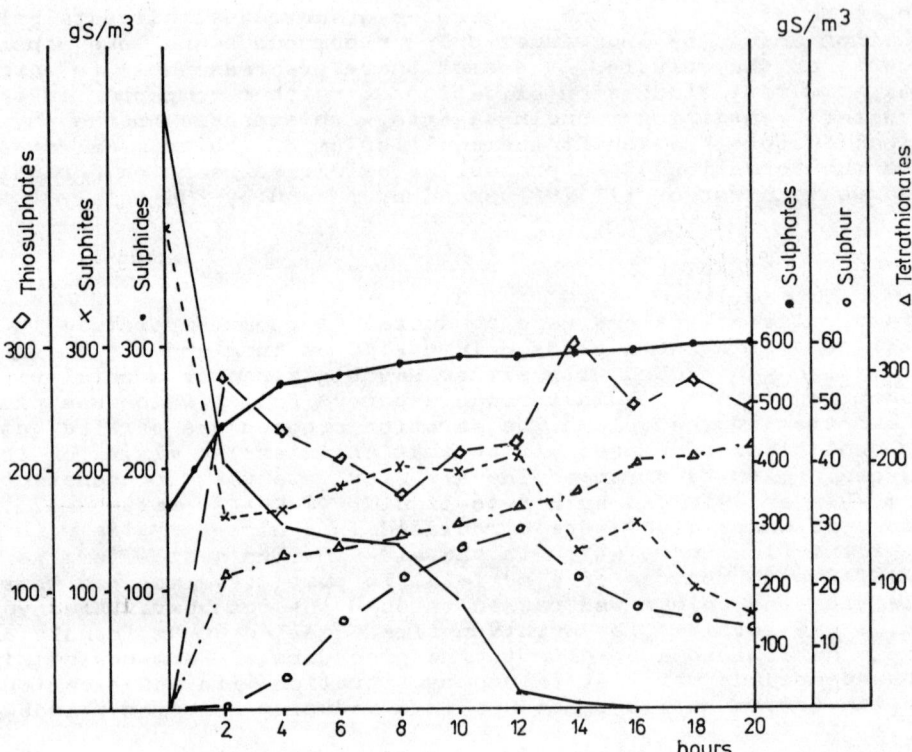

Fig. 1. Oxidation of sulphur compounds in artificial solution without catalyst at initial pH 9.5.

Figure 2 shows the dynamics of the conversion of sulphur compounds during aeration with the use of Hopkalit.

Discussion

The aeration of the model solution yielded complete oxidation of sulphides at an initial pH 7.0 and 9.5.At pH 7.0 sulphites were completly oxidized,whereas at pH 9.5 the removal of sulphites approached 80%.During aeration a periodic increase in sulphites concentration was observed /Figs.1 and 2/,which was due to oxidation of sulphides as follows /Avrahami, 1968/ :

$$S^{2-} + \frac{3}{2} O_2 \rightarrow SO_3^{2-} \qquad /a/$$

$$HS^- + \frac{3}{2} O_2 = HSO_3^- \rightarrow SO_3^{2-} + H^+ \qquad /b/$$

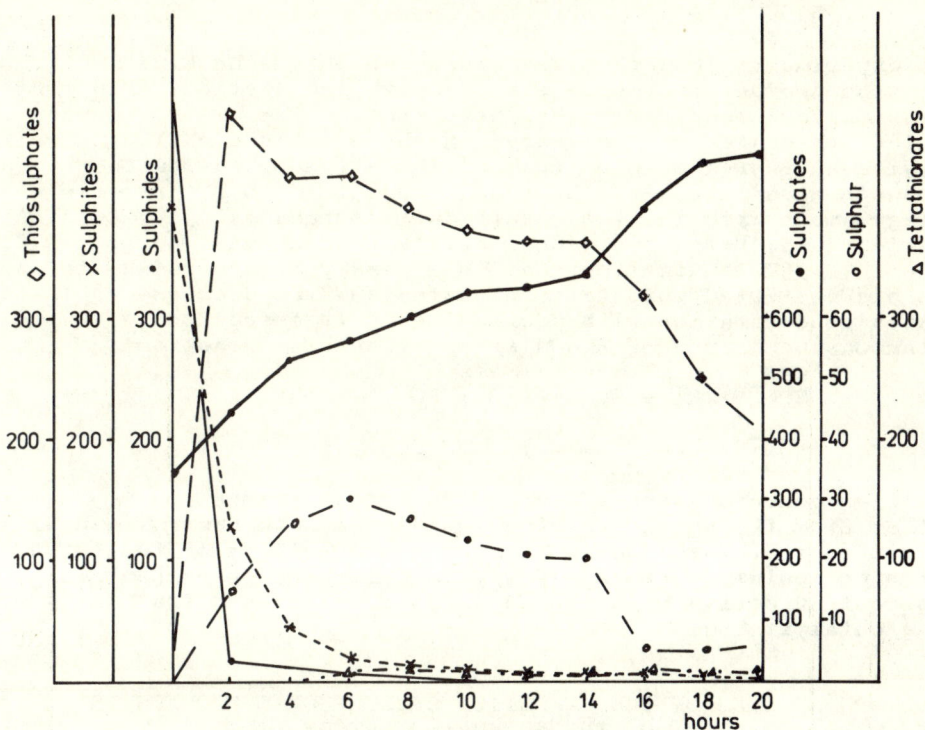

Fig. 2. Oxidation of sulphur compounds in artificial solution with Hopkalit at initial pH 9.5.

Sulphides, thiosulphates and phenols inhibit the sulphites oxidation rate /Cline and Richards 1969; Inale and Uzda, 1975/. The sulphites concentration were considerably decreased after sulphides removal. However, the process itself did not lead to an increase in sulphates concentration, but rather inceased the thiosulphates concentration. The decrease of colloidal sulphur in solution indicates the predominance of the reaction yielding thiosulphates /Avrahami, 1968/ :

$$SO_3^{2-} + S^o = SO_3^{2-} \quad /c/$$

$$HSO_3^- + S^o = S_2O_3^{2-} + H^+ \quad /d/$$

$$4\ S^o + 6\ OH^- = S_2O_3^{2-} + 2\ S^{2-} + 3\ H_2O \quad /e/$$

The aeration of sulphides and sulphites in the period studied did not lead to complete oxidation of the sulphur compounds to sulphates. A considerable amount of tetrathionates and colloidal sulphur remained in the solution at an initial pH 7.0, because they are less susceptible to oxidation in these conditions. At an initial pH 9.5 the concentration of colloidal sulphur and tetrathionates was twice as low as at pH 7.0. It can be expected that slow oxidation of thiosulphates will tend toward the formation of sulphates /Avrahami, 1968; Chen and Morris, 1970/ :

$$S_2O_3^{2-} + 5\ H_2O - 8\ e = 2\ SO_4^{2-} + 10\ H^+ \quad /f/$$

$$S_2O_3^{2-} + \tfrac{1}{2} O_2 = SO_4^{2-} + S^o \qquad /g/$$

The experiments show that the effect of pH on the kind and amount of reaction products formed was evidently more distinct than that of the catalyst used. In all the cases studied the highest concentration of tetrathionates were measured in the solutions of lowest initial pH. With an increase in ph the concentration of sulphates as reaction products also increased. The results obtained in the experiments are in agreement with the literature data /Avrahami, 1968; Bengtson and Bjerle, 1975; Berg, 1967; Charlot, 1976; Chen and Morris, 1970; Choppin, 1937/, indicating, that in a weakly acidic medium thiosulphates are converted into tetrathionates, and the decrease of thiosulphates concentration with decreasing pH is associated with the simultaneous occurence of the disproportionation reaction:

$$HS_2O_3^- + H^+ = SO_2 + H_2O + S^o \qquad /h/$$
$$S_2O_3^{2-} + H_2CO_3 \rightarrow CO_3^{2-} + H_2SO_3 + S^o \qquad /i/$$

The high concentration of thiosulphates at initial pH 9.5 may result either from the oxidation of sulphides in both neutral and alkaline solutions to thiosulphates or from the stability of the latter in the same medium /Choppin, 1937/. The oxidation of sulphur compounds was most effective with the Hopkalit catalyst and the blast-furnace slag catalyst bed.

STUDIES ON THE OXIDATION OF SULPHUR COMPOUNDS IN THE SCRUBBING WATERS

After 20 hrs of aeration at initial pH 7.0 and 9.5 the concentration of sulphides in the scrubbing waters decreased by 96% and 87% respectively, while sulphites decreased by 2.7% and 14%, respectively /Figs. 3 and 4/. The insignificant decrease of sulphites concentration in the effluent is due to a periodical rise in which sulphites have exceeded their initial concentration in the influent. Hence, it is easily seen that the amount of sulphites generated during sulphides oxidation was higher than the amount of sulphites which could be oxidized. The inhibition of the sulphites oxidation reaction can be attributed to the presence of sulphides, phenols and some other components of the scrubbing waters /Berg, 1967; Fuller and Crist, 1941; Roxburgh, 1962/. Aeration conducted in the presence of ferrosa, FC-2, Co-Mo, Hopkalit and blast-furnace slag yielded more complete oxidation of sulphides and sulphites to sulphates. The oxidation of phenols and reducing compounds /expressed in terms of COD/ was more efficient in the presence of the catalysts studied. Although complete oxidation of sulphides was not achieved /irrespective of the pH applied and the catalyst used/, the removals of these compounds were always very high /above 92%. The studies on the dynamics of sulphites oxidation indicate that the concentration of these substances increase in the initial stage of the process. This increase, however, is not as high as that observed in the initial stage of aeration with no catalyst. The determination of sulphates content after aeration process shows that the concentration of sulphates is higher than could be expected from the complete oxidation of sulphides and sulphites to sulphates. The literature data and the experiments on model solutions indicate that under the conditions presented here sulphates are not the only reaction product. As can be seen from these studies, other organic and inorganic sulphur compounds /which had been determined as sulphates after mineralization, see Table 1/ were also oxidized to sulphates.

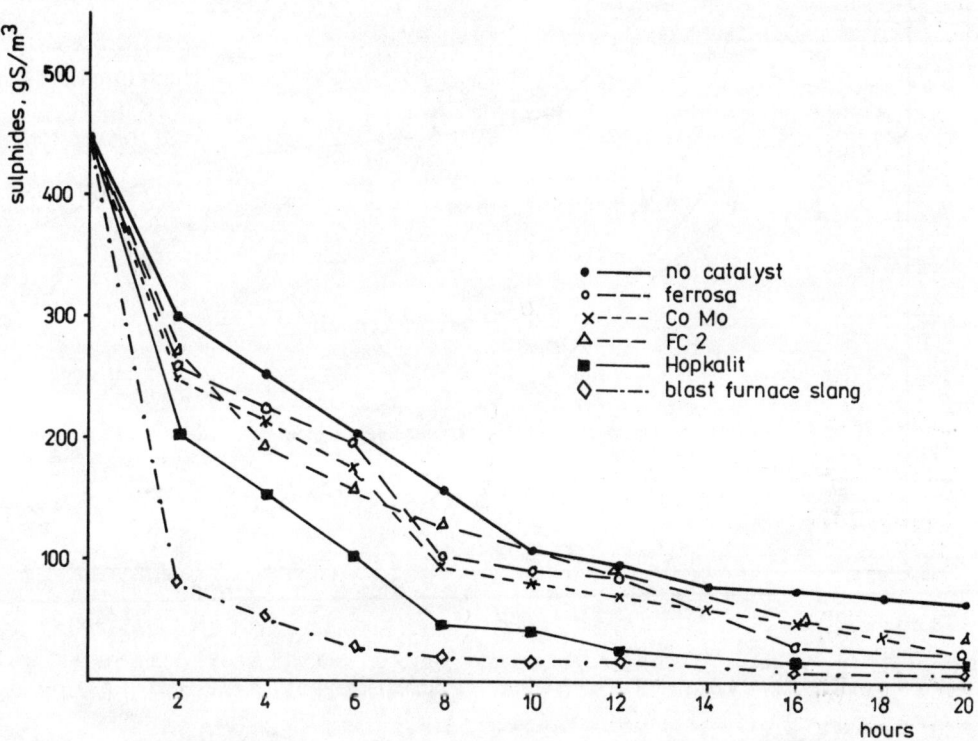

Fig. 3. Comparison of catalytic and non-catalytic oxidation of sulphides in pyrometallurgical scrubbing water at initial pH 9.5.

CONCLUSION

The studies on the oxidation of sulphur compounds in a strongly reducing solution /pyrometallurgical scrubbing water/ with and without catalyst indicate that the oxidation of sulphur compounds and the efficiency of the purification /expressed in terms of phenol and COD removals/ are the highest, when aeration is carried out in the presence of Hopkalit catalyst and blast-furnace slag at pH 9.5.

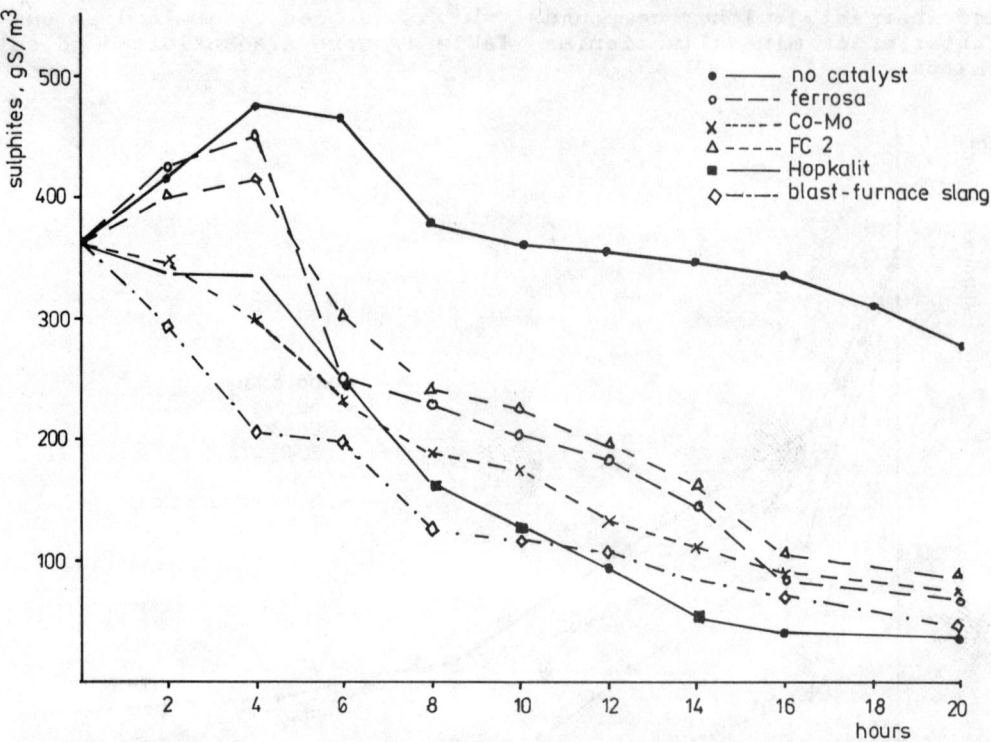

Fig. 4. Comparison of catalytic and non-catalytic oxidation of sulphites in pyrometallurgical scrubbing water at initial pH 9.5.

REFERENCES

Avrahami, M. /1968/. The oxidation of the sulphide ion of very low concentrations in aqueous solutions. J.Chem.Soc.,/A/, 647-651.
Bengtson, S. /1975/. Catalytic oxidation of sulfites in diluted aqueous solutions. Chem.Eng.Sci., 30, 1429-1435.
Berg, N. /1967/. Studies on the elimination of sulphide from tannery beamhouse effluents by manganese-catalysed oxidation. J.Am.Leath. Chem.Ass., 62, 684-693.
Cline, J.D., and F.A.Richards /1969/. Oxygenation of hydrogen sulphide in sea water at constant salinity, temperature and pH. Environ.Sci.Technol., 3, 838-843.
Charlot, G. /1976/. In PWN /Ed./, Analiza nieorganiczna jakościowa., 1rd ed. Warszawa.
Chen, K.Y., and J.C.Morris /1970/. Oxidation the aqueous sulfide by O_2 1. General characteristics and catalytic influences. Proc.5-th Int. Wat.Pollut.Res., III-32, San Francisco.
Choppin, R.A. /1937/. The oxidation of aqueous sulfide solutions by hypochlorites. J.Am.Chem.Soc., 59, 2203-2207.
Fuller, F.C., and R.H.Crist /1941/. The rate oxidation of sulfite ions by oxygen. J.Am.Chem.Soc., 63, 1644-1650.
Hermanowicz, W. /1976/. In Arkady /Ed./, Fizyko-chemiczne metody badania wody i ścieków., 1rd ed. Warszawa.
Inale, J., and K.Uzda /1975/. Antioxidation of sulfites. Jap.Kokai, 125, 993-999-ref.CA, 84, 107943w.

Jacobs, B.M. /1946/. The Analytical Chemistry of Industrial Poisons Hasards and Solvents,, New York.
Kłocek, E.M. /1978/. Chemical and biochemical oxidation of sulphur compounds in artificial solutions and industrial wastewaters. Ph.D.thesis., K-351/78, Technical Univ. Wrocław, Poland.
Roxburgh, J. /1962/. Catalyst effect on sulphite oxidation rates. Can.Jour.Chem.Eng., 6, 127-147.

SIMULATION OF THE FLOCCULATION PROCESS IN FILTER BEDS

A. L. Kowal and J. Maćkiewicz

Institute of Environment Protection Engineering, Technical
University, Wroclaw, Poland

ABSTRACT

Simulation of flocculation in filter beds is based on the filter-media density. Deterministic model of flocculation determined by the hydraulic parameters of the filter media was applied. Flocculation conditions for various filter media were compared in terms of the average velocity gradient. The filtration rate was found to depend on the density of the filter-bed media. Layer depth ratios were established, and the utility of the filters in the coagulation process was estimated.

KEYWORDS

Coagulation; filtration; velocity gradient; optimal filtration rate; filter-bed density; filter-bed depth.

INTRODUCTION

The coagulation process is conventionally employed to remove organic and inorganic pollutants occurring in the form of colloids or suspended solids, or sometimes in solution. Most of the water treatment plants involve coagulation in the purification of surface waters. Modern trends in wastewater engineering tend toward a wide application of the coagulation process combined with biological treatment to advanced wastewater treatment and to water renovation. Advanced treatment is generally employed to remove refractory compounds and nutrients. The removal of the latter prevents eutrophication. In highly industrialized and urbanized regions where water shortage is especially experienced, the available water resources can be enriched with water reclaimed from the second effluent. The water renovation process can involve several procedures-such as coagulation, filtration, sorption, desalination, disinfection and ammonia nitrogen removal. Among the various treatment techniques, coagulation deserves special attention, as this process is responsible for the treatment efficiency in both wastewater and water. A conventional system for volume coagulation includes installations for rapid mix, slow mix and sedimentation. Flocculation is often combined with sedimentation in multi-purpose settling tanks. A new solution to the coagulation process is the one involving single-or double media filter beds. In contact filters coagulation and filtration are carried out jointly on a filter bed with bottom supply. In multi-media filters the coagulated water can be passed from coarse to fine grains just as it goes in the contact filters with gravitational flow. The flocculation of water admixtures on coarse-grain single-layer filter beds may also be effective when aided with polyelectrolytes.

VOLUME COAGULATION

Flocculation depends on the number of particles and the probability of collision. Collision may result from the variable velocity of suspended particles and from micropulsation generated by mixing. The intensity of mixing can be defined by the variation in the velocity vector of fluid motion, which is described in terms of the average flocculation gradient

$$G = \sqrt{\frac{W}{\mu}} \quad \text{...............................(1)}$$

Energy dissipation depends on the kind and geometry both of the mixing basin and of the stirrer. To calculate energy dissipation for conventional flocculation tanks the following expressions will be used (Veytser and Lutsenko, 1975) :

(mechanical tank)

$$W = \frac{N \cdot 2 \pi s}{V} \quad \text{...............................(2)}$$

(turbulence and water torque)

$$W = \frac{Q v^2 \varrho g}{2V} \quad \text{...............................(3)}$$

and (compartment tank)

$$W = \frac{[n v_p^2 + (n-1)v_v^2] Q \varrho g}{2V} \quad \text{...............................(4)}$$

The efficiency of the coagulation process is also affected by the duration of the process itself. Micropulsation generated during mixing not only contributes to floc formation but causes floc damage as well. Floc formation and floc damage recur many times. If the velocity gradient exceeds the boundary value, floc damage proceeds faster than floc formation, which deteriorates the efficiency of the coagulation process. Extended mixing also multiplies the recurrence of floc formation and damage, leads to the screening of active centres, decreases the flocculation rate and reduces the size of flocs. Hence the coagulation criterion developed by Camp(Ca=Gt) characterizes the process in question for strictly defined ranges of either of the two values. For volume coagulation involving conventional hydrolysing coagulants the optimum flocculation time varies from 15 to 20 min. The application of polyelectrolytes (either as coagulation aid or as coagulants) cuts the flocculation time down to 10 min and below (Dzięgielewski, Dziubek and Kowal, 1978). Veytser and Lutsenko (1975) report that the optimum Camp numbers are 8,000 – 10,000 and 30,000 – 50,000 for alum coagulation and ferric chloride coagulation, respectively. This is due to the strength of $Fe(OH)_3$ flocs which is higher than that of the $Al(OH)_3$ flocs.

Alum flocs can be strengthened with polyelectrolytes. The Camp number will then approach the values that are typical of ferric chloride coagulation.

Veytser and Lutsenko (1975) have found that the optimum velocity gradient for the coagulation of wastewaters is approx $50\ s^{-1}$. At lower velocity gradients (about $10\ s^{-1}$), mixing can be applied only with aluminium sulphate as coagulant. In polyelectrolyte-aided coagulation with low velocity gradients the flocs formed do not exhibit the strength and structure required. Low velocity gradients are also inadvisable for the filtration process, as they cause a clogging of the surface layer.

FLOCCULATION IN FILTER BEDS

The requirements for the flocculation process in filter beds can be defined in terms of the expression for the average velocity gradient of fluid motion (Kowal and Maćkiewicz, 1977). The expression in question was derived on the basis of hydraulic parameters. Energy dissipation was determined from the increment of the head loss. For this purpose the Kozeny-Carman formula was used, which is in good agreement with the engineering practice

$$\Delta p = \frac{K\ L(1-\varepsilon)^2}{\varepsilon^3}\ \frac{v}{d^2}\ (\rho_m - \rho) \quad \dots\dots\dots\dots\dots\dots\dots(5)$$

The term ($\rho_m - \rho$) indicates the head-loss effect in a three-component mixture (filter media, water, suspended solids). By virtue of Eq. (5) the energy consumed in the water flow through the filter bed :

$$N = \frac{K(1-\varepsilon)^2}{\varepsilon^3}\ \frac{v^2}{d^2}\ \nu\ (\rho_m - \rho) \quad \dots\dots\dots\dots\dots\dots\dots(6)$$

Combining Eqs. (6) and (1) gives the expression for the average velocity gradient in the filtration process

$$G = \frac{1-\varepsilon}{\varepsilon^2}\ \frac{v}{d}\ \sqrt{\frac{\rho_m - \rho}{\rho}}\ K \quad \dots\dots\dots\dots\dots\dots\dots(7)$$

which is a function of the following parameters : grain size, porosity, filtration rate and media density. The velocity gradient justifies the choice of the proper filtration rate depending on the media density in order to assure the optimum conditions for flocculation. The density effect can be explained under boundary conditions, i.e., when the value the filter media density approaches that of the water density. Then G tends to zero. This resembles the phenomena that occur in sludge blanket clarifiers. The velocity gradient for sludge blanket coagulation approaches several s^{-1}, which is within the range of optimum conditions both for flocculation and sedimentation. If the density of the filter media increases, so does the velocity gradient. To maintain identical flocculation conditions in the filter beds it is advisable to employ such a filtration rate that decreases with increasing filter bed density (Fig. 1).

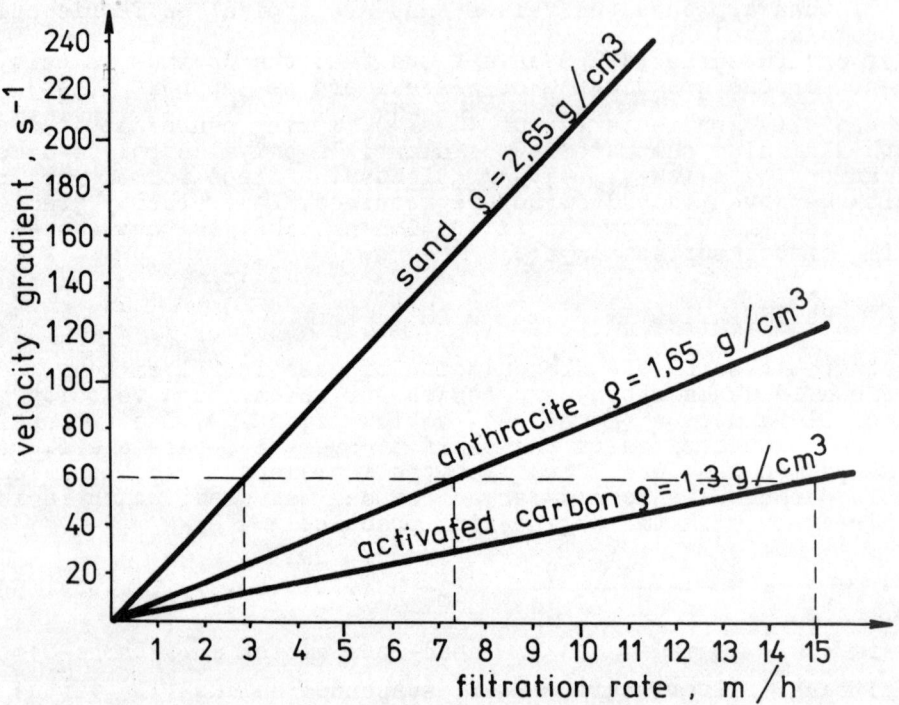

Fig. 1. The effect of filtration rate and filter bed media density on the value of velocity gradient.

If the density of the filter media is less than that of the water, the expression will change the sign (from plus to minus and vice versa) and the filtration velocity vector will change the direction. But then the filter bed begins to flotate so that some arrangements should be undertaken to prevent the outflow of the bed.

PARAMETERS OF FILTER MODELLING

The simulation of the filtration process can be carried out in terms of velocity gradient. For this purpose Eq. (5) has been transformed as follows:

$$G = \sqrt{\frac{\Delta h}{L} \cdot \frac{\rho_m - \rho}{\rho} \cdot \frac{v \, g}{\nu} \cdot \frac{1}{\varepsilon}} \quad \ldots \ldots \ldots (8)$$

In filter modelling the velocity gradient allows for determining the filtration rate and the filter bed depth as a function of the media density.
The relative filtration rate for verious filter beds was established by assuming the hydraulic gradients in the beds and the

flocculation conditions to be constant ($\Delta h/L$ = const; G = const).
Thus, the filtration rate becomes

$$\frac{v_2}{v_1} = \frac{\varrho_1 - \varrho}{\varrho_2 - \varrho} \cdot \frac{\varepsilon_2}{\varepsilon_1} \quad \ldots \ldots \ldots \ldots \ldots \ldots \ldots \ldots \ldots \ldots (9)$$

In multi-media filters the filtration rate for the individual layers is constant (v = const).
For a continuous flow (Δh = const) and for G = const the filter bed depth will be calculated in terms of

$$\frac{L_2}{L_1} = \frac{\varrho_2 - \varrho}{\varrho_1 - \varrho} \cdot \frac{\varepsilon_1}{\varepsilon_2} \quad \ldots \ldots \ldots \ldots \ldots \ldots \ldots \ldots \ldots \ldots (10)$$

The values of the filtration rates and the bed depths are shown in Fig. 2 for activated-carbon, anthracite, and sand-filter beds of a density of 1.3 g/cm³, 1.65 g/cm³ and 2.65 g/cm³, respectively.

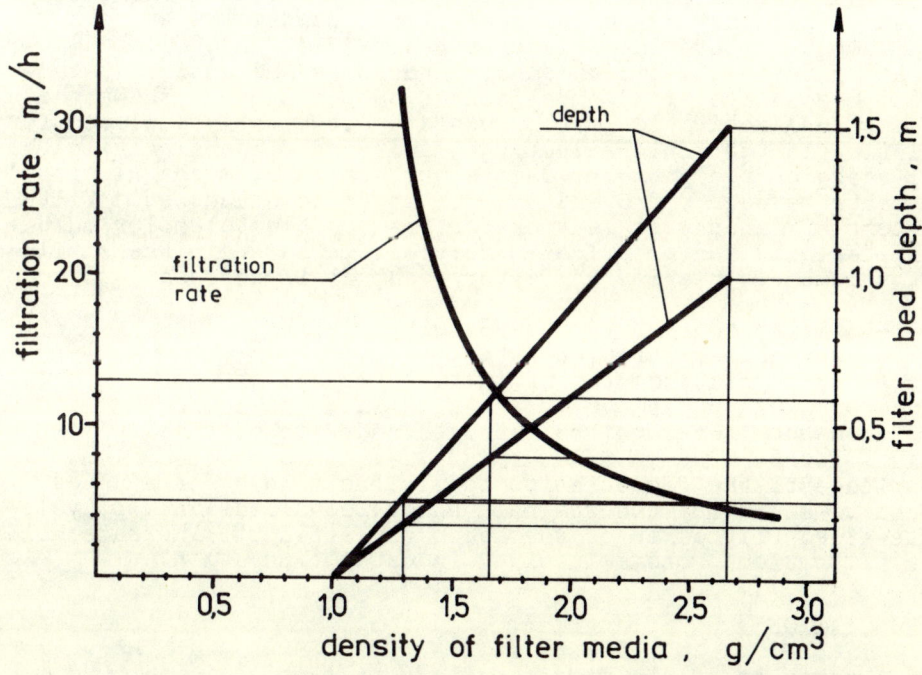

Fig. 2. The effect of density on the filtration rate and the filter bed depth.

The conventional filtration rate for a sand-filter bed is 5 m/h, and the optimum flocculation conditions for filter beds of decreased density can be kept at markedly increased filtration rates. To maintain optimum flocculation conditions it is necessary to decrease the filtration rate. The calculation of the layers in double-media filters is indispensible in the design of the flocculation process. The calculated depths of activated carbon beds are lower, whereas the calculated depths of anthracite beds are higher than those employed in the engineering practise. Activated carbon beds to be used in the filtration process are selected to comply with the requirements of sorption, which additionally improves the flocculation conditions. The depths of anthracite beds recommended for design (min. 0.3 m or 1/3 L) are too low.

CONCLUDING REMARKS

Velocity gradient and Camp's number have been selected as criteria for the simulation of the flocculation process. Camp's number is determined by velocity gradient and flocculation time and is considered a dimensionless similarity criterion for flocculation. The velocity gradient and flocculation time recommended for volume coagulation are $20 - 60 \, s^{-1}$ and 15 - 20 min, respectively, which corresponds to the Camp's numbers ranging from 18,000 - 24,000 to 54,000 - 72,000. Camp's number remains constant only in the stage of slow mixing. In filter beds Camp's number undergoes variations: decreases with bed depth and increases at each depth with the time of the filter run. This is due to the hydraulic conditions which affect the efficiency of the flocculation process. The average values of Camp's numbers in a filter run are 27,000-31,000; 11,500-39,000 and 5,000 - 30,000 for sand, anthracite and activated-carbon beds, respectively. A comparison of Camp's numbers shows that anthracite- and activated-carbon beds are efficiently employed for flocculation in the whole filter run, and sand filters only in the initial stage of the process. Camp's numbers in alum coagulation should range between 9,000 and 12,000. It follows that sand filters become less efficient. It is therefore reasonable to employ multimedia filter beds and or polyelectrolytes. To maintain the similarity of volume coagulation in filter beds, it is necessary to employ

- filter beds of various density,
- suitable filtration rates,
- appropriate depths of multi-layer filters,
- coagulants and coagulant-aids which will assure the required strength and size of the flocs.

The knowledge of the flocculation conditions allows for a wider use of multi-media and coarse-grain homogeneous beds in a polyelectrolyte-aided filtration of the secondary effluent. The same holds for the filtration process involved in water renovation.

NOTATION

ε = porosity of the filter media
μ = absolute viscosity
ν = kinematic viscosity
Q = flow rate
ρ_m, ρ = density of filter bed media and water, respectively
V = water volume
v = flow velocity and filtration rate
v_p, v_v = flow velocity in the direction parallel and vertical to the compartment of the mixing chamber
d = grain size of the filter media

G = average velocity gradient
Δ h = head less
K = Kozeny-Carman constant
L = filter bed depth
N = torque of the stirrer
n = number of compartment in the mixing basin
s = rotational speed of the stirrer
t = flocculation time
W = energy dissipation
1, 2 = subscripts denoting filter bed medium 1 and filter bed medium 2, respectively

REFERENCES

Dzięgielewski, B., A. M., Dziubek, and A. L. Kowal (1978). Stosowanie flokulantów do intensyfikacji procesu oczyszczania wody. Prace Naukowe Inst. Inż. Ochr. Srodow. P.Wr., 40, 17-28.

Kowal, A. L., and J. Maćkiewicz (1977). Sredni gradient prędkości w filtracji wody. Inż. Chem., VII, 3, 643-649.

Veytser, Y. I., and G. H. Lutsenko (1975). Optymalnyje usłowia obrazowania chłopiew pri koagulirowanii stocznych wod. Wodosnabżenie i Sanit. Techn. 9, 2-7.

TREATMENT OF WASTEWATER FROM HYDRAULIC TRANSPORT OF ASH IN POWER PLANTS

B. Dziegielewski, A. M. Dziubek and A. L. Kowal

Institute of Environment Protection Engineering, Wroclaw Technical University, Poland

ABSTRACT

The paper deals with the characteristics and neutralization methods of industrial wastewaters generated in hydraulic systems of ash transportation from brown coal power plants. The recarbonation with carbon dioxide and aeration process were applied to neutralization of these waters. It was found that during CO_2- recarbonation the largest quantities of calcium carbonate were precipitated within pH-range between 8.5 and 9.5. The prolonged aeration of the alkaline waters resulted in the pH-value decrease to the required level. Also the carbonate hardness of the water decreased due to precipitation of the calcium carbonate. The recarbonation by aeration did not reconvert the solid carbonates to bicarbonates, whereas recarbonation by carbon dioxide did.

KEYWORDS

Alkaline waters; neutralization; recarbonation; aeration; supernatant.

INTRODUCTION

Coal power plants which are equipped with hydraulic disposal and transportation of ash are often faced with difficulties in the supplementation of circulating water as well as releasing of surplus water back to the environment.

Ash-slurry usually contains 15 to 30 percent of suspended solids and is transported through a pipeline system into the exploited area where solid fraction of slurry is separated from liquid phase by sedimentation process. The supernatant from those sedimentation basins is pumped back to the ash handling plant.

Chemical composition of the fly-ash being caught on electrofilters depends on chemical characteristics of the coal used in the power plant. Fly-ash from Konin Power Plant contains about 35 % of CaO and only a small quantities of other metal oxides. In contrast, fly-ash from Turów Power Plant contains more than 30 percent of Al_2O_3 and only a small quantity of CaO (Kowal, 1968). Both plants use brown coal. Thus, the calcium oxide content in the ash depends on the carbon quality and ranges from several to 65 percent of CaO (Kumpf and co-workers, 1974). Because of permanent contact with ash, the chemical composition of supernatant water approaches to the saturated solution of calcium hydroxide. It is necessary to supplement periodically the recyrculating water with e.g. natural surface waters. It causes the plugging of the pipelines and ejecting devices due to deposition of calcium scale on surfaces which the water comes in contact with. The high pressure drops at the point where the ejecting nozzle is placed and increased temperature are additional factors resulting in acceleration of calcium carbonate precipitation. It is well known that calcium carbonate which is precipitated in water has a specific ability to deposite on solid elements. That phenomenon was adopted profitably in lime softening process with Virbos-type reactors. Granular dolomite, calcite or fine sand are used as a contact mass in that process. Plugging of the ejecting nozzles could be eliminated when carbonate hardness of supplemental water was considerably decreased, e.g. by addition of hydrochloric acid. The same chemical reactions take place in a river bed when surplus water from transportation system is mixed with natural river water. At a low flow of the river the water reaches an alkaline character.

We may assume that the chemical composition of the supernatant is constant. It depends on sedimentation effectiveness only.

TABLE 1 Characteristic of Supernatant from Sedimentation Basin

Parameters	Average Values
pH-value	11.6
Hardness, g/m^3 $CaCO_3$	625.0
Alkalinity, g/m^3 $CaCO_3$	350.0
TSS, g/m^3	150.0
Chlorides, g/m^3 Cl^-	73.0
Sulphates, g/m^3 SO_4^{2-}	247.0
TDS, g/m^3	1380.0
Sodium, g/m^3 Na	31.0
Potassium, g/m^3 K	6.0

Table 1 presents average physical-chemical composition of supernatant during one year period (Kowal and co-workers, 1978). It can be seen that only pH-value and carbonate hardness are increased. So, surplus water from the system must be neutralized before its disposal. Increased pH-value of the river water results in leaching of clay from the river bed, and what follows, water turbidity will increase. Particles of calcium carbonate aglomerate with clay particles forming strongly dispersed suspension. These suspended solids do not settle because of turbulent river flow and cause secondary pollution of water, which is harmful to fish and causes problems for users of river water. Direct neutralization of supernatant in the river needs large quantities of natural water and is intolerable with respect to water protection.

The paper deals with neutralization of alkaline water in physical-chemical processes including carbon dioxide recarbonation and prolonged aeration. The neutralization with mineral acids was not taken into consideration because it is not economically justified and maintainces high concentration of TDS.

RECARBONATION OF SUPERNATANT WITH CARBON DIOXIDE

Neutralization of strongly alkaline water in recarbonation process consists in carbon dioxide saturation. The recarbonation is a process which has been used for many years in water treatment for downward of pH following lime-soda softening. The hydroxides and carbonates - predominant forms in supernatant - react with carbon dioxide and are reconverted to bicarbonates. In the first stage calcium carbonate is formed and this reaction runs quantitatively towards calcium carbonate precipitation, due to low solubility of calcium carbonate. This reaction is the first stage of two-stage recarbonation process. Theoretically, this reaction is completed when pH is decreased to a value of 8.3. In the second stage - calcium carbonate is dissolved to bicarbonate with substantial decrease of the pH value.

The investigations dealing with supernatant water recarbonation with carbon dioxide were carried out in a batch process. It was found that pH range of minimum solubility of calcium carbonate is between 8.5 and 9.5 pH (Fig. 1). If the recarbonation process is stopped within that pH range and calcium carbonate removed, the carbonate hardness of water may be decreased up to 50 g/m^3 $CaCO_3$. The point of minimum solubility of calcium carbonate was reached after 3 minutes saturation with carbon dioxide and for the next 3 minutes solid calcium carbonate, previously formed, was dissolved again to bicarbonates in 80 percent. This is important to note, that the recarbonation's reactions in supernatant are not instantaneous, in spite of the fact, that the CO_2 gas may enter rapidly the water as dissolved CO_2.

In designing the reactors it has to be taken into account a considerable time required for recarbonation reactions to be completed and the pH to be lowered to the desired value. If detention time is too

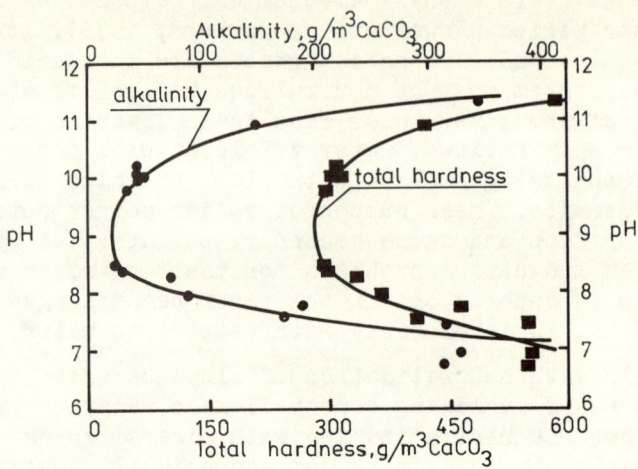

Fig. 1. Recarbonation of the supernatant from sedimentation basin with carbon dioxide.

short the reaction will proceede out of the reactor and the calcite scale will be formed on the surfaces of downstream units and piping. If two-stage recarbonation is used, sufficient detention time must be allowed to provide both reactions and settling in the first stage. Required dosage of carbon dioxide for alkaline supernatant water to be neutralized, generally ranges from 200 to 400 g/m^3 fitting the stoichiometric values. Recarbonation equipment consists usually of reaction basin for recarbonation and diffusion system to introduce the carbon dioxide to the water. The construction and technological parameters of reactor depend on how recarbonation process is designed. In two-stage recarbonation it is recommended to design combined reactor and settling basin assuming detention time at least 30 minutes with an overflow rate not more than 100 $m^3/m^2 d$ (Parker and co-workers, 1975). The equipment required to deliver the carbon dioxide to the reaction basins will depend on the source of carbon dioxide. Liquid CO_2 can be fed in either the gaseous or liquid form. For liquid-feeding, an equipment similar to solution-feed chlorinators may be used but it may be made from less resistant materials. If carbon dioxide is produced from fuel then underwater burners may be used. With underwater burners, air and natural gas are compressed and then burned at the point of application. For dispersing gaseous carbon dioxide into the water, an absorption system consisting of a grid of perforated PCV pipe is recommended. Gas transfer efficiencies of 85 percent or greater can be obtained with 5 mm diameter holes discharging 0.03 to 0.05 m^3 of gas per minute. The holes should be spaced along the pipe at least 7.6 cm apart, and the pipes should be spaced 0.46 m. The absorption system must be submerged 2.4 m in the water (Parker and co-workers, 1975).

RECARBONATION OF SUPERNATANT BY THE PROLONGED AERATION

The concentration of CO_2, which can be dissolved in water, depends on the partial pressure of CO_2 in gaseous phase. If stack gas is used, the equilibrium concentration of CO_2 in the water, which may be obtained, is lower as compared with pure CO_2, and what follows, it may be lower than the required dosage for hydroxides to be neutralized. Hence, it may pointed out, that partical pressure of CO_2 in gaseous phase and the time required for recarbonation are closely correlated.

In further investigations dealing with alkaline supernatant water from hydraulic transportation of ash we used an atmospheric air as a source of carbon dioxide. The air was fed by compressor to the chamber of 20 dm^3 volume equipped with a grid of perforated pipe. Aeration water was sampled and analyzed after 30 minutes of sedimentation. Figure 2 presents the changes in composition of aerated water in relation to the time of aeration. The carbonate hardness and pH-value were lowered to the desired level after 6 hours of aeration. The carbonate hardness of the water was decreased due to precipitation of calcium carbonate. The chemical reactions which take place in the course of prolonged aeration are analogous to the reactions in the first stage of two-stage recarbonation. However, when pH-value was lowered to the point of minimum solubility of calcium carbonate, further aeration did not cause solubilization of the carbonates to bicarbonates.

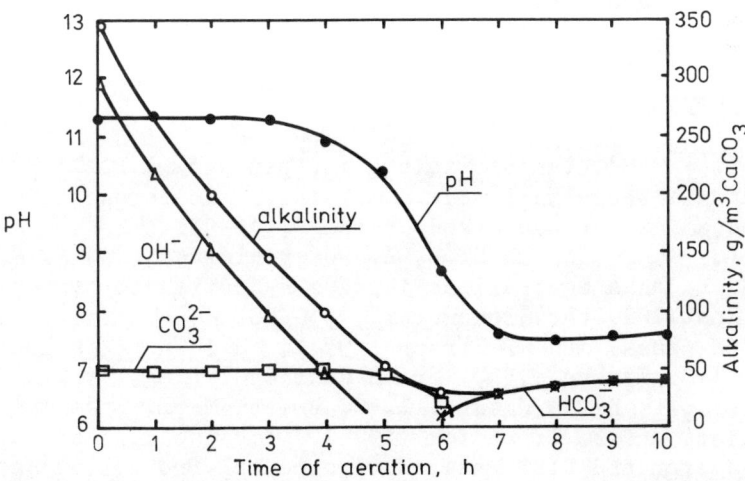

Fig. 2. Chemical composition of the supernatant versus time of aeration.

In spite of the fact, that particles of calcium carbonate were present in the water and pH was lowered to 8.3 and below, it was only insignificant increase of carbonate hardness of water. For that reason the control of recarbonation process is not so difficult like two-stage recarbonation with carbon dioxide. In addition, it was found, that the particles of calcium carbonate which are formed in the course of aeration are better floculated and settle much faster as compared with small, crystalic particles obtained in carbon dioxide recarbonation.

CONCLUSIONS

The supernatant water from hydraulic ash-disposal system contains significant concentration of dissolved solids including calcium hydroxide which causes its alkaline character. It is required for those waters to be treated before they are mixed with natural river water. The supernatant must be neutralized and partially desalined. The choice of neztralization method depends on awailable source of neutralization agent. The recarbonation with pure carbon dioxide or stack gases is usually used for that purpose. In this case, the recarbonation should be carried out in two steps rather the pH be brought down to about 7.0 in a single step. Two-stage recarbonation allows for carbonate hardness to be decreased.

The investigations concerning neutralization of supernatant indicated that atmospheric air might be effective agent to lowere pH to the desired value. Application of prolonged aeration is very clean technological method because no chemical means must be used. Use of atmospheric air to neutralization of alkaline waters requires further investigations to consider these problems from both theoretical and experimental viewpoints. At present, an investigation concerning water stabilization in prolonged aeration process and diffusion rate of carbon dioxide from air are carried on.

REFERENCES

Kowal, A. L. (1968). Zastosowanie glin, popiołów i żużli pylistych do wspomagania koagulacji wody. Materiały z konferencji "Koagulacja, filtracja i fluorowanie wody." PZITS, 46-53.
Kowal, A. L., M. Świderska-Bróż, B. Dzięgielewski, and A. M. Dziubek (1978). Badania neutralizacji wód z hydrotransportu popiołu. Raporty Inst.Inż.Ochr.Środow.PWr., 330/78.
Kumpf, W., K. Mass, and H. Straub (1974). Müll-und Abfallbeseitigung, teil III-Schlacken, 7530. Erich Schmidt Verlag, Berlin 1964-1977.
Parker, D. S., E. de la Fuente, L. O. Britt, M. L. Spealman, R. I. Stenquist, and F. I. Zadick (1975). Lime use in Wastewater Treatment: Design and Cost Data. Environmental Protection Agency. Brown and Caldwell, Walnut Creek, California. Report No. EPA-600/2-75-038. October, 1975, pp. 63-66.

ESTIMATION OF CONVENTIONAL METHODS FOR THE TREATMENT OF URBAN STORM WATER

M. Swiderska-Bróz

Institute of Environment Protection Engineering, Technical University, Wroclaw, Poland

ABSTRACT

In the paper the results of the research on pollutants removal from urban storm water are given. The following conventional processes were investigated: volume coagulation, sedimentation, coagulation in filter beds, filtration through sand-beds for raw and pretreated water. In volume coagulation alum and ferric salt were used. This process was carried out with pH adjustment. The influence of pollutants concentration in urban storm water and technological parameters on effectiveness the investigated processes was determined. Sedimentation and raw water filtration should not be as the only processes in urban storm water treatment. For urban storm water contained a lot of colloidal pollutants, coagulation was necessary. The following technological system appeared to be succesful in urban storm water treatment volume coagulation at the optimum pH, a 0,5 h sedimentation and filtration with the rate 5 m/h.

KEYWORDS

Urban storm water; volume coagulation; sedimentation; coagulation in filter beds; filtration through sand beds.

INTRODUCTION

The concentration of pollutants carried by urban storm water depends on the location, the kind of the catchmant area and its management. This concentration is also influenced by the frequency and intensity of both rainfall and snowfall. However, the basic factors that affect the concentration of interest are the geological structure, urbanization and industralization level, intensification of agriculture, air pollution intensity of the traffic, kind and method of cleansing. Chojnacki (1970) studied the coastal area and noticed that the rain waters sampled had contained chlorides concentrations four times as high as those contained in the water from other areas. Rain waters sampled in highly industrialized areas showed an increased pollution level which is due to the process gases emitted into the atmosphere as well as to dust emissions. Fertilization processes and chemical interventions in agriculture are also responsible for the increase of rain water pollution. The highest concentration of nitrogen and phosphorus compounds is recorded in the fertilization (spring and summer) season, whereas the highest biocides concentrations are measured in spring (Weibel, 1972). The physico - chemical composition of the urban storm water is also affected by the kind of the road surface. Szigarin (1966) found that run-off waters from asphalt roads carrying heavy traffic exhibited 1.1 - 1.3 times higher pollution levels than run-off waters from paved roads. It follows that traffic is the direct source of organic (BOD, COD) and inorganic pollution (heavy metals) (Brunner, 1977). Studies on heavy metals concentration in urban storm water were carried out by Roberts (1976). He determined the Pb, Cr, Zn, Cu and Cd levels 0.14 g/m^3, 0.003 g/m^3, 0.16 g/m^3, 0.015 g/m^3 and 0.001 g/m^3 respectively. Penkert (1974) determined the concentration of the some heavy metals in rainfall and snowfall. His results are listed in Table 1.

TABLE 1 Maximum Concentrations of the Heavy Metals in Rainfall and Snowfall (g/m^3)

	Mn	Fe	Cu	Zn	Cd	Pb
Rainfall	0.64	0.40	0.01	0.70	0.35	0.14
Snowfall	0.06	0.54	0.08	0.48	0.88	0.06

Lead, not completely oxidized hydrocarbons, nitrogen oxides, smoke - black and tar - pitch products contained in the process gases emitted to the atmosphere create a serious hazard to the environment (Rybicki, 1978). Run - off water and thaw water pollution is contributed by road cleansing and glaze prevention. The composition the rain water varies from season to season. The most pronounced fluctuations of the suspended solids concentration is observed during spring. The highest BOD_5 concentrations appears in late summer and autumn. The water in question also contains some the nondegradable organic pollutants. Heilmann (1976) recorded the hydrocarbon concentration in the rain water, which was 0.8 g/m^3 whereas the concentration of TOC ranged from 4.0 to 14.0 g/m^3 (Roberts, 1976). The concentration of aromatic polycyclic hydrocarbons in the sewage during storm and intensive rainfall was measured by Harrison (1975). He demonstrated that rain water carried large amounts of aromatic hydrocarbons. Golwer (1973) who had studied the concentration of pesticides in the rain water, found that maximum concentrations of pesticides are recorded in the spring months. The concentration of pollutants in run - off water varies with the time of discharge to the sewer. The first "wave" carries the highest pollution load. However if the intensity of precipitation is small (up to 8 dm^3/s ha) or if the run-off area is characterized by high concentration time then the pollution load will vary (Field, 1973 ; Osuch, 1972).

Nowadays, when water shortage becomes a serious problem, the studies on rain-water treatment raise more and more interest.

URBAN STORM WATER TREATMENT

Samples were taken at three sampling sites located in various quarters of the city of Wrocław, and were analysed for physical - chemical composition.

Physical - Chemical Composition of Urban Storm Water

The analyses showed that this water was characterized by a turbidity of up to 1100 g/m^3 and intensive opalescent coloured matter which was an indication that colloidal pollutants were present. TSS ranged from 28 g/m^3 to 1523 g/m^3. Permanganate value, COD and BOD_5 were 335 g O_2/m^3, 765 g O_2/m^3 and 260 g O_2/m^3 respectively. Total hardness fell below 3.57 eqv/m^3 and pH varied from 6.4 to 7.8. During the investigation period (15 June 1977 - 9 June 1978) the organic and inorganic pollution level underwent variations.

TECHNOLOGICAL STUDIES

Sedimentation, volume coagulation, coagulation in filter beds and filtration were employed to study the effectiveness of these processes in the removals of the individual pollutants. A technological system involving volume coagulation, sedimentation and filtration, as well as its efficiency in the treatment of urban storm water were also studied.

Sedimentation

Sedimentation was carried out in Imhoff canes. Two - hour sedimentation evidently decreased the pollution level. TTS, permanganate value, BOD_5 and COD were removed in 22.3 - 91.4 %, 6.9 - 33.5 %, 4.0 - 50.0 % and 5.0 - 31.9 % respectively. The efficiency of the sedimentation process was found to depend on the pollution level in the raw water. When the raw water was highly polluted, the efficiency of treatment process was low. After 2 hrs of sedimentation the following concentrations were recorded: TSS up to 394 g/m^3, permanganate value up to

268 g O_2/m^3, BOD_5 up to 250 g O_2/m^3 and COD up to 727 g O_2/m^3. As the pollution level in urban storm waters is generally high, sedimentation alone appears to be insufficient. It seems therefore advisable that this process be employed prior to other treatment procedures so as to decrease both the concentration of pollutants and the coagulant doses required.

Volume Coagulation

Alum and ferric sulphate were used as coagulants. As the water under treatment displayed a low alkalinity, volume coagulation was carried out with pH adjustment.

Efficiency of alum coagulation. At natural pH. The alum doses applied varied from 40 to 160 g/m^3. The effectiveness of the coagulant depended on the level of the raw water pollution. When the concentrations of pollutants were low, the coagulant doses ranged from 40 to 60 g/m^3 $Al_2(SO_4)_3 \cdot 18 H_2O$. With these doses turbidity was decreased down to 10 g/m^3, coloured matter down to 20 g Pt/m^3, permanganate value down to 4.0 g O_2/m^3 and TSS down to 60 g/m^3. When the raw water was characterized by a high concentration of organics, the coagulation process was less effective even with a dose of 150 g/m^3 $Al_2(SO_4)_3 \cdot 18 H_2O$. A 0.5 h sedementation did not improve the treatment effeciency either.

With pH adjustment. pH was changed with $Ca(OH)_2$, NaOH and HCl. Small variations of H^+ concentration have signifficently increased the effectiveness of the process and decreased the coagulant dose required. In Fig. 1, pH is plotted versus optimum alum and ferric sulphate doses. From Fig. 1 it is easily seen that pH adjustment decreased the coagulant doses needed. The removals of individual pollutants versus pH are presented in Fig. 2. As shown by this figure, an improved pH does not improve the effectiveness of alum coagulation.

Effects of ferric sulphate coagulation. At natural pH. Ferric sulphate doses varied up to 140 g/m^3. This coagulant was also effective when the concentrations of pollutants were low, and its efficiency decreased with increasing organic matter concentration. A comparison of results (alum and ferric salts versus turbidity, coloured matter and permanganate value removals) is given in Fig. 3. From Fig. 3 it is evident that either of the coagulants tested yielded similar turbidity removals. Permanganate value removals were higher when ferric sulphate was as coagulant, coloured matter removals were better when alum was employed.

With pH adjustment. The efficiency of the treatment process is pH dependent. The best effects were obtained with a pH of 4.5.

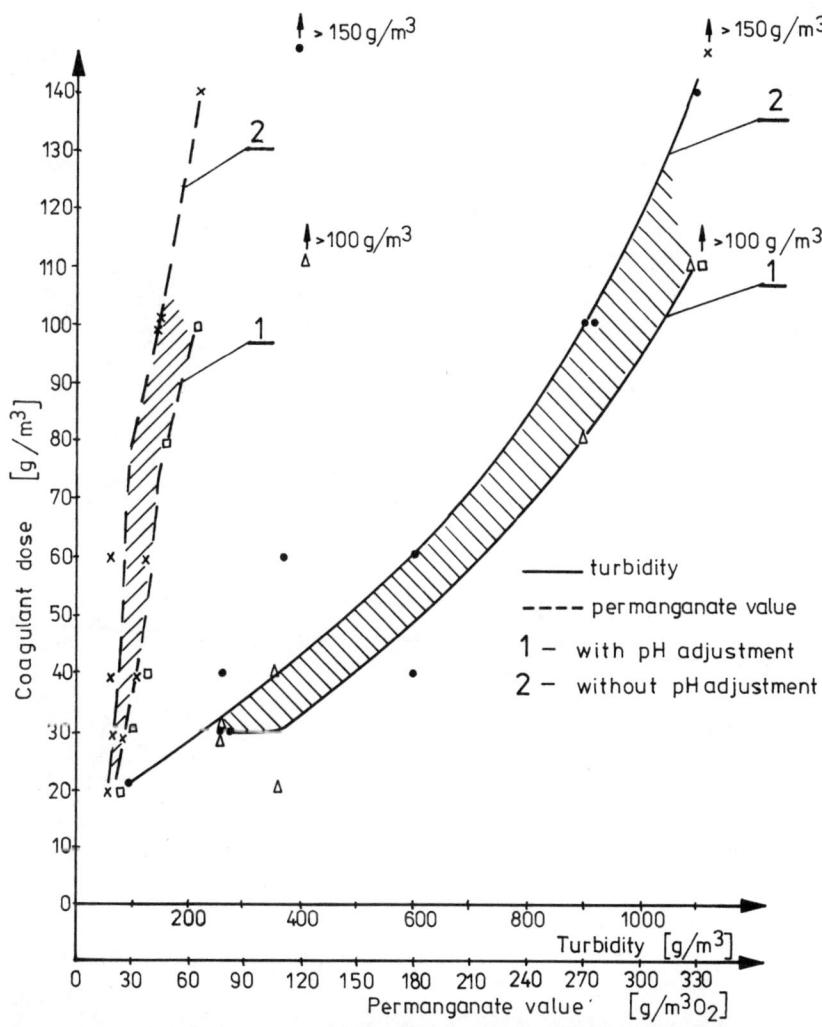

Fig. 1. Coagulant dose require versus pH.

A slight increase in OH⁻ concentration also improved the effectiveness of ferric sulphate coagulation. This improvement was not achieved in alum coagulation (Fig. 2).

Filtration and Coagulation in Filter Beds

Filtration and coagulation in filter beds were carried out using a system of two different model filters. The parameters of the sand beds were as follows: bed I; d_{10} = 0.6 mm, WR = 1.57, bed II; d = 0.75 - 1.02 mm, the depths of the beds were respectively, 0.7 m and 1.1 m. The filtration rate was 5 and 10 m/h. When coagulation was carried out in the filter beds, the coagulant dose was the same as the one used in volume coagulation.

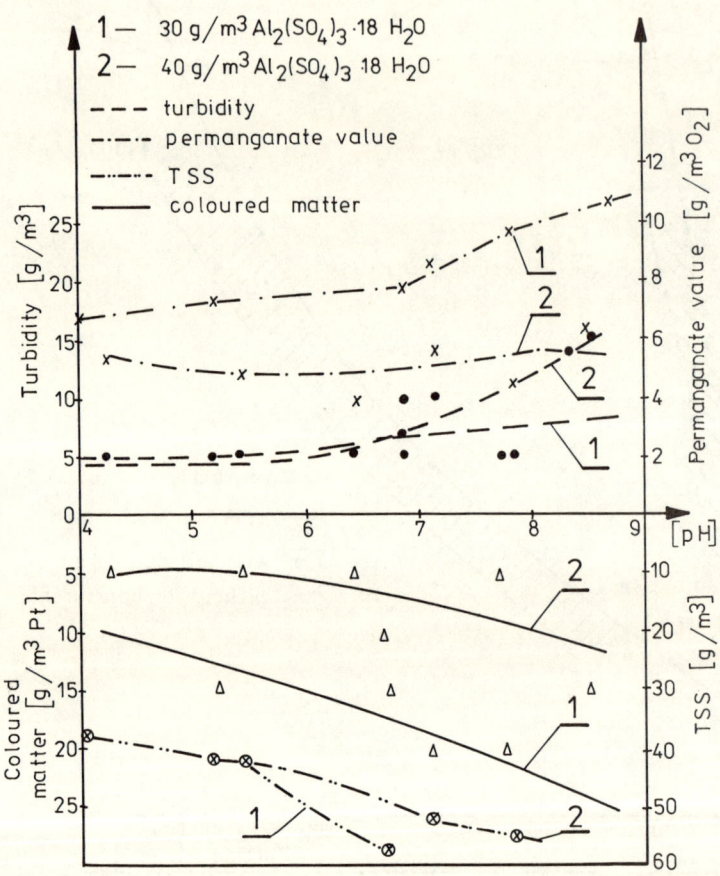

Fig. 2. Coloured matter, turbidity, permanganate value and TSS in the effluent versus pH.

<u>Raw water filtration.</u> Raw water filtration was found to be ineffective. The effluent showed high turbidity (up to 290 g/m^3), opalescent coloured matter (up to 50 g Pt/m^3), high permanganate value (up to 21.0 g O_2/m^3) and TTS up to 237 g/m^3. During filtration a rapid increment of head-loss was observed. A comparison shows that volume coagulation is evidently more effective even with low coagulant doses (10 - 20 g/m^3) than the filtration of raw urban storm waters.

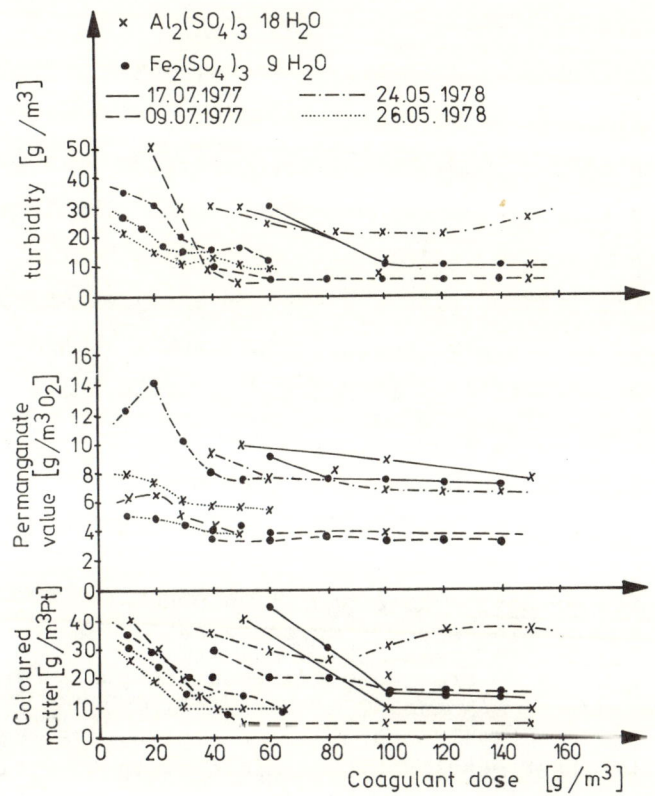

Fig. 3. Efficiency of aluminium and ferric sulphate in removing coloured matter, turbidity and permanganate value.

Filtration of urban storm water after pretreatment. Before entering the filter bed, the water is subjected to pretreatment. This is a combined process involving either 2 h sedimentation and filtration or alum volume coagulation wit pH adjustment, 0.5 h sedimentation and filtration. When a pretreatment conducted via the above route a considerale abatement of pollutant concentrations is achieved on one hand and on the other hand the filter run is extended. The water subjected to experiments was characterized by a high concentration of organics. The experiments were carried out in a technological system which involved presedimentation to decrease the pollution level, and filtration with the rate 5 m/h to further improve the quality of the water. In this way turbidity was removed by 75 %, coloured matter by 30 %, permanganate value by 80 % and TSS by 80 %. However, the effluent was still containing organic matter, which had the disadvantage of stabilizing the colloids present in the water. The pretreatment of the urban storm water under test has considerably improved both the conditions of filtration and the effectiveness of the treatment process. When the urban storm water was less polluted, the filter run lasted 10 h, and the purity of the effluent was high (turbidity 5 g/m^3, coloured matter 10 g Pt/m^3 and permanganate value 3.6 - 8.0 g O$_2$/m^3). When the urban storm water tested showed a high pollution level (opalescent coloured matter 100 g Pt/m^3, turbidity 400 g/m^3 and permanganate value 335 g O$_2$/m^3), the technological system involved volume coagulation (40 g/m^3 Al$_2$(SO$_4$)$_3$ · 18 H$_2$O) with pH adjustment, 0.5 h sedimentation and filtration with a rate of 5 m/h yielded the following results: coloured matter was decreased down to 40 g Pt/m^3, turbidity was removed to a level of 40 g/m^3 and permanganate value was reduced to 67 g O$_2$/m^3. From among other technological systems, the one employed in our experiments was found to be the most effective especcially in the treatment of urban storm water exhibiting high concentrations of organic matter.

<u>Coagulation in filter beds.</u> The coagulation process was carried out with aluminium and ferric sulphate as coagulants. The coagulant doses varied from 25 g/m^3 to 50 g/m^3, depending on the pollution level. The filtration rates were 5 m/h and 10 m/h. At a filtration rate of 5 m/h the coagulation process was found to be more effective than at a rate of 10 m/h. But the best results were achieved with an optimum coagulant (usually 50 g/m^3) and a filtration rate of 5 m/h. A comparison of the two coagulation processes - volume coagulation and coagulation in a filter bed - shows that with the same coagulant dose the latter yields higher removals of coloured matter and turbidity than the former. A serious disadvantage of filter bed coagulation is the rapid increment of the head loss. It is to be emphasized that coagulation in filter beds should be carried out with pH adjustment.

CONCLUSIONS

The effectiveness of the treatment process depends on the pollution load carried by the urban storm water to be treated.

Sedimentation alone is found to be insufficient, but can be successfully applied as a pretreatment step.

Because of the considerable amounts of colloids present in the storm water, the treatment process should involve coagulation.

Volume coagulation combined with pH adjustment improves the effectiveness of the treatment process. If alum is used as coagulant, pH is to be decreased; on the contrary, pH may be slightly increased, if ferric salts are used.

Filtration of raw water is found to be ineffective.

Filter-bed coagulation yields satisfactory removals of coloured matter and turbidity nevertheless, the rapid increment of the head loss makes the usefulness of this process in storm water treatment doubtful.

Based on the experimental results it is suggested to use multi-media filter beds.

The treatment process should also include sorption to improve the removal of organic compounds present carried in large amounts by urban storm waters.

Among the technological systems studied the following one is found to yield the best results: volume coagulation with pH adjustment, 0.5 h sedimentation and sand filtration (at a rate of 5 m/h).

The coagulation process is carried out with pH adjustment, as storm waters are characterized by a low alkalinity.

REFERENCES

Brunner, P.G. (1977). Strassen als Ursuchen der Werschmutzung von Regen-wasserabflüssen - Uberlick uber den Stand der torschung. Wasser-Wirtschaft, 4.
Chojnacki, A. (1970). Zawartość składników mineralnych w opadach atmosferycznych na tle warunków przyrodniczych i gospodarczych kraju. Polish Journal of Soc. Science, 3.
Field, R., Seeley, D. (1974). Urban runoff and combined sewer overflows. JWPCF, 6.
Field, P., and D. Weigel (1973). Urban runoff and combined sewer overflows, JWPCF, 6.
Golwer, A., and W. Schneider (1973). Belastung des Bodens des Unteridiesches Wassers durch Strassenverkehr. GWF, Wasser-Abwasser, 4.
Harrison, R., R. Perry, and R.A. Wellington (1975). Polynuclear aromatic hydrocarbons in raw potable and waste water. Water Research, 9.
Heilmann, H., M.Holeczek, and H.Zehle (1976). Organische Stoffe im Regenwasser. Vom Wasser.
Osuch, E., and R.Miłaszewski (1972). Charakterystyka zanieczyszczeń ścieków pochodzących od wód opadowych. GWTS,4.
Penkert, V. (1974). Untersuchungen des Einflusses atmospharischer-Verunveinigungen auf die Wasserbschaftenheit der Gewässer. Wasserwirtschaft und Wassertechnik, 5.
Roberts, P.V.(1976). Pollutant loadings in urban storm water. I.A.W. P.R., 7, Workshop, Wien.
Rybicki, S. (1978). O pewnych możliwościach zmniejszania zanieczysz-

czeń w obszarach stref ochronnych ujęć wody. Materiały Konferencyjne pt. Zagadnienia zaopatrzenia w wodę miast i wsi, Poznań, Poland.
Weibel, S.R. (1972). Characterization treatment of urban storm water. JWPCF, 7.

THE RECOVERY OF AMMONIUM NITRATE FROM FERTILISER FACTORY WASTES

L. D. Roland

Foster Wheeler Ltd., Reading, Berks., U.K.

KEY WORDS: Effluent Treatment, Continuous Ion Exchange, Ammonia, Ammonium Nitrate, Nitric Acid, Fertiliser Production.

ABSTRACT: The discharge of ammonia or nitrate in liquid effluents from a factory to non tidal waves is undesirable for several reasons. Numerous ways of recovering it have been considered but the most efficient and now widely used in fertiliser factories in the U.S.A. is a Chemical Separations Continuous Counter Current Ion Exchange loop using nitric acid and ammonia as regenerants.

This paper gives a brief history of early work in this field. It gives details of the basic ion exchange system to recover the ammonia and nitrate as better than a 16% ammonium nitrate solution and some of the alternatives and modifications that have since been made. The analysis of the liquors treated at ten such plants, an indication of costs and a cost comparison with the only possible practical alternative - distillation - are also included.

The problem of using concentrated nitric acid in an ion exchange system is also discussed.

INTRODUCTION

Nitrogenous fertilisers can cause water pollution problems both in manufacture and use. Although a prime one is ammonium nitrate other nitrogen containing fertilisers can ultimately produce the same effect.

Ammonia in natural waters will cause a reduction of the dissolved oxygen content with its subsequent effect on fish and plant life whilst excessive nitrate will upset the ecology of inland water ways and more important still be a potential danger if it is subsequently used as a source of potable water.

There is now considerable circumstantial evidence to suggest that if high nitrate water is drunk by pregnant women and children under six months there is a risk of methaemoglobinaemia (blue babies). It is also belived that it leads to an increased incidence of stomach cancer in adults

When fertiliser is applied to land - particularly where high levels are used - there is always the danger that it will leach out into rivers, streams and lakes raising the nitrate content temporarily to unacceptable levels or into underground supplies giving a more gradual but permanent increase.

Treatment of nitrate waters containing 10-50 mg/dm^{3}N as NO_3 to make them fit for human consumption is possible by either biological or ion exchange methods but is not the subject of this paper.

However, the waste waters from factories manufacturing ammonium nitrate and allied fertiliser can contain nitrogen present as ammonia or nitrate one or two magnitudes greater than this. Effluents with 0.5% ammonium nitrate plus considerable free ammonia are not unknown.

These liquors come essentially from the neutraliser overheads which are generally the most concentrated but are supplemented by general leakages, spillages and wash down. Where there is significant leakage of powder as for example prilling tower exhausts and this settles on nearby land and buildings then even rain water run off can be a significant source of polluted liquid effluent.

If the fertiliser factory is situated on a coast or tidal estuary then dumping of the untreated waste liquors can possibly be tolerated but where it is well inland and must discharge into a natural water course or even into a public sewer then nowadays treatment if not statutorily obligatory is almost a moral responsibility for the manufacturer.

Many inland fertiliser factories exist in the U.S.A. and in the late 1960's one such establishment decided to treat its waste waters. At that time there was no recognised method available to treat high nitrogen waste waters so that a research programme was instituted.

A first step was to investigate and then limit and segregate the various sources of contamination. This produced some improvement but ultimately an irreducible volume of effluent was reached which required further treatment. All known possible means were then considered and these included:

(1) Biological nitrification/denitrification
(2) Concentration by distillation
(3) Reverse Osmosis
(4) Precipitation of ammonia as magnesium ammonium phosphate
(5) Recovery by ion exchange

For various reasons the first four were eventually abandoned and ion exchange - utilising a form of demineralising - offered the most promising solution with distillation being the only practical alternative.

Discussions with resin manufacturers and equipment suppliers led to the conclusion that if regeneration with nitric acid and ammonia was utilised in a continuous counter current ion exchange (CCIX) system then the ammonium nitrate could be recovered in a strong solution for return to the product stream whilst the treated water would be available for re-use or suitable for discharge.

Now in 1968 the Simplot Chemical Company located at Brandon Manitoba Canada had installed a Chemical Separations CCIX plant to demineralise water for boiler feed. This comprised a strong group cation followed by a weak base anion unit. It treated a water containing 1200-1800 mg/dm^3 total dissolved solids and used nitric acid and ammonia as regenerants since these were both in plant chemicals.

With this background of experience Cham Seps were contacted by the company and carried out pilot plant work on a 45 cubic metre sample of actual waste water from the fertiliser factory and from this developed the basis flow sheet that has now been used on many plants in the U.S.A.

The Chem Seps System

In order to explain the working of a typical unit as for example when used as a softener a simplified flow sheet is shown in Fig. 1. Contaminants are removed from the feed liquid in the contacting section and are stripped from the resin in the regenerating section during the process cycle whilst the resin is moved around the loop in the pulse cycle.

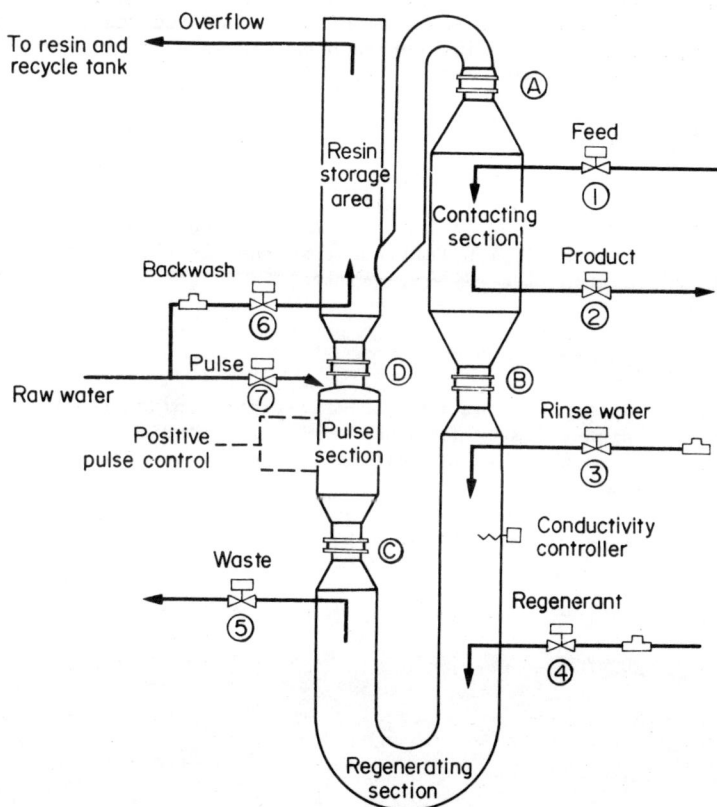

Fig. 1. Diagram of the Ion exchange operation.

In more detail in the process cycle feed (1) enters the contacting section through a set of distributors and passes downflow and counter current through the resin which is in the packed bed form. The product (2) leaves through a set of collectors.

In the process cycle valves A. B and C are closed thus isolating all the sections. At the same time a predetermined quantity of regenerant enters through (4) and leaves through valve (3) and its duration is controlled by the interface conductivity controller.

Valve D is also open to begin with and this permits back washed resin from the resin storage area to drop into the pulse section. When it is refilled to the upper pulse control valve D is closed.

The duration of the process cycle can vary from 5 minutes upwards depending on the level of contaminants to be removed and is controlled either by a timer, if the contaminants are fixed, or a controller in the product stream if they fluctuate.

At the end of the process cycle all valves on lines going into and out of the unit are automatically closed as is valve D if it is open. Valves A, B and C are then opened and a slug of water enters the pulse section through valve (7). This moves a predetermined volume of exhausted resin into the regenerating section which in turn displaces an equal volume of regenerated and rinsed resin into the bottom of the contacting section whilst an equal amount of exhausted resin leaves the top of this section and goes to the resin storage area for subsequent backwashing.

The resin movement is plug flow with Reynolds' number being in the laminar region of the curve and no fluidisation occurs during the pulse cycle.

The pulse cycle is ended when the resin level in the pulse section reaches the lower pulse control. The pulse valve (7) closes followed by valves, A, B and C and the plant reverts to the process cycle. All valves are sequenced and interlocked, whilst the entire plant is design fail safe, so that in the event of a malfunction it immediately shuts down with all valves closed.

Ammonium Nitrate Recovery

The original flow sheet developed from the first pilot plant tests is shown in Fig. 2. Here contaminated water from the collection pond is pumped through sand filters and then the cation unit containing a strong group resin in the hydrogen form.

Here any free ammonia will be absorbed and, any neutral salts converted to their equivalent mineral acid.

$$R1H + NH_3 \longrightarrow R1NH_4$$

$$R1H + NH_4NO_3 \longrightarrow R1NH_4 + HNO_3$$

$$R1H + NaCl \longrightarrow R1Na + HCl$$

R1 = Cation Resin

Thus the liquor leaving the cation unit will be a dilute acid equivalent to the anions present in the original effluent.

When the exhausted resin is treated with 22% nitric acid it is converted back to the hydrogen form whilst the spent regenerant which forms the product consists of nitrates plus excess nitric acid.

$$R1NH_4 + HNO_3 \longrightarrow RH + NH_4NO_3$$

$$R1Na + HNO_3 \longrightarrow RH + NaNO_3$$

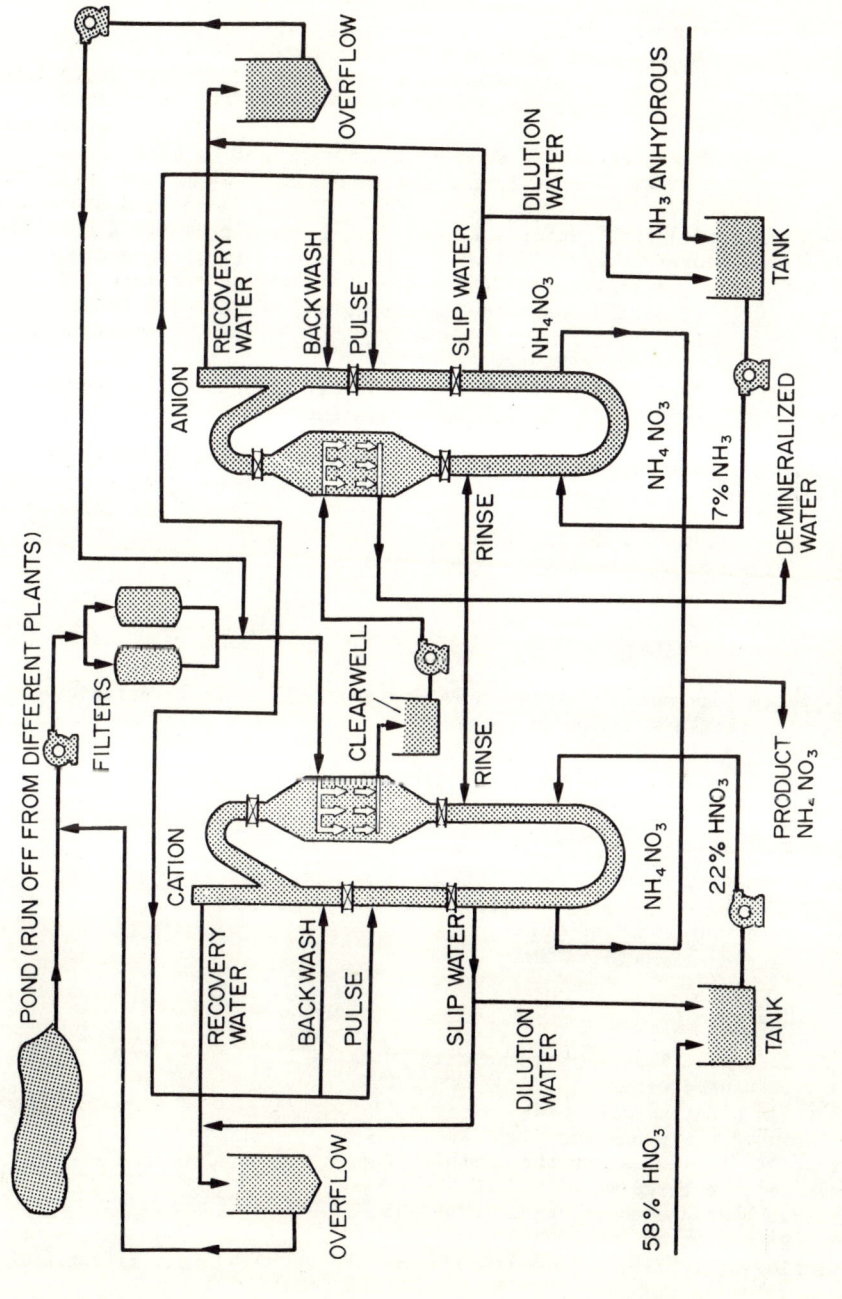

Fig. 2

It will be noticed that a slip water outlet line is included below the pulse section. Immediately after pulsing the regenerant/water interface below the pulse section will have been depressed so that if the product valve opens at the beginning of the process cycle water will flow out thus diluting the product.

Instead a valve on the slip line opens. Eventually the regenerant/water interface rises above the product outlet line and will be detected by a conductivity probe (not shown) situated between the product and slip lines. The slip valve then closes and full strength product leaves at the lower outlet.

The slip water which may contain traces of regenerant can be used to make up more regenerant or else recycled back via the overflow tank.

It will be appreciated that because of the counter current flows in the regeneration section fully regenerated resin is last in contact with fresh regenerant thus ensuring the highest utilisation of the resin whilst spent regenerant is last in contact with fully exhausted resin thus ensuring maximum utilisation of the regenerant.

The decationised water is pumped from the clearwell through the anion unit which contains weak base anion resin in the free base form.

Here the mineral acids it contains are absorbed by the resin so that the water leaving the anion unit is essentially demineralised.

$$R2 + HNO_3 \longrightarrow R2\ HNO_3$$

$$R2 + HCl \longrightarrow R2\ HCl$$

$$R2 = \text{Anion Resin}$$

In the regeneration section the resin is stripped with a 7% solution of ammonia and the product is then a solution of ammonium nitrate in the presence of excess ammonia.

$$R2\ HNO_3 + NH_4OH \longrightarrow R2 + NH_4NO_3 + H_2O$$

$$R2\ HCl + NH_4OH \longrightarrow R2 + NH_4Cl + H_2O$$

If the product streams from the cation and anion loop are combined there will generally be a slight excess of nitric acid. If so desired this can be neutralised with additional anhydrous ammonia.

Closed Circuit

It will be seen from Fig. 2 that the only incoming flows are
    a) Untreated water
    b) 57% plant nitric acid
    c) Anhydrous plant ammonia
whilst the only flows leaving the circuit are:
    a) demineralised water
    b) recovered ammonium nitrate stream.

Any other flows occurring within the circuit are recycled thus maintaining a closed loop.

"The original plant during the course of erection at Chattanooga"

Later Developments

This basic flow sheet was used on the first two plants installed for C.F. Industries Tennessee and North Carolina but as developments took place subsequent plants included various modifications or alternatives. These include:

a. Single Loop Plants

Most nitrate fertiliser factories make their own nitric acid on site and require a supply of water to the absorption towers. Now if the waste liquor is essentially ammonia and ammonium nitrate and does not contain chlorides or sulphates then after passing through the cation unit it is converted to a dilute solution of nitric acid. This can be used directly in the nitric acid plant and so eliminate the need for further treatment. Thus the ion exchange plant is cheaper and produces a smaller volume of recovered ammonium nitrate for return to the main circuit.

Such a plant as this has been in operation at Illinois Nitrogen since 1974.

b. Regeneration Recycle

With the original flow sheet it was found that the flow of resin through the regeneration section was greater than the flow of regenerant. This resulted in diffusion of pulse water into the product - despite the provision of slip. It was found advantageous to increase the flow of regenerant in the last part of the regenerant loop by recirculation of spent regenerant back just below the product take off. This resulted in a 1-2% increase in product strength.

c. Weak Group Resin Loop

If the untreated liquor contains large quantities of free ammonia then it is probably worth while considering a third loop containing a weak group acid resin ahead of the other two. This is because it will remove the free ammonia more efficiently than a strong group resin which will lead to a smaller volume of recovered product.

Whether or not it is decided to utilise the third loop will depend on such things as the additional capital cost and complexity with the third loop, the volume and strength of the liquor to be treated, and the utilisation of the product.

Nitric Acid

Cation regeneration systems must be designed carefully to prevent the possibility of extra strong nitric acid coming in contact with the resin. For this reason continuous blending of the strong (57-60%) plant acid with water to the required 22% which will shut down automatically on dilution water failure is considered preferable to batch make up of regenerant so common in normal ion exchange practice. Stronger acid gives no additional chemical advantage but merely increases the rate of deterioration of the cation resin. Equally it is advisable to limit the temperature of the diluted acid to less than $40°C$ in order to minimise resin breakdown by thermal shock.

It is now well known that even moderately dilute nitric acid - one or two percent - should not come in contact with an anion resin due to the possibility of an uncontrolled reaction or even an explosion. For this reason therefore the treated waste liquid cannot be polished through a mixed bed regenerated by nitric acid - only two bed systems are considered safe.

The Chem Seps system utilizes 22% nitric acid as regenerant for the cation unit. This is best prepared on a continuous basis with automatic shut down in the event

Revised Flow Sheet

Some of the later developments in this flow sheet have been incorporated into version shown in Fig. 3.

Recovered Ammonium Nitrate

Since the ion exchange plant acts as a demineraliser most trash ions as well as the ammonium nitrate in the waste liquid will appear in the concentrated product stream. Consequently thought has to be given as to where this is returned. Should there be any possibility of chlorides being present then it must not go back into a main ammonium nitrate plant stream that is ultimately prilled. In this case it is better disposed of into a liquid or mixed fertiliser.

However, where it is a pure stream then prilling presents no problems, and at least two companies dispose of their recovered liquid this way.

One further cause for concern is the possible pick up of organic matter through the ion exchange plant.

Here tests have shown no detectable increases of the carbon content after passage through an installation but nevertheless it is advisable to install fine screens in the product stream to guard against the possibility of broken resin contaminating the recovered ammonium nitrate.

Analysis and Results

Table 1 gives details of the waste water being treated through ten of the nitrogen recovery systems.

It will be noted that there is a very considerable variation in the strength and volume of the liquid being treated. Generally the smaller volume plants have stronger waste waters. This is an indication of the segregation that has been possible at the various factories. In one case for example the contaminated rain water treated by the plant represents about 20% by volume of its throughput.

An ion exchange system is a percentage remover of total dissolved solids. On a once through basis better than a 97% removal of nitrogen is to be expected. For example on a plant containing 1000 mg/dm^3 N as NH_3 and 400 mg/dm^3 N as NO_3 the total nitrogen in the treated effluent is in the 10-15 mg/dm^3 range, whilst on a 23 day test at Illinois Nitrogen the mean influent was 1700 mg/dm^3 NH_3 and the mean effluent was 9 mg/dm^3 NH_3, which is better than a 99% removal. A recovered neutralised product of at least 16% is to be expected but this can be influenced by many factors so that it is not unusual to get strengths in excess of 20%.

Comparison with Distillation

As mentioned at the beginning of this paper the only possible practical alternative to the ion exchange route is distillation.

A prime candidate for using distillation was the plant whose analysis is given in column F of Table 1 since this had a total dissolved solids of more than 1%. The

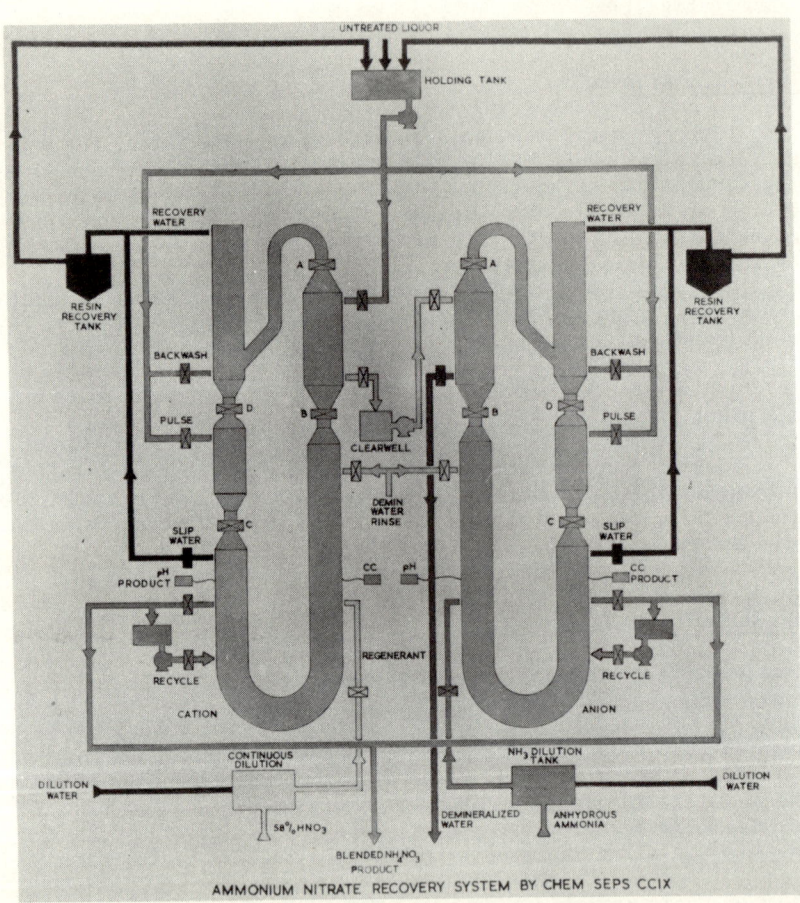

Fig. 3

total weight of liquid to be treated daily was 219 tonnes containing 2.38 tonnes of ammonium nitrate.

To obtain a 25% solution of ammonium nitrate, the same concentration as obtained by the Chem Seps CCIX system, would have required a triple effect evaporator using 126 tonnes per day of steam or 80,000 Kwh. At a cost of 0.68c per Kwh the total annual cost would be $201.480.

Even so to attain the low levels of 3-10 mg/dm^3 required in the treated water an ion exchange polishing unit would still have been required. In other words, two unit operations instead of one.

The cost of operating the CCIX system, including electric power, resin replacement and utilities would be less than $20,000 per annum. It is assumed that labour, maintenance and amortization and all other costs involved would be essentially equal in both cases. The chemicals are also not considered in the operating cost of the ion exchange since these are returned back to the product.

It will be noticed that plant F has the highest dissolved solids of the plants quoted so that comparison of this type for the other cases would have been even more favourable for ion exchange. This is because essentially operating costs for distillation are directly proportional to weight of water evaporated whilst for ion exchange they are proportional to the weight of solids removed.

Operating Costs

There are two possible ways of considering this:

a. The cost per tonne of the ammonium nitrate recovered

b. The cost increase per tonne of ammonium nitrate produced.

Now the Chem Seps process is a pollution control system and not a fertiliser production plant. Without it the factories would not be allowed to operate so that the real economic significance of the process is what cost does it add to the business of making fertiliser just the same as services, depreciation or other associated running costs.

In order to provide actual costs for the system it is necessary to depend on users and the following cost figures are supplied entirely by them. The interpretation, the mathematics and the values of the ammonium nitrate and water have been added. The figures given for both the chemical and the water covered a wide range but those used are a reasonable concensus and the economic analysis can easily be adjusted by whatever local prices and costs happen to be.

Table 2 lists operating costs as provided be C.F. Industries personnel for a one year period ending June 30 1975 at the Chattanooga complex and similar annual figures from the North Carolina complex based on actual operating costs over the nine months period between July 1 1973 to March 31 1974 as published by Van Moorsel. The greatest cost difference in the two plants is under "utilities" which includes an extremely high figure for evaporation costs at Tunis. All other cost categories seem reasonably comparable. The different nature of the two plants' operations is shown under credits where the Tunis complex recovers twice as much ammonium nitrate reflecting a much higher concentration feed, and the Chattanooga complex recovers twice as much water reflecting a more dilute waste water. This difference is also reflected in the cleanup costs per ton produced and the cost of cleanup per 1000 gallons even though the Tunis complex produces almost twice the annual volume of ammonium nitrate.

In brief, this cost data shows a net cost of £131,000 to operate the system at Chattanooga and £325,650 to operate at the Tunis complex after credits are taken for recovered ammonium nitrate and water. This represents a per ton cost ot 54¢ and 71¢ per ton of ammonium nitrate produced and represents in both cases approximately 0.5% of the value of a ton of ammonium nitrate at £125 per ton. The relative concentration of the waste of both complexes results in the cost of treatment per thousand gallons being almost 5 times greater at Tunis where concentration is excessive.

In summary, after taking credits for product and water recovered, the Chem Seps nitrogen recovery process costs 50-70¢ per ton of ammonium nitrate produced at these two complexes or adds about 0.5% to the sales price of the ammoniun nitrate. Is this to be considered as reasonable or outrageous for pollution abatement?

TABLE 1.

FLOW AND ANALYSIS OF WASTE WATERS

CHEM-SEPS ION EXCHANGE NITROGEN RECOVERY SYSTEMS

CONSTITUENT	Analysis mg/dm³ as	PLANT A	PLANT B	PLANT C	PLANT D	PLANT E	PLANT F	PLANT G	PLANT H	PLANT I	PLANT J
CATIONS											
Calcium	Ca^{++}	60	60	–	1	80	–	–	81	0.671	6
Magnesium	Mg^{++}	4.8	4.8	–	–	48	–	–	11	1	3
Sodium	Na	0	0	–	20	130	–	–	22	154	2
Potassium	K	–	–	–	–	–	–	–	1500	–	–
Ammonia	NH_3	340	300.800	1100–2500	563	600	10212	2425	–	–	899
TOTAL CATIONS	$CaCO_3$	1170	1170–1600	3200–7400	1700	2448	30,000	7150	2200	2029	2674
ANIONS											
Carbonate	CO_3	–	–	–	–	–	–	–	–	10	–
Bicarbonate	HCO_3	–	–	–	–	–	–	–	–	–	–
Hydroxide	OH	–	–	–	608	–	10,000	2000	–	164	293
Sulfate	SO_4	72	72	–	3	1300	–	–	75	2	21
Chloride	Cl	53	53	–	1	55	–	–	14	202	5
Nitrate	NO_3	1240	1200–1600	2930	1315	458	775	1550	2392	1493	2214
TOTAL ANIONS	$CaCO_3$	1170	1170	2363	1703	1800	30,000	7130	2227	2029	2674
Total Hardness	$CaCO_3$	170	170	–	215	400	–	–	248	1693	27
Silica	SiO_2	12–50	12–20	–	–	–	–	–	–	–	–
TOTAL DISSOLVED SOLIDS		1770	1700–2500	4000–5500	2537	2671	11,000	3975	4100	2777	3470
pH		–7	–7	7–10	7–11	7	7–11	7	7	10–11	8–11
Flow m³/h		159	159	9	227	159	9	227	20	70	141

TABLE 2

ECONOMIC ANALYSIS

CHEM-SEPS NITROGEN RECOVERY ANALYSIS

	July 1, 1974– June 30, 1975 CF CHATTANOOGA	July 1, 1973– June 30, 1974 CF TUNIS
ANNUAL OPERATING COSTS		
Operating Labour & Payroll Costs	$ 98,280	$ 86,400
Maintenance-Labour & Materials	65,520	84,000
Operating Supplies-Chiefly Resin	65,520	52,800
Lease Payments	56,160	51,000
Engineering & Laboratory	23,400	26,000
Utilities (Incl. Evaporative Heat)	47,120	259,700
Overhead (Incl. Supervision & Depreciation	122,000	104.000
TOTAL COSTS PER YEAR	$473,000	$663,900
RECOVERY CREDITS		
Tons per year NH_4NO_3 Recovered $ Value per year at $125/Ton	870 Tons $109,000	1650 Tons $206,000
Million Gallons Water Processed $ Value per year at $1/1000 gal	233 MGY £233,000	132 MGY $132,000
TOTAL CREDITS PER YEAR	$342,000	$338,250
TOTAL COSTS PER YEAR LESS TOTAL CREDITS PER YEAR	$473,000 342,000	$663,900 338,250
NET COSTS PER YEAR	$131,000	$325,650
COST RELATED TO FULL PLANT PRODUCTION		
TONS PER YEAR AN PRODUCED	240,000	456,000
Net Cost Chem-Seps Process/Ton AN	$0.54/Ton	$0.71/Ton
Net Cost as % of AN at £125/Ton AN	0.4%	0.6%
Net Cost Chem-Seps Process/1000 gas Treated	$0.56/1000 gal	$2.47/1000 gal

EXPERIENCES WITH PRIMARY WATER PURIFICATION AND WASTE WATER TREATMENT PLANTS IN NUCLEAR POWER PLANTS WITH PRESSURIZED WATER AND BOILING WATER
New Development in this field

N. Buser and J. P. Ghysels*

Theodor Christ AG, Aesch, Switzerland
*Kernkraftwerk Mühleberg, Switzerland

ABSTRACT

Powdered resin precoat filtration with the aid of structured filter media permits the adaption of water treating techniques to even the most demanding problems in nuclear power plants, particularly in the field of primary water and waste water treatment.

With the aid of available actual experiences, the authors demonstrate the advantages of this technique over other techniques, taking into account the various objectives that are pertinent for the operation of nuclear power plants.

KEYWORDS

Nuclear power plants; condensate purification; primary water purification; reactor water purification; personnel irradiation; precoat filter; powdered resin precoat filtration; liquid waste discharges; boiling water reactors;

Next to condensate purification, primary water purification (reactor water purification) and waste water purification are the most important water treatment processes in nuclear power plants.

The suitability of various processes for use in these areas is determined by:

- personnel irradiation
- liquid discharge volume
- purity of the reactor water and thus formation of deposits on the fuel elements, production of activation products and transparency of the reactor water.

These are important criteria in determining the feasibility and economy of energy conversion through nuclear engineering. The fulfilment of these water treatment tasks can and must be measured by available empirical values of these criteria.

Figure 1 shows how the water treatment processes, essentially coined by the operation of the plant, influence these criteria.

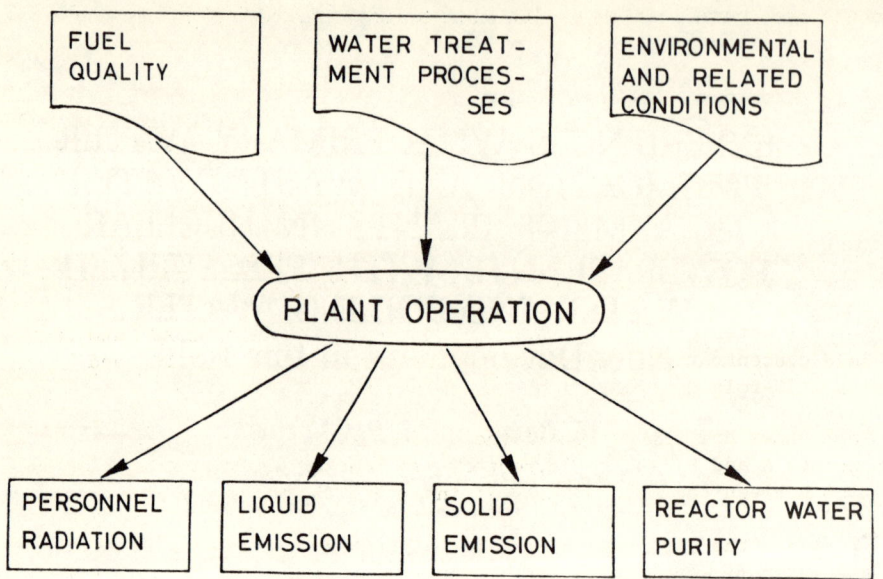

FIG. 1: CONDITIONS AND OBJECTIVES FOR THE MANAGEMENT IN NUCLEAR POWER PLANTS

It further shows that these criteria are also dependent on fuel quality and environmental and related conditions.

The fields of application of condensate purification and reactor water purification can be graphically presented in the reference field TDS - TSS, i.e. the actual ratio of dissolved and undissolved substances in the water before and after these purification stages.

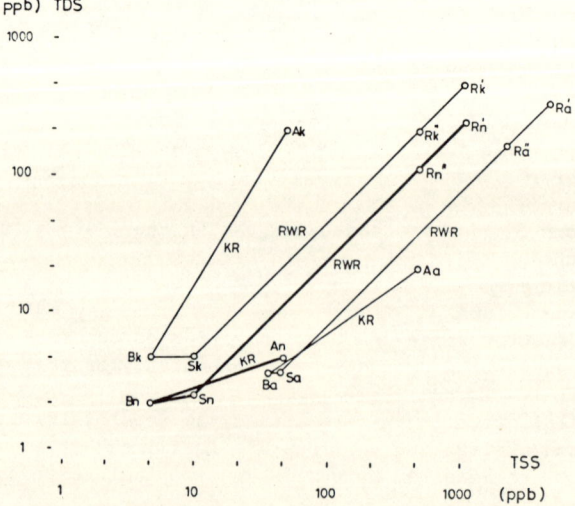

FIG. 2: CONCENTRATION CURVES IN THE CONDENSATE PURIFICATION AND REACTOR WATER PURIFICATION IN BWR NUCLEAR POWER PLANTS

The condensate purification plants are required to treat raw condensate with these typical compositions:

A_n for normal operation

A_k for coolant water leakage into the condensate

A_a for start-up operation

Pure condensate compositions B_n, B_k and B_a must be produced wich through absorption of corrosion products change to the feed water compositions S_n, S_k and S_a.

With reactor water purifying plants which treat 1 % of the feed water quantity (100-fold concentration) the reactor water compositions R'_n, R'_k and R'_a result, and with a 50-fold concentration, the values R''_n, R''_k and R''_a.

The graph shows how the action of the condensate purification plant (sections between points A and B) and the effect of the reactor water purification system, (section between the points S and R) influence the purity of the reactor water.

In both plants the amount of dissolved and undissolved substances that is removed from the water (i.e. the products of the treated quantity of water times the concentration differences before and after the plants) determine the reactor water quality.

Since in the condensate purification plant already the maximum possible quantity of water, i.e. the entire feed water quantity, is treated, the solution of the objective of "condensate purification" can be improved only by raising the pure condensate quality. The illustration shows both qualitatively and quantitatively how the reactor water quality improves with a rising pure condensate quality.

The proportionality factor corresponds to the degree of concentration, i.e. to the ratio between the solids in the feed water and the solids completely removed in the reactor water purification system. An incomplete reactor water purification can be compensated by an increase of the mass flow. The same effect is achieved, if from 10 t/h all contaminations are removed or if from 20 t/h half of all contaminations are removed.

The essential difference between the demands to be made on "condensate purification" and "reactor water purification" consists in that the condensate purification plant operates at a contamination level that lies about 10...100 times below the operating concentrations of the reactor water purification, and in that these concentrations lie in the order of magnitude of the values optimally achievable with all known water purification processes.

It follows that:

1. the demands to be made on the condensate purification plant are considerably higher than those made on the other water treatment plants in a nuclear power plant, particularly the waste water plants.

2. the required degree of efficiency of condensate purification depends on what the other water treatment plants, viz. primary water purification and waste water purification, can achieve.

Evaluation of the experiences gained in operation shows the advisability of distinguishing between the following two water treatment principles as used in nuclear power plant operation:

Principle A: Precoat filter for suspended solids and bead resin mixed-beds for a dissolved solids purification, used together or individually.

Principle B: Powdered resin precoat filtration or special structured filter layers with suspended solids and dissolved solids removal capability.

A report of the American Nuclear Regulatory Commission of June 1976 (1) lists the following personnel radiation levels for several comparable reactor plants:

	Years	MW . yr	Man-Rem	$\frac{\text{Man-Rem}}{\text{MW . yr}}$
Personnel radiation in power plants with water treatment plants according to principle A:				
Pilgrim	73...75	1026	1233	...
Millstone Point	72...75	1498	4668	...
Total		2524	5901	...
Average				2.3
	Years	MW . year	Man-Rem	$\frac{\text{Man-Rem}}{\text{MW . yr}}$
Personnel radiation in power plants with water treatment plants according to principle B:				
Monticello	72...75	1508	1917	...
Quand Cities 1+2	73...75	2001	2078	...
Vermont Yankee	73...75	954	440	...
Total		4463	4435	...
Average				1.0

TABLE 1 - Personnel irradiation in nuclear power plants

Furthermore, the Mühleberg nuclear power plant attained in the last two operating years irradiation levels of only 1 man-rem per MW year. Of this, about half is from shut-down periods, particularly cobalt (Co-60) radiations on surfaces outside the core playing a role. The second half was largely caused by nitrogen (N-16-radiations), viz. during routinge controls or small-scale repairs in the turbine house. But precisely this quantity depends to a considerable degree on the totality of the objectives in the plant and therefore will vary considerably from one power plant to another as within different operating periods of the same power plant.

Dissolved and undissolved iron, mainly from the steam generation, are incidentally the main reason for the irradiation of the personnel - particularly at inspections - and this should probably be the main cause for the lower personnel irradiation in powdered resin condensate purification plants, since such contaminations are more effectively removed from the condensate with powdered resin purification systems.

When per MW year the personnel irradiation is 1.3 man-rem lower in American power plants with systems that use principle B, then according to a study of the Canadian Atomic Energy Commission (2) which estimates the total costs of irradiation damage of 1 man-rem to be $ 9 million/year would result for a 1000 MW power plant.

It may be expected that solid and liquid waste discharge volumes, with otherwise equal working conditions and equal management, will be the smaller, the less water-soluble substances are introduced into the plant. For the water treatment plants operating by principle B neither acid nor base, but practically only fully regenerated ion-exchange resings with large absorption capacity for active substances and inert waterinsoluble filter aids are brought into this balance sheet.

Table 2 shows the values for the liquid waste discharges of comparable power plants taken form a report of the U.S. Agency (3). It is worth noting that here too conspicuously low values for the power plants with water treatment plants operating according to principle B are registered.

Power Plants with water treatment systems according to principle A

Power Plant	Output (MWe)	1972	1973	1974	1975	1976
Dresden 1	215	6.76	9.24	6.89	0.84	0.353
Big Rock	75	1.09	2.65	1.07	2.02	0.77
Humboldt Bay 3	65	1.40	2.37	4.4	3.47	1.08
La Crosse	55	48.5	35.9	13.1	14.1	5.68
Oyster Creek	670	10.0	4.15	0.66	0.41	0.2
Nine Mile Point 1	610	34.6	40.8	25.7	21.0	2.14
Millstone 1	690	51.6	33.4	198.0	199.0	9.65
Pilgrim 1	685	1.45	0.91	4.22	2.06	2.34
Dresden 2,3	1666	22.0	25.9	33.2	0.81	1.21
Fitzpatrick 2	821	-	-	0.01	9.39	6.01
Curies/GWe		37.50	32.83	51.74	49.59	5.30

Power plants with water treatment systems according to principle B

Power Plant	Output (MWe)	1972	1973	1974	1975	1976
Monticello	580	<0.01	0	0	0	0
Vermont Yankee	540	0	<0.01	0	<0.01	<0.01
Cooper Nuclear	801	-	-	1.42	1.73	0.65
Duane Arnold	565	-	-	<0.01	0.01	<0.01
Hatch - 1	813	-	-	<0.01	0.03	0.04
Quad Cities 1,2	1666	2.41	21.4	38.7	17.1	6.99
Browns Ferry 1,2,3	3294	-	-	0.72(2)	2.79(2)	<4.0
Peach Bottom 1,3	2196	-	-	0.95	0.93	2.3
Brunswick 1,2	1698	-	-	-	1.92(1)	3.28
Curies/GWe		0.87	7.68	4.47	2.40	1.46

Table 2. Total activity (excluding tritium) in the waste water

The literature (4) lists for boiling-water reactors with powdered resin condensate purification systems, solid, concrete embedded wastes of 206 m3 per GWe and year and for those with mixed-bed filter systems 650 m3 per GWe and year. In the Mühleberg nuclear power plant, 58 m3 solid, not concrete embedded material with 60 to 70 % water per GWe are emitted annually.

The differing results of the various water purification processes can be visualized in a coordinate system in which the total content of a water in dissolved and undissolved substances relative to their particle size is plotted.

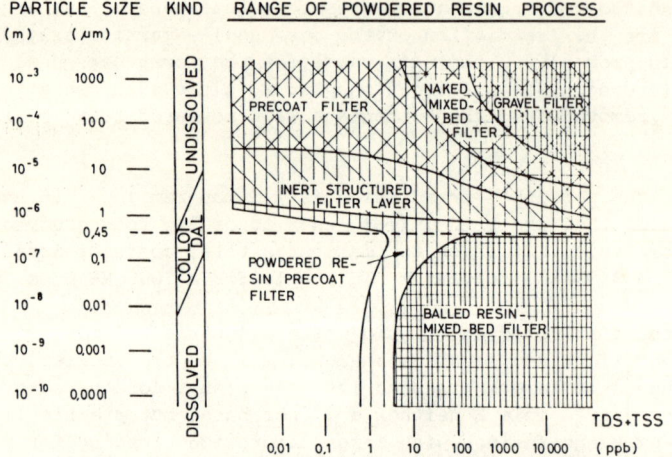

FIG. 3: THE OVERLAPPING RANGES OF WATER TREATMENT PROCESSES ARE COVERED BY THE POWDERED RESIN PRECOAT CAPABILITY

The ranges within which the various water purifying processes operate with an effectiveness of e.g. 90 % are given in this qualitative manner of presentation. The techniques employed in nuclear power plants are illustrated in Figure 3. The historical developement of condensate purification systems proceeded in the following way:

- Gravel filters and bead resin mixed-beds
- Precoat filter and bead resin mixed-beds
- Naked bead resin mixed-beds
- Powdered resin precoat filters

The powdered resin precoat filter process, due to synergistic removal of dissolved and suspended solids, covers the major part of the illustrated reference field of dissolved and undissolved substances. The process, through its beneficial effect on purity of the reactor water and thus the production of activation products, surely influences environmental irradiation, i.e. human irradiation and emission activities.

Concerning the formation of deposits on the fuel elements as a function of the water treatment techniques, no empirical values are known, however, the higher transparency of the reactor water and the fuel element tank water cleaned with the powdered resin precoat systems has brought important advantages during the manipulation of the fuel.

Was it not obvious to use the expected and now also proven features of the powdered resin precoat filters likewise for the treatment of the waste water from nuclear power plants? Here also the higher solid capacity of the powdered resin preacoats as opposed to conventional precoat filter materials of inert material is of importance. The powdered resins with a higher filtration effect, offer special advantages in this application.

Indeed, this technique brings good decontamination results particularly for activities that are bound to undissolved particles. For isotopes whose decontamination is subject to retardation on the ion-exchange resin, the prescribed maximum permissible values are less easy to adhere to than with bead-resin mixed-beds. The decontamination effect achieved in the Mühleberg nuclear power plant by powdered resin precoat filters is illustrated in Figure 4.

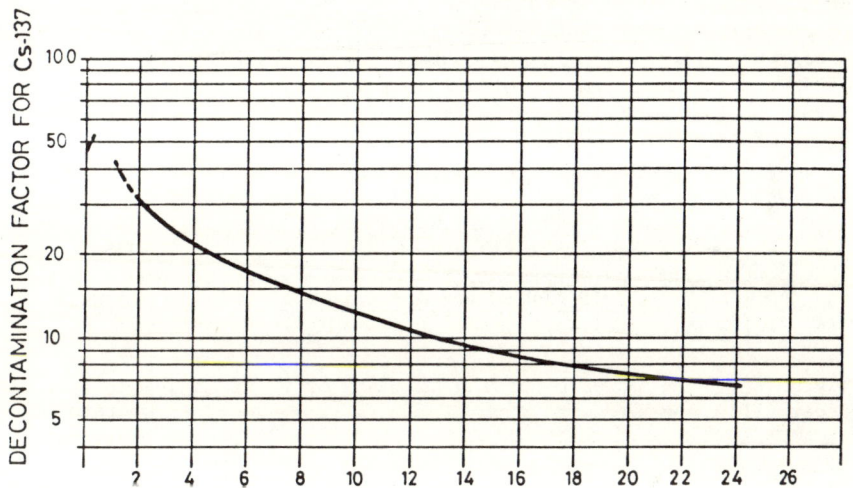

FIG. 4: DECONTAMINATION IN THE FUEL ELEMENT-STORAGE TANK AT THE MUEHLEBERG NUCLEAR POWER PLANT AS A FUNCTION OF FILTER RUNNING TIMES

Selective precoat layers that are effective for various isotopes, e.g. cesium, are in the process of development and are already being tested.

If the results achieved in the Mühleberg nuclear power plant are compared with those from the evaporation in the Würgassen nuclear power plant which is equipped with the same condensate purification system the following data for the year 1976 with the values published by Hepp (5) result:

Solid discharges for 1976 per m3 of treated waste water (sump water, unwashed):

- in Mühleberg   0.3. l resin with 60 to 70 % water content

- in Würgassen   4.1. l evaporator-concentrate with 75 to 85 % water content

Liquid discharge rate per m3 of treated waste water:

- in Mühleberg   ca. $20 \cdot 10^{-5}$ Ci (particularly cesium)

- in Würgassen   ca. $5 \cdot 10^{-5}$ Ci (particularly iodine)

Today, structured filter layers with any desired ratio of ion exchange to mechanical purification action are available. Also layers without ion exchange material are structured by controlled surface charge of the inert material. Common feature of all structured filter layers is their better filtration action and their higher solid capacity, because their charge and flow diversion results in depth filtration rather than surface filtration, as is almost exclusively the case with conventional precoat filters. When filter layers contain ion exchange material, a considerable synergistic effect out of the combination of mechanical filtration and ion-exchange action results. Precoat filters with structured filter layers will therefore find many more profitable applications in the future. They can aid in optimally solving all water treatment tasks met within nuclear power plants. Either alone, with a ratio of mechanical to ion exchange action specifically adapted to the particular problem, or in combination with bead resin mixed beds.

Sipp (6) reported that the service life of a waste water mixed bed filter could be increased 500 % when the upstream precoat filter of silica gel was converted to the use of structured precoat materials.

Currently more than 20 power plants in the USA, of which 13 are nuclear power plants, use precoat filter systems with structured precoat materials in more than 30 applications.

LITERATURE

1. U.S. DEPARTMENT OF COMMERCE
   National Technical Information Service

   PB - 257 054

   Occupational Radiation Exposure at
   Light Water Cooled Power Reactors, 1969 - 1975,
   Nuclear Regulatory Commission, June 1976

2. Stickert, R.J. "Incentives and Objectives for Decontamination of Nuclear Plants" Presented at Seminar on Decontamination of Nuclear Plants, Columbus, Ohio, May 1975

3. U.S. Environmental Protection Agency,
   Office of Radiation Programs.
   Summary of Radioactivity released in Effluents from
   Nuclear Power Plants from 1972 through 1975. / Document No EPA 520/3-77-006

4. U.S. DEPARTMENT OF COMMERCE
   National Technical Information Service

   ERDA-76-43 VOLUME 1

   Alternatives for Managing Wastes from Reactors and
   Post-Fission Operations in the LWR Fuel Cycle, May 1976

5. H.Hepp: Waste Water Treatment in Nuclear Power Plants:
   Practical Experiences Omnibus Volume VGB-Conference
   "Kraftwerk und Umwelt 1977" (Nuclear Power Plant and Environment)

6. Joseph R. Sipp, "Product Evaluation of Graver ECODEX and ECOCOTE"
   Presented at the April, 1977 meeting of the Edison Electric
   Institute (EEI) in Columbus, Ohio

This conference has been published in german language by the periodical "VGB Kraftwerkstechnik" in november 1978.

AUTHOR INDEX

Barcicki, J. 237
Bartkiewicz, B. 221
Bień, J. 205
Błażejewska, G. 255
Boari, G. 41
Bunch, R. L. 3
Buser, N. 313

Calmon C. 27
Cichocki, A. 229, 237

Drabent, Z. 107
Dziejowski, J. 107
Dzięgielewski, B. 283
Dziubek, A. M. 283

Ennet, P. 65

Ghysels, J. P. 313
Gomółka, E. 193
Gorzka, Z. 153, 163, 175
Grossman, A. 211

Hannus, M. 65
Hermanowicz, S. W. 141

Jaroszyński, T. 185
Jasińska, K. 163
Jóźwiak, A. 153

Kaźmierczak, M. 175
Kępiński, J. 87
Klocek, E. M. 265
Korczak, M. 113
Kowal, A. L. 123, 265, 275, 283
Kowalska, E. 205
Kozłowska, B. 153
Kucharski, B. 211
Kusznik, W. 211

Leszczyńska, D. 123
Liberti, L. 73
Limoni, N. 73
Łanowy, T. 113

Maćkiewicz, J. 275
Mika-Gibała, A. 255
Molder, H. 65

Niewęgłowska, Z. 221

Passino, R. 73
Pawłowski, L. 1, 229, 237
Petruzzelli, D. 73
Pielichowski, H. 113
Poranek, A. 245

Roland, L. D. 299
Roman, M. 141
Rutkowski, M. 113
Rybka, J. 193

Search, W. J. 15
Smoczyński, L. 107
Socha, A. 163
Steiner, M. P. 55
Sykut, K. 99
Sztark, W. 133
Świderska-Bróż, M. 289

Tiravanti, G. 41

Winnicki, T. 245, 255
Wiśniewski, J. 245

Zagulski, L. 229, 237
Zielewicz, E. 205

SUBJECT INDEX

Adsorption
  activated carbon   141, 211, 227
  BIOT number   211
  breakthrough curves   145, 148
  expanded-bed   146
  film transfer coefficient   146, 147
  filter optimization   211
  fluid dispersion coefficient   146, 147
  model   142
  oil   199
  packed-bed system   141
  powdered activated carbon   189
Analytic method
  electrochemical   100
  wastewater   99
  water   99
Anion exchanger
  acrylic   33
  fouling   31, 83
  inorganic   35
  macroporous   32
  silica removal   27
  stability   80, 237
  thermally regenerable   35
  vanadium compounds recovery   229
Coagulants
  aluminium hydroxide   113
  aluminium sulphate   117, 134, 292
  basalt detritus   115
  dose versus pH   293
  ferric sulphate   292
  ferrous salts   134
Coagulation
  in filter beds   296
  process   275
  theory   276
  volume   276, 289, 292

Concentration
  degree of   315
Demineralization
  brackish water   41
  economic evaluation   51, 238
  electrodialysis   41
  ion exchange   41, 55, 299
  reverse osmosis   55
  wastewater   21, 229, 237, 299
Desalination
  distillation   88
  electrodialysis   90
  equipment   89
  hyperfiltration   90
  reverse osmosis   90
  saline mine waters   87, 92
Electrodes
  analysis error   106
  characteristics   103
  detection limit   101
  ion selective   99
  liquid-membrane   101
  solid state   101
  types   104
Electrodialysis
  demineralization   41, 260
  water desalination   90
Evaporation   319
Filtration   275, 320
  sand-beds   289
Flocculation
  in filter beds   277
  sludge   205
  sludge from metal plants   207
  sludge from municipal sewage   207
  sludge from the cellulose   208

Ion exchange
  ammonia recovery 21, 73, 237, 299
  clinoptilolite 79
  condensate polishing system 58
  deionization of water 27, 55
  demineralization 41
  membranes 256
  mechanical purification 328
  mixed-bed system 60
  nitrate recovery 21, 237, 299
  water recovery 21, 229, 237, 299
  vanadium compounds recovery 229
  Irradiation levels 317
Oxidation
  catalytic 175, 265
  cetyl sulphate 156, 158
  detergents 175
  electric energy consumption 170
  electrochemical 153, 164
  inorganic sulphur compounds 265
  nonionic surface-active substances 163
  ozone 3, 123
  rokaphanol N-6 163
  sulphides 265
  sulphites 265
Ozonation
  catechol 126
  contactor 6
  dihydroxybenzenes 123
  hydroquinone 129
  pH 123
  resorcinol 128
  wastewater 3
Ozone
  contactors 6
  disinfection 3
  dissolution 3
Particles
  undissolved 319
Polysulphone membranes
  separation properties 245, 252
Powdered resin
  precoat filters 319
Precipitation
  glulam sludge 224
  phosphorus 65
Precoat filter 316, 318
Recovery
  acid 41
  ammonia 21, 73, 237, 299
  nitrate 21, 237, 299
  phosphate 73
  water 21, 229, 237, 299
  vanadium compounds 229
Reduction
  COD 108, 265
  colloid 108
  BOD 108

Removal
  acetonitrile 188
  alkalinity 285, 287
  ammonia 20, 21, 299
  chlorobenzene 188
  coliform 10
  detergents 175
  dioxane 188
  heavy metals 290
  nabam 134
  nitrate 22, 299
  NMP 188
  oil 193
  organic compounds 119
  organic solvents 185
  pesticides 133
  phenol 222
  phormaldehyde 222
  phosphorus 65, 73, 111
  rokaphenol N-6 163
  silica 27
  urea 21
Reverse osmosis
  desalination 90
  water demineralization 55
Sorbent
  for oil removal 193
  polyurethane resin 193
  resin for oil sorption 199
Urban storm water 289
Urea
  biological treatment 21
  hydrolysis 21
  nitrogen industry 16
Wastewater treatment
  activated carbon 211
  activated sludge 118
  aeration 190, 283
  alum coagulation 188
  biological electrochemical method 65
  coagulation 109, 113, 115, 133
  dialytic processes 255
  disinfection 12
  electrochemical oxidation 163, 175
  electrolysis 107
  from hydraulic transport of ash 283
  ion exchange 21, 229, 237, 299
  neutralization 283
  nitrogen fertilizer industry 15, 237, 299
  ozonation 10
  piezodialysis 261
  pig farm 107
  piggery 113
  pilot plant test 67
  recarbonation 283
  sedimentation 291
  sorption 189
  ultrasonic field 205

Water
  analysis 99
  conditioning 211
  electronic industry 55
  pharmaceutical industry 55
  purity demands 60
  pyrometallurgical scrubbing water 265
  reactor 318
  renovation 255
  ultrapure 55
Water recovery
  ion exchange 21, 229, 237, 299